In-situ Mechanics of Materials

Pranjal Nautiyal · Benjamin Boesl · Arvind Agarwal

In-situ Mechanics of Materials

Principles, Tools, Techniques and Applications

Pranjal Nautiyal
Dept of Mechanical and Materials Eng
Florida International University
Miami, FL, USA

Benjamin Boesl
Dept of Mechanical and Materials Eng
Florida International University
Miami, FL, USA

Arvind Agarwal
Mechanical and Materials Eng
Florida International University
Miami, FL, USA

ISBN 978-3-030-43322-2   ISBN 978-3-030-43320-8   (eBook)
https://doi.org/10.1007/978-3-030-43320-8

© Springer Nature Switzerland AG 2020

This work is subject to copyright. All rights are reserved by the Publisher, whether the whole or part of the material is concerned, specifically the rights of translation, reprinting, reuse of illustrations, recitation, broadcasting, reproduction on microfilms or in any other physical way, and transmission or information storage and retrieval, electronic adaptation, computer software, or by similar or dissimilar methodology now known or hereafter developed.

The use of general descriptive names, registered names, trademarks, service marks, etc. in this publication does not imply, even in the absence of a specific statement, that such names are exempt from the relevant protective laws and regulations and therefore free for general use.

The publisher, the authors, and the editors are safe to assume that the advice and information in this book are believed to be true and accurate at the date of publication. Neither the publisher nor the authors or the editors give a warranty, expressed or implied, with respect to the material contained herein or for any errors or omissions that may have been made. The publisher remains neutral with regard to jurisdictional claims in published maps and institutional affiliations.

This Springer imprint is published by the registered company Springer Nature Switzerland AG
The registered company address is: Gewerbestrasse 11, 6330 Cham, Switzerland

Preface

Seeing is believing—this age-old adage holds the key to decipher the response of materials—more so in the present times, with the design and development of materials taking place at multiple length scales, from the molecular and atomic level all the way up to the macroscopic level. The advancements in nanotechnology, nanomanufacturing, and additive manufacturing have resulted in an explosion of research activity to develop superior load-bearing materials, which are stronger, stiffer, harder, and tougher. One of the early strategies to accomplish this was to engineer microstructures with ultrafine, sub-micron-sized grains in a narrow window where the Hall–Petch relationship predicts peak material strength. In the 1960s, 1970s, and 1980s, the discovery and development of high-strength, micron-sized fibers, such as Kevlar fiber, boron fiber, glass fiber, and high-strength carbon fiber popularized the development of composite materials. 1991 saw the breakthrough discovery of carbon nanotubes—and it was subsequently established that these nanotubes are the strongest materials known to humankind. A decade later, graphene, a single atomic layer of carbon was isolated—with tensile strength hundreds of times higher than that of steel, replacing carbon nanotube as the world's strongest known material. Inspired by the remarkable physical and chemical characteristics of carbon nanotubes and graphene, the discovery and fabrication of alternative 1D and 2D nanomaterials have received immense attention in the last decade. These nanomaterials with exceptional mechanical properties have been extensively employed to engineer stronger, stiffer, and tougher nanocomposites. Manipulating the dimensions of materials can also impact their mechanical properties. The first article on cylindrical pillar compression technique—using focused ion beam machining to fabricate micropillars with varying diameters and applying compressive forces using a nanoindenter was published in 2004. The strength of the metallic pillars was seen to scale with their diameter—there was an order of magnitude jump in strength compared to the bulk material strength, as the pillar diameter dropped below 10 µm—a phenomenon known as *"smaller is stronger."* This finding inspired nanomechanical interrogations of matter at smaller length scales—a wave of articles on micropillar compression, microbeam bending, and miniature tensile testing were published in the decade that followed—and now we know that the early observation made for Nickel pillars is not universal and *smaller can also be weaker*. More recently, the advent of nanomanufacturing and high-resolution additive manufacturing techniques has produced an interest in mechanical metamaterials, which have well-defined, multi-scale architecture down to the nanometer length scale. Unlike conventional foams or cellular materials that experience loss in stiffness with decreasing density, the metamaterials display near-constant stiffness even at ultralow densities because of a well-defined network of unit cells. By employing ultrafine individual features, carefully considered architectural design for load transfer, and using materials with suitable intrinsic properties, 3D metamaterials provide an unprecedented opportunity to translate the exceptional properties observed at the nanoscale for real world applications at the macroscale—albeit with some challenges that need to be solved.

These marvelous results of manipulating matter provoke so many questions: *Why do the materials behave so differently at different length scales? How does adding a strong fiber or a nanosheet strengthens an otherwise weak polymer? How should the unit cell design and dimensions be tweaked to achieve superior flexural strength in a metamaterial? What is the effect of the alignment and distribution of nanofibers in a metal matrix on the resultant mechanical properties?* These questions are not just for curiosity's sake, but the answers to these questions also hold the key to design superior load-bearing and synergistically functional materials. As scientists and researchers, we rely on theories, measurements, and observations to explain the unknown. But there are some challenges with this approach of understanding matter–mechanics relationship:

- As we foray into the world of nanomaterials, the classical laws of physics often do not hold true anymore. The mechanisms we observe at the macro do not necessarily explain the phenomena seen in the nano- or microworlds.
- The complexity of material systems we deal with has increased manifold—material microstructures now consist of multiple elements, phases, and components, with different kinds of physical and chemical interactions active between them. It is rather difficult to model and decipher all these effects that may be influencing the mechanics of materials.
- The conventional evidence-collection approach relies on postmortem evaluation of materials, that is, looking at the deformation or failure mechanisms in a microscope after a mechanical test has been performed. These observations lead to speculations and conjectures—but a lack of hard evidence *(what actually transpired during the test)* leads to skepticism, disagreements, and many times, contradicting theories and explanations.

Failure to understand actual mechanisms, or an incorrect understanding can lead to wasted time, misspent resources, and a missed opportunity for engineering better materials. It can also lead to catastrophic failures. Therefore, a lot is at stake when we engineer advanced materials. Hence, we see pervasive inertia in the industry to accept new materials—there is a huge time gap in translating exciting developments in research laboratories to actual engineering applications.

We believe the solution to this challenge lies in the capability to gather hard evidence on the suitability of novel, advanced materials for critical load-bearing applications. The ability to observe and record the deformation and failure of materials *in action, as it is happening* can provide us a precise understanding of load-bearing mechanisms and how they are related to the material microstructure. Different mechanisms can be activated at multiple length scales: for instance, the macroscale fracture is caused by microscale crack initiation and propagation, which in turn is affected by atomic-level dislocation activity. In composite materials, nano-sized filler materials influence the crack propagation pathways at the micron length scale, and they create obstructions to dislocation motion at the atomic length scale. These dynamic phenomena cannot be effectively captured or understood just by *postmortem* imaging. Seeing them *in action* is fascinating—but it is also challenging. How do we trigger and observe these muti-scale mechanisms, all the way from a macroscopic point of view, down to the atomic-level phenomena?

This book is written with the intention to explore the answers to all these questions. *In-situ* Mechanics approach couples mechanical measurements with real-time imaging of material deformation. This is accomplished by performing mechanical tests inside or under a microscope. We have written this book with the aim of tackling this subject in great depth and breadth. This book aspires to serve a beginner as well as an expert in the field. Some of the salient features of the book are listed below:

- *In-situ* testing often requires specialized instrumentation and sample preparation. Combing mechanical and imaging instrumentation can introduce complexity in the test setup. We have devoted a chapter to different tools and techniques for in-situ characterization of materials.
- Deformation mechanisms are influenced by the stress-state induced in the material. A dedicated chapter delves into different in-situ test methods, such as indentation, scratch, micropillar compression, microbeam bending, tensile testing, and double cantilever testing.
- The environment is a dominant factor in the mechanics of materials. There is a strong focus on the development of advanced engineering materials that can perform in extreme temperatures. There is an exclusive chapter on in-situ elevated and cryogenic temperature testing, with case studies on real-time examination of deformation mechanisms in a wide temperature window of 130–1000 K.
- In-situ mechanical characterization is a versatile approach for examining a wide diversity of samples. We have included case studies on testing and imaging of nanomaterials, composite materials, metamaterials, smart materials, biomaterials, and biological samples. Due to the visual nature of the work, we have included 27 supplementary videos to complement the discussion in the chapters.
- The in-situ mechanics approach can inform and inspire the development and improvisation of mechanics models. Chapter 6 focuses on image correlation and molecular dynamic simulations to complement in-situ interrogations.
- We have discussed some practical limitations, challenges, and considerations during in-situ testing of materials. Like every other approach, in-situ character-

ization has its own constraints. But this is an evolving field and many more innovations in instrumentation and methodology are anticipated. We have elucidated different areas where the in-situ approach holds promise for future applications. The usefulness of the technique transcends disciplines—it is promising for structural, functional, as well as biological and biomedical applications.

- Last, but not the least, we lay out our vision for education and training in this field. As an evolving field with a high impact on future engineering developments, a concerted approach for curriculum design and workforce development is the need of the hour.

This is the first comprehensive and dedicated book on the subject of In-situ Mechanics. We wrote this book with the intention of educating students, equipping engineers, aiding researchers, and informing leaders in the field. We humbly acknowledge that this is not a perfect book, and we will be delighted to hear the opinions, criticisms, and suggestions from our readers. We hope this book brings value to you.

Miami, FL Pranjal Nautiyal
Miami, FL Benjamin Boesl
Miami, FL Arvind Agarwal

Acknowledgments

We would like to acknowledge some key actors without whom this project could not have been realized. First and foremost, the authors are grateful to Ms. Noemie Denis, an undergraduate student and researcher, who assisted us with schematics, figures, and copyright permissions, since we started working on this book two years ago. We commend her painstaking attention to detail.

We are extremely grateful to the funding agencies, the program managers, and Florida International University for providing the resources and infrastructural support to carry out our research in this field for nearly 15 years. The inception of nanomechanics research activities in our laboratories dates back to a 2006 DURIP award (Defense University Research Instrumentation Program by the Office of Naval Research), which was utilized to acquire a state-of-the-art nanoindenter with nanotribology and scanning probe imaging capabilities. In 2017, an in-situ SEM nanoindenter was acquired through another DURIP award. These notable acquisitions set off a flurry of research activities on the mechanics of nanomaterials, nanocomposites, advanced coatings, ultrahigh-temperature ceramics, mechanical metamaterials, shape memory materials, multifunctional composites, and cold-sprayed materials in our group. We acknowledge the Office of Naval Research (ONR), the Army Research Office (ARO), the National Science Foundation (NSF), the Air Force Office of Scientific Research (AFOSR), the Department of Energy (DOE), and the National Nuclear Security Administration (NNSA) for providing financial support for these endeavors.

Since the Fall of 2017, we are working with the nanomechanics thrust of an NSF Engineering Research Center, CELL-MET (EEC-1647837). CELL-MET has an ambitious goal of engineering cardiac *tissue patches* to repair the human heart damaged by heart attacks. Our group has been working with bioengineers, mechanical engineers, and chemists to use the in-situ characterization approach for interrogating the mechanics of soft materials, micro-scaffolds, and cardiac tissues. This is yet another example of the versatility of the in-situ mechanics approach. The authors acknowledge the support of NSF for this endeavor, and we are appreciative of the partnerships we have nurtured in the Center, particularly with Professors Alice White and Chris Chen at Boston University.

We place on record our sincere thanks to Mr. Lance Kuhn and his colleagues at Bruker Nano Surfaces (erstwhile Hysitron) for their exceptional support over the last 14 years.

Pranjal Nautiyal acknowledges the FIU University Graduate School for the Presidential Fellowship and the Dissertation Year Fellowship awards.

The authors acknowledge Drs. Chris Rudolf and Andy Nieto, former graduate students in our groups, for their early work and contributions on in-situ mechanical characterization of advanced nanocomposites.

We are indebted to the pioneering scientists and researchers for their valuable contributions to this field. We have drawn from their work, and this book would not have been written without the knowledge they created.

We thank our teachers and mentors, who taught us science, engineering, research, and life.

Most importantly, we are filled with gratitude for our families and loved ones. The realization of this project would not have been possible without their unconditional love, support and inspiration. We derive courage, determination, and above everything else, our humanity from them.

Contents

1 In-Situ Mechanics: Introduction and Importance... 1
  1.1 What Is In-Situ Mechanics?... 1
  1.2 Historical Perspective ... 4
    1.2.1 Low-Load Indentation ... 5
    1.2.2 Microfriction ... 6
    1.2.3 Dislocation Motion ... 6
    1.2.4 Studying Crack Propagation in High Resolution... 8
    1.2.5 Real-Time Image Analysis... 9
  1.3 Why In-Situ Mechanics?... 10
    1.3.1 Certainty Over Speculation ... 10
    1.3.2 Validation and Development of Theories and Models... 11
    1.3.3 Testing Local Mechanical Properties... 12
    1.3.4 Small-Volume Testing ... 13
    1.3.5 Environmental Effects ... 16
  1.4 Summary ... 17
  Questions and Assignments ... 17
  References... 20

2 In-Situ Mechanics: Experimental Tools and Techniques ... 25
  2.1 Miniature Sample Preparation ... 26
    2.1.1 Focused Ion Beam for Machining Specimens... 26
  2.2 Imaging Tools and Techniques for In-Situ Observation ... 32
    2.2.1 Optical Techniques for In-Situ Imaging... 32
    2.2.2 Electron Microscopy for High-Resolution Real-Time Imaging ... 39
    2.2.3 Electron Backscatter Diffraction During In-Situ Mechanical Testing ... 48
    2.2.4 In-Situ Tomography for 3D Internal Imaging ... 53
  2.3 Mechanical Instrumentation for Multi-Scale Characterization... 55
    2.3.1 In-Situ Atomic Force Microscope Inside Electron Microscopes... 55

		2.3.2	Instrumented Nanoindenter Inside	
			Electron Microscopes.	58
		2.3.3	Microelectromechanical Systems	
			for Mechanical Testing.	61
		2.3.4	Micromechanical Stage for In-Situ Testing	63
		2.3.5	Macroscale In-Situ Characterization	65
		2.3.6	Summary	67
	Questions and Assignments			67
	References.			70
3	**Test Methods for In-Situ Mechanical Characterization**			75
	3.1	Indentation for Localized Deformation Study		75
	3.2	In-Situ Tribology to Study Surface Interactions		84
	3.3	In-Situ Nano- and Micro-Pillar Compression		90
	3.4	In-Situ Beam Deflection		93
	3.5	In-Situ Tensile Characterization.		99
	3.6	In-Situ Double Cantilever Testing for Studying Fracture		106
	3.7	Summary		107
	Questions and Assignments			107
	References.			110
4	**In-Situ Mechanical Characterization**			
	as a Function of Temperature			113
	4.1	High-Temperature In-Situ Mechanical Testing		114
		4.1.1	In-Situ High-Temperature Nanoindentation	114
		4.1.2	In-Situ High-Temperature Micro-pillar Compression	119
		4.1.3	In-Situ High-Temperature Microbeam Bending	126
		4.1.4	In-Situ High-Temperature Tensile Testing.	128
	4.2	Cryogenic In-Situ Mechanical Investigations		132
	4.3	Summary		136
	Questions and Assignments			137
	References.			139
5	**Application of In-Situ Mechanics**			
	Approach in Materials Science Problems.			141
	5.1	In-Situ Characterization of Nanomaterials		143
		5.1.1	0D Nanomaterials	143
		5.1.2	1D Nanomaterials	149
		5.1.3	2D Nanomaterials	160
	5.2	In-Situ Characterization of Composites.		165
	5.3	In-Situ Characterization of 3D		
		Architectures and Metamaterials		177
	5.4	In-Situ Mechanics of Smart Materials.		186
	5.5	In-Situ Mechanics of Biological Materials		188
	5.6	In-Situ Mechanics of Irradiated Materials.		194
	5.7	Summary		197

		Questions and Assignments	197
		References..	200
6	**Interfacing In-Situ Mechanics with Image Correlation and Simulations**		205
	6.1	Digital Image Correlation Analysis.........................	207
	6.2	Interfacing with Molecular Dynamic Simulations..............	217
	6.3	Summary	221
		Questions and Assignments	224
		References..	225
7	**Challenges During In-Situ Mechanical Testing: Some Practical Considerations and Limitations**		227
	7.1	Sample Misalignment and Slippage.........................	228
	7.2	Effect of Environment	229
	7.3	Effect of Electron Beam Exposure..........................	231
	7.4	Summary	235
		Questions and Assignments	237
		References..	237
8	**Future Outlook for In-Situ Mechanics Approach**		239
	8.1	Research and Innovations	239
		8.1.1 New Material Development........................	240
		8.1.2 Novel Manufacturing Processes.....................	241
		8.1.3 Extreme Environment Performance..................	243
		8.1.4 Coupling Mechanics with Functional Properties.........	245
		8.1.5 Interfacing with Machine Learning	246
	8.2	Education and Training	247
		References..	249
Index...			255

Chapter 1
In-Situ Mechanics: Introduction and Importance

This chapter introduces the concept and capabilities of in-situ mechanical characterization of materials. The importance of multi-scale assessment of mechanical properties and deformation mechanisms is highlighted. Some pioneering works from past decades are reviewed to highlight the historical circumstances and scientific problems that led to the development of the in-situ mechanics approach. The indispensable role played by in-situ characterization in modern materials research and development is highlighted by discussing several characterization scenarios where conventional techniques are insufficient.

1.1 What Is In-Situ Mechanics?

Mechanical characterization is a quintessential step in the development of new materials and processes. Depending on the application and material type, strength, stiffness, Young's modulus, ductility, toughness, adhesion, and wear are some of the most widely studied mechanical properties of materials. In addition to the mechanical measurements, qualitative information about the deformation mechanisms and failure characteristics is also of interest to the metallurgists and mechanical engineers for real-world applications. Traditionally, this is done by observing the material in a microscope after it is subjected to mechanical testing (Hull 1999; Hayes and Edwards 2015; Straffelini 2015). The observed regions of interest could be a fractured surface, cracked regions, plastically deformed areas, delaminated interfaces, worn surfaces or a combination of these (Höche and Schreiber 1984; Huang and Hall 1991; Könönen et al. 1995; Dubnikova et al. 1997; Chandrasekaran and Loh 2001; Klimanek and Pötzsch 2002; Nautiyal et al. 2016a, c; Antillon et al. 2018;

Electronic supplementary material The online version of this chapter (https://doi.org/10.1007/978-3-030-43320-8_1) contains supplementary material, which is available to authorized users.

Bustillos et al. 2018; Thomas et al. 2018; Fontoura et al. 2018). The qualitative examination provides critical insights into the nature of mechanical deformation and the causes of material failure. As modern engineering efforts are directed toward advanced applications requiring high-performing materials, comprehensive knowledge of material behavior at multiple length scales is vital. In-situ mechanics is a specialized approach to capture and decipher the mechanical behavior of materials with extraordinary range and resolution of length scales, microstructure, time, mechanical conditions, and environmental factors (Nili et al. 2013; Wheeler and Michler 2013; Rudolf et al. 2016a; Jiang et al. 2017; Chang 2018). *In-situ Mechanics can be defined as a group of mechanical characterization techniques in which deformation is visually observed in real time while the material is subjected to mechanical loading.* This is accomplished by performing mechanical testing of materials under/inside a microscope, facilitating the recording of deformation events and their correlation with the microstructure.

Depending on the nature of materials being tested and the desired information from the test, in-situ studies are conducted in a wide diversity of microscopes. Optical microscopy (OM), scanning electron microscopy (SEM), and transmission electron microscopy (TEM) are the three most broadly defined imaging techniques based on the length scales of resolution that can be achieved (shown in Fig. 1.1). Real-time imaging throws light on the deformation mechanisms that are key determinants of the mechanical performance of the material. Deformation videos provide the opportunity to identify the factors that trigger material failure. The nature of loading during the in-situ tests can be tensile, compressive, flexural, shear, penetration or surface abrasion, depending on the mechanical property of interest. The load–displacement response captured from the test can be correlated with the 'instantaneous snapshots' of the material microstructure captured by the microscope. In other words, in-situ testing provides the ability to see transformations

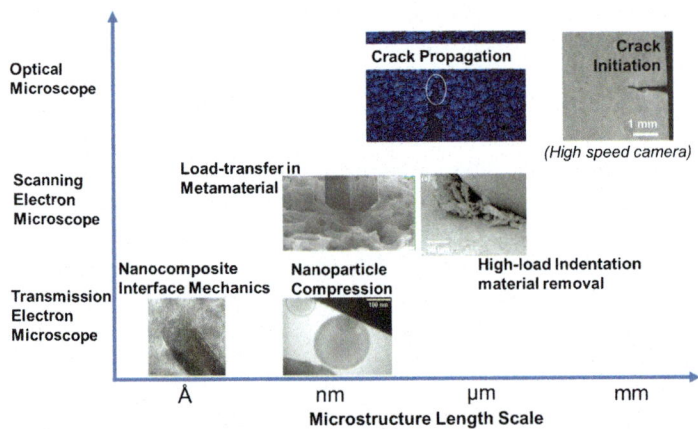

Fig. 1.1 Multi-scale imaging of mechanical deformation during in-situ tests. (Some of the images are reproduced/adapted with permissions from Nautiyal et al. (2018c), Deneen Nowak et al. (2007), Nautiyal et al. (2018b), Rudolf et al. (2016a), Embrey et al. (2017), Nautiyal et al. (2019b))

1.1 What Is In-Situ Mechanics?

corresponding to each load/displacement step. By analyzing the load-displacement data after the test is concluded, stresses and strains associated with key microstructure deformation events can be quantified. For instance, the critical stress required to activate the crack initiation in a ceramic material can be precisely computed because of the availability of microstructure snapshots corresponding to load–displacement response. Therefore, in-situ mechanics approach is not only confined to the qualitative assessment of material performance, but also aids in the quantification of minute transformations in the material when it is subjected to the external load.

Mechanical properties can be highly sensitive to material and environmental temperatures. The deformation mechanisms that are activated upon mechanical loading can be very different for the same material, but at different temperatures. The wreckage and sinking of the Titanic in 1912 is believed to be because of the ductile-to-brittle transition of the steel used for making the hull of the ship. Steel, which is a ductile material in ambient conditions, fails in a brittle fashion at very low temperatures. As an example, ASTM A36 has a ductile-brittle transition temperature of -27 °C (Felkins et al. 1998). However, the steel used in the Titanic at that time had a ductile-brittle transition temperature exceeding $+30$ °C (Foecke 1998). The temperature of the seawater was -2 °C at the time of the collision, making the steel fail catastrophically. Another relatively recent event of significance to material scientists is the failure of NASA space shuttle Columbia in 2003. The failure initiation site is believed to be the left wing leading edge, made of reinforced carbon-carbon composite (RCC) material. Extreme temperature and stress conditions in the upper atmosphere caused the erosion of RCC (Bykowski et al. 2006). The leading edges of the wings experience temperatures exceeding 1400 °C during re-entry, and mechanical properties of materials degrade rapidly at such extreme thermal conditions. These two classical material failure examples highlight that suitability of a material can only be evaluated by probing its mechanics at/around the *true in-service conditions*. Moreover, some material classes are more sensitive to minute temperature changes. Considering these real-world engineering challenges, there is a rising interest in the in-situ mechanics of materials as a function of temperature (Wheeler et al. 2015). The materials can be subjected to elevated or very low temperatures during mechanical testing, while the deformation is observed in real time in a microscope. In-situ investigations unravel how the mechanisms responsible for failure change as a function of temperature, which can be extremely useful information for material scientists to tweak the processing and constitution of the materials to suit the desired service conditions.

Over the past decades, we have seen the development of superior alloys and composites with remarkable mechanical performance. The material chemistries are becoming increasingly complex, as new materials being developed are composed of far more elements, compounds and/or a mixture thereof. The spatial distribution of the microstructure constituents is carefully controlled for imparting unprecedented mechanical and functional properties. Feedback about the effect of chemical composition, elemental distribution or phase formation during material processing on deformation behavior is vital. To address this pressing requirement of material scientists, in-situ mechanical characterization is being coupled with different spectroscopic techniques. As an example, real-time testing of materials in an SEM equipped with electron dispersive spectroscope (EDS) can facilitate the mapping of material microstructure at different stages of mechanical loading (especially just before/during/after failure) to understand how deformation mechanisms can be altered by tweaking the chemical makeup of the material. With the advancements in in-situ mechanical testing in recent years, it is now possible to precisely map the failure mechanisms associated with different constituents in the microstructure. Therefore, in-situ mechanical characterization techniques provide the ability to unravel the behavior of the material, starting from individual building blocks to the aggregate material response. Figure 1.2 schematically summarizes the concept and capabilities of the in-situ mechanics approach.

1.2 Historical Perspective

The in-situ mechanics approach has gained widespread attention and popularity for material characterization in the last two decades. However, some of the early works date back to the 1950s and 1960s. In this section, some of the seminal works on the development and application of in-situ mechanics approach are highlighted. This section provides an overview of the historical circumstances and scientific problems that led to the development and adoption of in-situ mechanical characterization techniques.

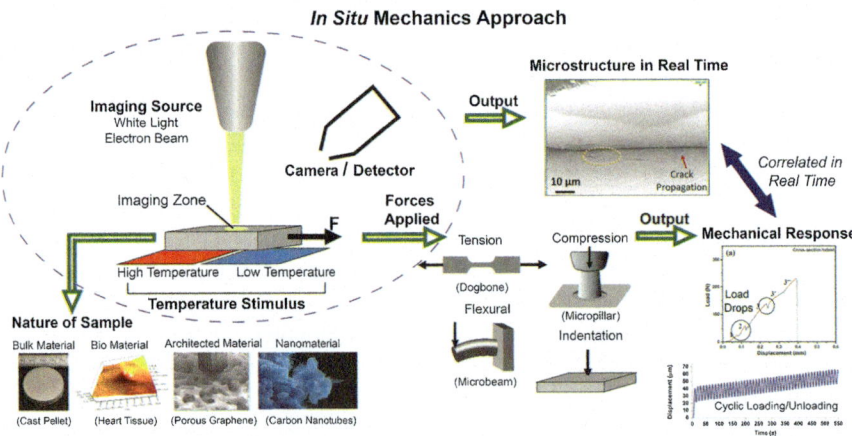

Fig. 1.2 Schematic representation of the concept and capabilities of in-situ mechanics approach. (Some of the images are reproduced/adapted with permissions from Balani et al. (2009a), Nautiyal et al. (2018b), Boesl et al. (2014), Nautiyal et al. (2019a), Loganathan et al. (2017), Agarwal group (FIU))

1.2 Historical Perspective

1.2.1 Low-Load Indentation

One of the first studies on low-load indentation was published in 1968 by Gane and Bowden, who investigated the small-volume deformation of metal crystals by using a very fine stylus (~100 nm in diameter) inside SEM (Fig. 1.3a). It was noticed that no deformation occurred upon pressing the stylus against the sample surface, until a *critical load* was reached (Gane and Bowden 1968). The pressure required for initiating deformation was significantly higher than bulk material hardness. The shear strength was found to approach the theoretical shear strength. This is one of the first visually demonstrated evidence of size effect, which is a very well-known and established concept now for nanoindentation of materials (Fisher-Cripps 2005). Bangert and Wagendristel developed a fully programmable ultramicrohardness tester for using inside the SEM (Bangert and Wagendristel 1985). The developed instrumentation had a wide operating range of 50 μN to 20 mN, allowing in-situ SEM testing of specimens with different morphologies and dimensions. The application of the tester was demonstrated for probing a mouse hair and thin films. The creep effect during indentation-hardness measurement was demonstrated by varying dwell time from 1 s to 100 s, which resulted in enhanced plastic deformation or lower hardness value (Fig. 1.3b). Over past two decades, the low-load indentation has emerged as a popular alternative technique for analyzing creep behavior of a variety of soft and viscoelastic materials (Syed Asif and Pethica 1997; Fischer-Cripps 2004; Oyen 2007; Wang et al. 2010; Nautiyal et al. 2015, 2016b; Lee et al. 2016). The modern advanced instrumentation can capture creep response for thousands of seconds and over a wide range of temperatures. In-situ creep characterization provides the opportunity to visually observe the progression of plastic deformation due to loading for an extended duration of time.

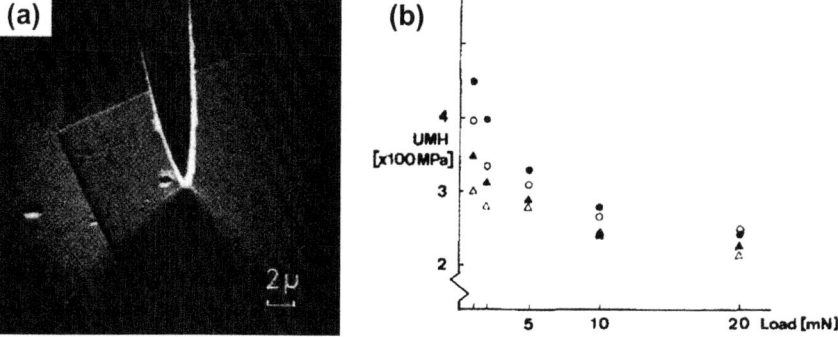

Fig. 1.3 (a) Indentation of gold surface inside SEM, and (b) plot showing dependence of ultramicrohardness of Al single crystal on load, loading rate and dwell time during indentation ($t = 1$ s, $v = 10$ μN/s (○); $t = 1$ s; $v = 1000$ μN/s (●); $t = 100$ s; $v = 10$ μN/s (Δ); $t = 100$ s; $v = 1000$ μN/s (▲). (Reproduced with permission from Gane and Bowden (1968). Purchased for reproduction from Bangert and Wagendristel (1985) American Institute of Physics)

1.2.2 Microfriction

Gane and Bowden, in their seminal study on micro-deformation of metal crystals inside SEM, also probed the sliding action of a sharp stylus on metal crystals (Gane and Bowden 1968). Some interesting deformation mechanisms were observed due to scratch using a chisel-shaped stylus (Fig. 1.4). The nature of tip–surface interactions was varied during the sliding process by orienting the stylus parallel and non-parallel to the direction of motion. This produced remarkably different mechanisms: cutting of the surface with the material pile up for former orientation, and formation of a spiral metal chip for the latter case. Slip lines emanating from the scratch-track were observed in the SEM. Although there were no lateral force or friction measurements extracted by the authors, but this early work is significant and paved way for further developments in the in-situ nanotribology of materials. Today, sliding/scratch experiments are more sophisticated with real-time lateral force/displacement measurement capabilities (Rabe et al. 2004; Balani et al. 2009b; Bakshi et al. 2010; Lahiri et al. 2011, 2015; Pitchuka et al. 2014). It is now possible to extract friction plots as a function of sliding distance, ramp loading and repeated multi-cycle scratch.

1.2.3 Dislocation Motion

In 1956, Hirsch and coworkers accidentally discovered moving dislocations in TEM while removing the condenser aperture (Hirsch 1986). The researchers observed cross-slip, dislocation bowing and pinning at the surface oxide film, and took cine films of the images on the fluorescent screen (Hirsch et al. 1956). This was the first

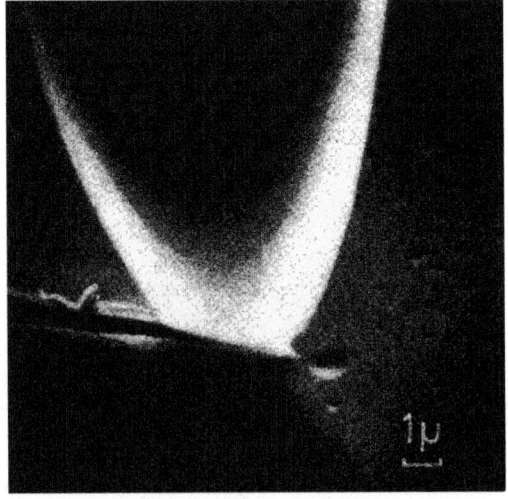

Fig. 1.4 Ploughing mechanism observed during sliding of stylus on the gold surface. (Reproduced with permission from Gane and Bowden (1968))

1.2 Historical Perspective

visual evidence of the glide of dislocations on slip planes. In subsequent years, heating, cooling and straining stages were developed and used for in-situ observation of dislocation activity. Figure 1.5a shows a custom-built straining stage for testing inside TEM from a study published on bcc metals in 1973 (Takeuchi 1973). In a separate study, Sumino and Sato used a "hot tensile stage" for conducting in-situ testing of silicon crystals between 650 and 800 °C. The researchers observed the lines of moving dislocations followed the directions of the Peierls valleys (Sumino and Sato 1979). Low temperature investigations also received great interest in the 1970s. Louchet and coworkers used a helium-cooled in-situ straining stage in a high-voltage TEM to *quantitatively* assess dislocation activity during tensile deformation of bcc crystals at low temperatures (Louchet et al. 1979). Most dislocations were observed to be of screw character at very low temperatures (~100 K), which exhibit slow movement. With the increase in temperature, dislocations were seen to become more flexible as evidenced by equilibrium curved shape (Fig. 1.5b–d).

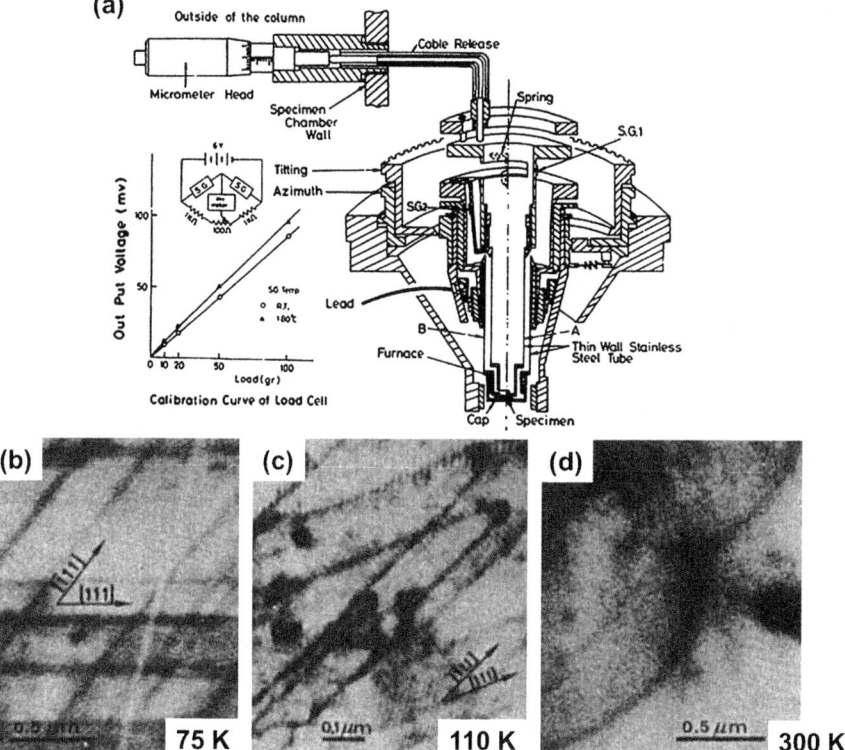

Fig. 1.5 (a) Schematic representation of straining holder custom fabricated for high-voltage TEM, and (b–d) TEM imaging of dislocations at 75, 110, and 300 K, respectively. (Reproduced/adapted with permissions from Legros (2014), Louchet et al. (1979))

Dislocation activity associated with *crack tip* has also been studied by in-situ investigation (Ohr 1985). Emission of dislocations from the crack tip takes place during crack propagation, creating a dislocation-free zone. Crack propagation takes place by a combination of elastic and plastic mechanisms: while dislocations emitted from the crack tip were observed to be responsible for the plastic crack opening, a brittle fracture was noticed in the dislocation-free-zone (elastic phenomenon).

1.2.4 Studying Crack Propagation in High Resolution

In 1980, Mindess and Diamond used a device with the ability to apply controlled forces and which can fit inside the SEM chamber, to induce slow crack growth in Mortar (Mindess and Diamond 1980). Wedge-loaded samples were tested inside SEM, to be able to observe crack initiation and propagation. This loading configuration also allows to compute K_{IC} from wedge displacement measurements. Figure 1.6a

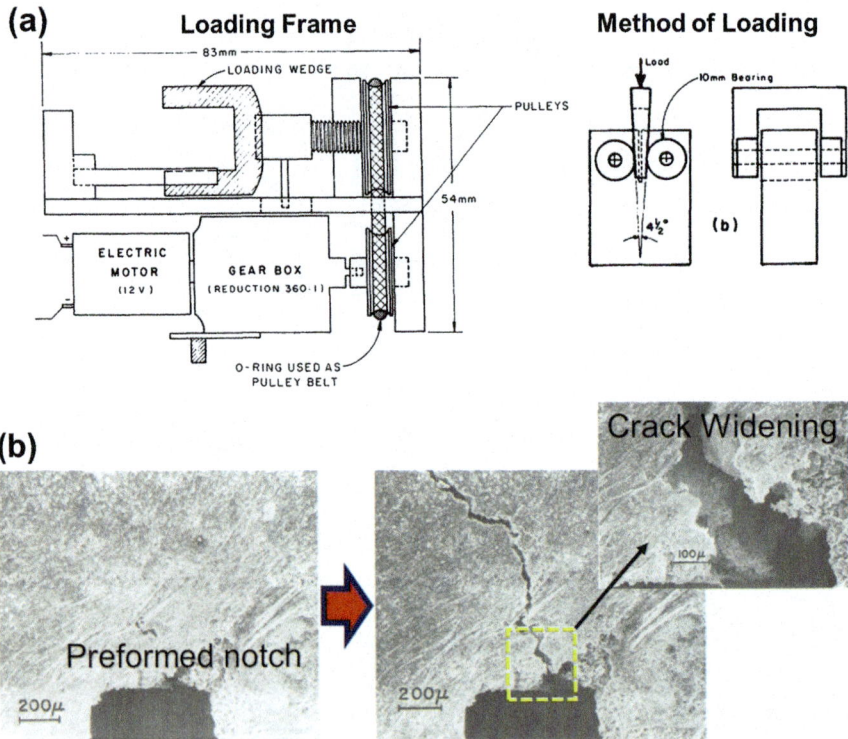

Fig. 1.6 (a) Loading frame for testing inside SEM and illustration of the wedge loading method for inducing crack growth, and (b) SEM images showing crack propagation and widening after it initiates from the pre-notch in the Mortar sample. (Reproduced/adapted with permission from Mindess and Diamond (1980))

1.2 Historical Perspective

shows the loading frame and test setup used. Preformed notch acted as the crack initiation point. SEM imaging captured crack propagation and crack widening (Fig. 1.6b). In-situ testing also showed the crack branching mechanism, where only one *active* branch of the crack experienced further propagation and opening, and the passive branch did not experience any visible change. A key insight obtained from the in-situ testing was that simple fracture mechanics models do not account for the complex geometric features of crack propagation in the material. Today, the study of crack formation and growth in different material systems is an area of active interest and in-situ investigations have been at the core of this research area (Rudolf et al. 2015, 2016a; Zhang et al. 2016; Sernicola et al. 2017).

1.2.5 Real-Time Image Analysis

The above sections highlight the development and evolution of in-situ techniques for observing deformation mechanisms. However, real-time imaging can also be useful for extracting the properties of the constituents of the sample being tested. Mulvaney and Halpern, in their 1976 study on intrinsic mechanical properties of vascular smooth muscle cells (SMC), monitored the mechanical deformation of arterial resistance vessels from rats using a Nomarski interference microscopy (Fig. 1.7) (Mulvany and Halpern 1976). The vessel specimens used in the study consisted of ~1000 SMCs (shown in the SEM micrograph in Fig. 1.7c). The specimens were connected to force transducer and displacement device using two wires

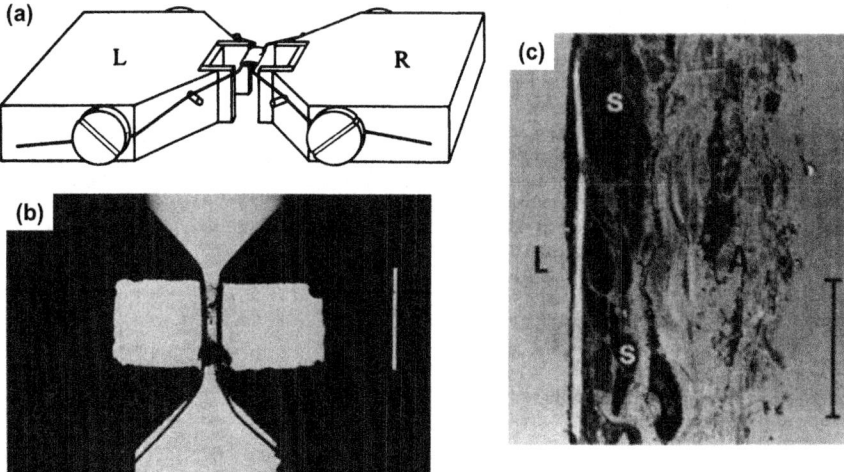

Fig. 1.7 (a) Schematic showing the assembly of arterial vessel specimen threaded with wires, (b) micrograph of the mounted vessel (from Nomarski interference light microscope), and (c) electron micrograph of the vessel showing smooth muscle cells. (Reproduced/adapted with permission from Mulvany and Halpern (1976))

(Fig. 1.7a). The microscope images of the changes in vessel circumference, imposed by the displacement device, were monitored in real time (Fig. 1.7b). The comparison of the frames revealed that the length change of the constituent SMCs was in the same proportion as the length change of the vessel specimens. Therefore, force–strain characteristics for SMCs are the same as the vessel specimens. This early study shows the application of in-situ characterization for understanding the mechanics of much smaller units or microstructure building blocks, which are otherwise difficult to probe. The in-situ *frame analysis* technique has now evolved for more complex microstructural strain evolution (Embrey et al. 2017; Nautiyal et al. 2018b, d), and can be applied for in-situ characterization at multiple length scales. The in-situ image correlation and its applications are discussed in detail in Chap. 6.

1.3 Why In-Situ Mechanics?

1.3.1 Certainty Over Speculation

Classical approach of investigating material deformation mechanisms relies on postmortem observation of samples in a microscope after the tests are performed. For instance, fracture surface after tensile failure, residual impression after an indentation test, or wear track after a tribology test can be observed under optical, atomic force and/or electron microscopes to decode the plasticity and failure mechanisms. Post-failure microscopy can indicate potential mechanisms, such as ductile vs. brittle failure, ploughing vs. cutting wear, twinning vs. slip dominated indentation, or activation of reinforcement mechanisms such as crack bridging, crack deflection or sword-in-sheath failure (Lahiri et al. 2010, 2013; Nautiyal et al. 2015, 2016c; Antillon et al. 2018; Thomas et al. 2018). However, the emergence of new materials with complex microstructures has made this approach insufficient for understanding mechanical behavior. Some instances or cases include composites with multiple nano-microfillers, materials with multiple phases formed under the reactive synthesis conditions, or complex 3D architectures (metamaterials) with multi-scale hierarchy (Nautiyal et al. 2017c, 2018c; Zhang et al. 2017b; Idowu et al. 2018; Wu et al. 2018). Different constituents can deform in remarkably different fashion and display mechanical interactions, which may be difficult to capture by post-failure microscopy. In such situations, researchers have to rely on load–displacement curves or computational modeling to complement the post-failure microscopy for speculating different mechanisms that can possibly be governing deformation behavior. However, this approach has limitations: load–displacement behavior of complex material systems depends on a host of factors during mechanical testing, and modeling these systems is challenging and computationally impossible in some cases. Lack of understanding of the deformation behavior could be a major limiting factor for adoption and applications of new materials. The in-situ approach is the most effective way to observe the deformation behavior of these new and complex material systems in real time as they are subjected to mechanical loading. Simultaneous load-displacement curves also enable the experimentalists to quantify the *critical stresses/strains* required to trigger different mechanisms such as crack initiation, dislocation nucleation, de-bonding and interface sliding, nano-filler fracture or material pileup. Mechanical feedback, in turn, can be useful for tailoring the processing/fabrication conditions to suppress or enhance specific mechanisms. The *combined qualitative and quantitative assessment* is a powerful characteristic of in-situ mechanics approach.

1.3.2 Validation and Development of Theories and Models

In-situ mechanical characterization captures material deformation with remarkable emporal and spacial resolution. When coupled with quantitative mechanical data, the set of information extracted from in-situ tests can be useful to validate or negate theories on mechanics of materials, and also inspire scientists to improvise and advance the existing theories that explain deformation behavior. The most classic example of this is the concept of slip dislocation. The cause of plasticity in metals was not well understood in the early twentieth century. Although the concept of dislocations or defects was developed in 1907, it was not until 1934 when Orowan, Polanyi and Taylor proposed that the m*ovement of dislocations* can be responsible for plastic deformation. However, the glide of dislocations on their slip planes was experimentally demonstrated by in-situ observation in TEM in 1956 (Hirsch et al. 1956), confirming the theory of dislocation motion. Last two decades have seen the rise of new advanced materials, such 2D materials, advanced high-temperature ceramics, new classes of high entropy alloys and complex nanocomposite systems comprising of a combination of different nanomaterials with metals/polymers/ceramics (Choi et al. 2007; Agrawal et al. 2013; Yue et al. 2013; Rudolf et al. 2016b, 2017; Zhang et al. 2017a; Nautiyal et al. 2017a, b, 2018a; Gao et al. 2018; Thomas et al. 2019). Material scientists and metallurgists are still trying to understand how these new materials behave mechanically. While there are different works on analytical and computational modeling, the validation and acceptability of these theories (and the underlying assumptions) is contingent on experimental evidence. In-situ assessment is central to verify the theories that explain how these advanced materials deform and inform/inspire material design for superior performance. Figure 1.8 shows tensile deformation of a 3D graphene foam architecture investigated by a mesoscopic model (Pan et al. 2017). Stretching, bending, and alignment of graphene flakes constituting the 3D foam were observed. These findings were in sync with the observations made during in-situ tensile deformation in a prior study by our group (Nieto et al. 2015a). The comparison of mesoscopic modeling outcomes and real-time SEM images is shown in the figure. The correlation with in-situ experimentation validates the application of a 2D mesoscopic model for a 3D numerical sample. This example highlights the importance of in-situ experiments

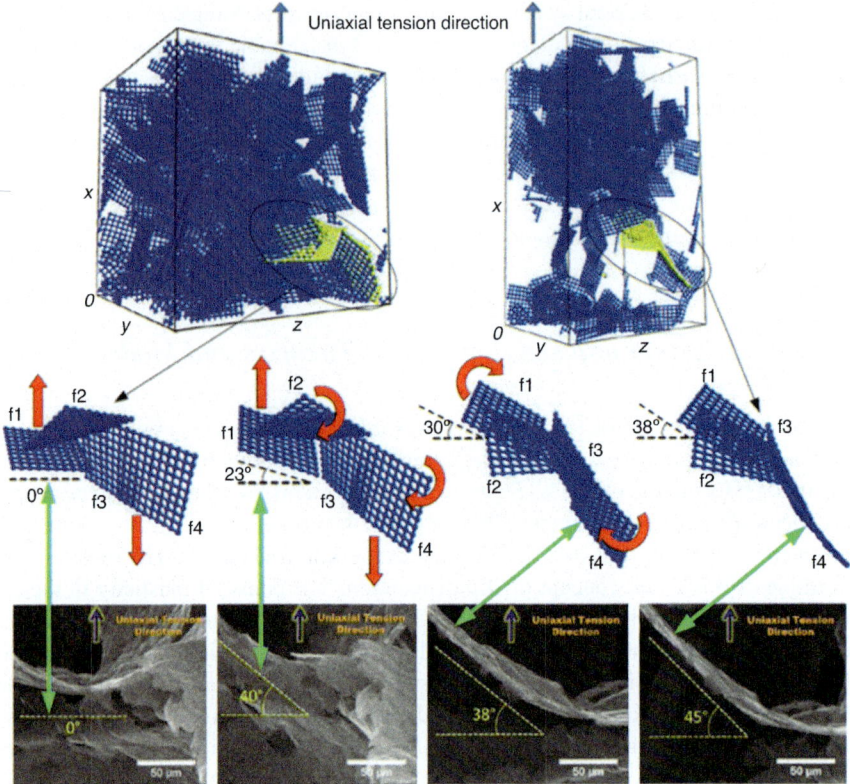

Fig. 1.8 Tensile deformation of a graphene foam branch flakes by mesoscopic model, and verification of the flake configurations from in-situ SEM tensile images. (Reproduced/adapted with permissions from Pan et al. (2017), Nieto et al. (2015a))

for development, application, and validation of computational tools/techniques to understand deformation behavior of complex hierarchical materials.

1.3.3 Testing Local Mechanical Properties

In order to engineer materials with desired mechanical attributes, understanding the characteristics of *individual constituents* in the microstructure can be insightful. As an example, advanced nanocomposites have complex microstructures with different material phases, such as matrix, filler and interfacial products (Nautiyal et al. 2017c, 2018c). The mechanical properties of these new materials are greatly influenced by the deformation behavior of the constituents. The feature size of nanofillers requires nanometer-size resolution capability (Lahiri and Chen 2012). More recently, mechanical metamaterials have attracted great attention. Metamaterials can display

1.3 Why In-Situ Mechanics?

unprecedented strength, stiffness, and flexibility, by careful control of 3D architecture. This design process requires an understanding of deformation behavior of individual features, such as nodes and struts that constitute the metamaterials. The size of these features can vary from nanometer to millimeter length scale. Conventional bulk-scale mechanical testing techniques are not suitable to understand how these hierarchical structures behave. The material microstructures are also influenced by processing techniques. The last decade has seen a massive upsurge of interest in additive manufacturing approach. Additive techniques can introduce complex microstructures. For instance, cold-sprayed metal deposits consist of splats with severe work hardening. The bonding along splat interfaces is of prime importance for structural applications (Chen et al. 2009; Nautiyal et al. 2018d). 3D printing of polymer structures can introduce anisotropy in mechanical properties due to the nature of layer-by-layer deposition (Chen et al. 2017). It is important to probe these *local mechanical effects* to develop comprehensive process maps for the new, upcoming manufacturing techniques. These examples highlight the importance of localized mechanical properties from the standpoint of material development, 3D architecture design as well as better understanding of new processing techniques. In-situ mechanics approach is indispensable for resolving and characterizing microstructural features. An example of resolving and indenting different local features of a honeycomb-shaped scaffold structure is shown in Fig. 1.9. Readers are referred to Supplementary Video, Video 1.1 showing the localized indentation of the walls in a honeycomb scaffold fabricated by direct laser writing.

1.3.4 Small-Volume Testing

For applications in nano-microelectronics, robotics and biomedical technology, miniature materials and systems are gaining prominence. Thin films in optics, magnetics and electrical devices, micro-scaffolds for tissue engineering, targeted drug delivery devices and nano/microelectromechanical systems (NEMS/MEMS) are examples of the technologies that are moving toward miniaturization. It is a well-known fact that materials can display strong *size effects* in mechanical properties.

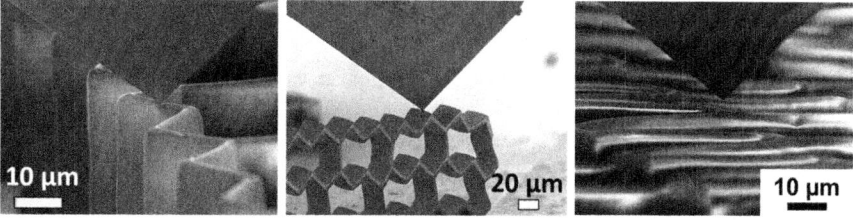

Fig. 1.9 Targeted indentation-based mechanical characterization of different *individual features* in a honeycomb scaffold architecture. (Courtesy: Agarwal group (FIU))

For instance, nucleation and propagation of dislocations or shear bands can be tailored in surface-dominated geometries like nanopillars (Greer et al. 2011). The preexisting flaws in the ceramic material are sites for failure initiation and limit their fracture strength. However, for small material volume, the probability of finding flaws is very low (Jang et al. 2013). As a result, the strength of nano-sized solids is much higher than bulk materials, and approaches the theoretical strength. Flow strength of metal crystals depends on the sample diameter. Annihilation of dislocations at free surfaces (dislocation starvation) results in enhanced strength for smaller samples. The variation can span an order of magnitude. Miniaturization demands understanding how small volumes of materials behave mechanically. There has been great interest in probing nano and micropillars, microbeams, and nanometer-thin films to understand mechanisms at play at these length scales. This is possible only by in-situ mechanics approach, as resolving nano-micrometer sized samples essentially requires the assistance of microscopes. Readers are referred to Supplementary Video, Video 1.2 showing the in-situ deflection of a polymer microbeam.

As the development of new nanomaterials has gained pace in recent decades, there is a growing interest in characterizing their mechanical properties. Nanomaterials can have different geometries, such as spherical particles, nanotubes or nanowires, nanosheets or nanoplatelets, or multifaceted particles (or polygons). These nanomaterials have at least one dimension in "nanometer" length scale. For example, nanotubes or wires have nanometer-scale diameter, nanoplatelets or sheets have nanometer-thickness, and nanocubes have nanometer edge length. This makes the placement and clamping of the samples challenging. Conventional bulk-scale mechanical testers cannot be used for gripping these extremely fine nanomaterials. Severe forces during gripping the samples can induce stress-concentration or local deformation, creating a weak link for failure initiation during mechanical loading. As a result, the mechanical strength would be underestimated and not an accurate measure of "intrinsic" material strength. Contrary to this, poor clamping causes slippage and overestimated strains. Incorrect strains can then be carried over to elastic modulus calculations, resulting in the underestimation of material stiffness. Some of the nanomaterials, such as nanotubes or nanowires also need to be aligned parallel to the loading axis. Significant deviation in their orientation can result in erroneous measurements. Therefore, it is vital to observe and verify the sample alignment during clamping these nanomaterials for mechanical testing. Nanomaterials can only be resolved in electron microscopes. Typically, electron-beam-induced deposition of C/Pt/Cu/W (inside the microscope chamber) is a useful strategy for clamping the samples for mechanical tests (Tang et al. 2011; Espinosa et al. 2012). These deposition techniques are employed in a vacuum environment, while observing and manipulating the sample. An example of the sample manipulation and clamping is shown in Fig. 1.10. As a result, use of in-situ SEM/TEM techniques has become indispensable for testing of nanomaterials.

During mechanical testing, the response of the sample/material is captured as load-displacement profile. Mechanical stresses can be computed only if the

1.3 Why In-Situ Mechanics?

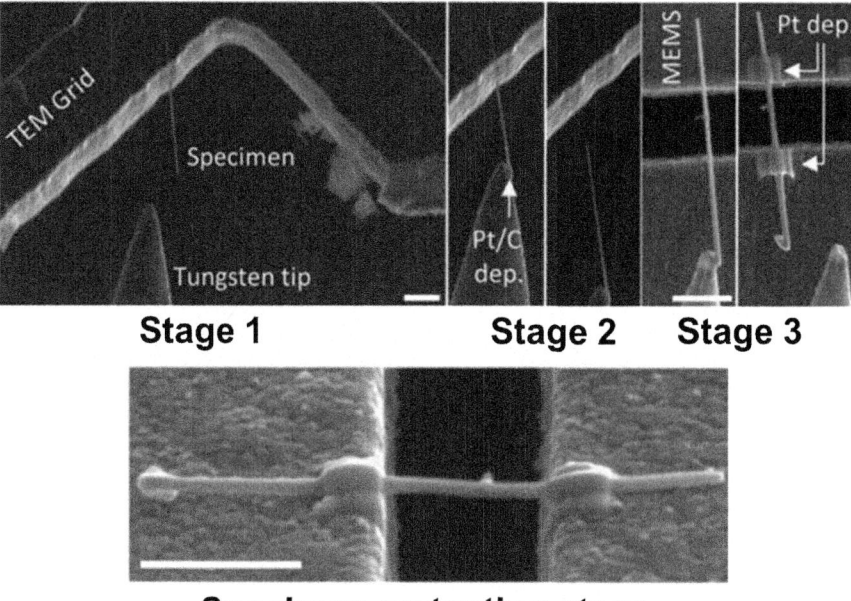

Fig. 1.10 Steps showing the sample clamping technique for tensile testing of a nanotube. (Reproduced/adapted with permission from Espinosa et al. (2012))

cross-section area that bears the mechanical load is known. Conventional mechanical testing is performed on standard geometries, such as dogbones, cylindrical rods, or blocks, which have well-defined and easy to calculate cross-section areas. However, it is difficult to accurately determine the area of low-dimension materials (like nanotubes, nanosheets) or materials with complex nano-microscale architecture (metamaterials). High-resolution imaging of samples is necessary to determine their dimensions, such as the thickness of the nanotube wall, thickness and width of multilayer nanoplatelets, or size of hollow cells in a nanoporous material. Often, these dimensions can vary in the same batch of fabricated materials. For example, nanotubes in a batch display variation in the number of walls, outer/inner diameters and length. Therefore, for accurate determination of stresses, it is important to measure the dimensions of each *individual sample* that is being tested (shown in Fig. 1.11). This makes in-situ testing in microscopes necessary, so that each sample can be resolved and the cross-section area can be computed. The choice of microscopy technique depends on the feature size of the samples. For instance, a microporous foam can be resolved under an optical microscope, but an electron microscope would be required for resolving features in a nanotube buckypaper.

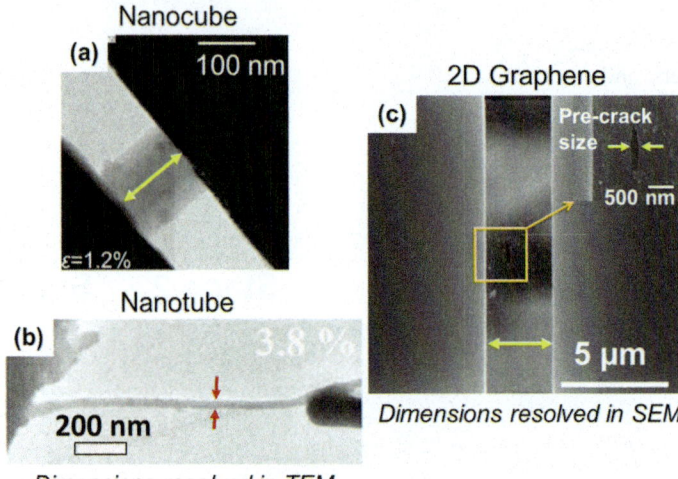

Fig. 1.11 Resolving low-dimension materials subjected to mechanical test inside electron microscopes: (**a**) the height/thickness of a nanocube under compression inside TEM, (**b**) the diameter of a nanotube undergoing tensile test, and (**c**) the gage length and the pre-crack width for a graphene sheet under tension. (Reproduced/adapted with permissions from Issa et al. (2015), Liu et al. (2013), Zhang et al. (2014))

1.3.5 Environmental Effects

The loads and displacements associated with mechanical testing of low-dimension materials can be extremely small. Environmental noise/vibrations can introduce errors. Isolation from environmental noise is also important for characterizing specific events during deformation, such as dislocation activity, twin formation, crack initiation/propagation/deflection, etc. These events manifest as strain bursts or load drops in force–displacement curves. Vibrations can limit the ability of mechanical equipment to capture these events. Some materials are highly sensitive to temperature and moisture (air humidity), and creating isolation is vital for measuring their true properties. Environmental exposure can also induce chemical changes, such as surface oxidation. Minimal surface oxides, although insignificant for bulk testing, can be problematic during extremely localized mechanical testing such as modulus mapping of an alloy microstructure. In-situ mechanical testing in microscope chambers, such as SEM or TEM (which have vacuum environment), can address these challenges. Microscope chambers provide isolation from air flow, temperature fluctuations, surrounding vibrations, humidity as well as oxygen. In-situ testing in isolated microscope chambers provides the opportunity to create controlled conditions for mechanical characterization. For instance, temperature-dependence of mechanical properties of materials can be studied by heating/cooling the sample inside an electron microscope chamber without worrying about sample oxidation.

1.4 Summary

In-situ mechanics has become an indispensable approach for material scientists, researchers, and engineers. Real-time imaging is useful to decipher deformation in materials with complex microstructures. The deformation behavior captured by in-situ testing can inform the development of mechanics theories and validation of analytical/computational models. High-resolution imaging enables testing of nanomaterials, since they have ultrafine dimensions and cannot be gripped/tested in conventional mechanical testers. In-situ imaging also allows targeted testing of the specific features in a material's microstructure or examination of small volumes of material. The application of knowledge derived from in-situ studies is vital for a wide diversity of material applications: aerospace, marine and automotive structures, electromechanical devices, robotics, biomedical devices, scaffolds and other accessories, miniature sensors and actuators, etc.

Questions and Assignments

1. Identify the types of microscope suitable for in-situ visualization of deformation in the following samples:

    (a) Tensile testing of a single nanofiber (5 nm in diameter and 500 nm in length).
    (b) Tensile testing of a nanocomposite rectangular bar (5 cm × 50 cm), consisting of graphene dispersed in epoxy polymer.
    (c) Compression of a microparticle (50 μm in diameter) deposited on a substrate.

2. Which of the following microscopes can resolve dislocation motion during plastic deformation?

    (a) Optical microscope.
    (b) Nomarski interference microscope.
    (c) Scanning electron microscope.
    (d) Transmission electron microscope.
    (e) Both b and d.

3. Which of the following statements is incorrect regarding in-situ mechanical characterization of materials?

    (a) In-situ characterization provides information about deformation mechanisms in real time.
    (b) In-situ tests may require more sophisticated sample preparation than conventional testing.

18 1 In-Situ Mechanics: Introduction and Importance

(c) In-situ characterization focuses only on real-time observation, without any load–displacement data for the tests.
(d) In-situ characterization can be performed as a function of temperature.

4. Identify any low-dimension nanomaterial not mentioned in this chapter and read an article on in-situ mechanical characterization of that material. Based on your reading, prepare a summary with the following sections:

   (a) Tools used (microscope and mechanical tester).
   (b) Characterization technique adopted (tension/compression/tribology/bending/indentation etc.)
   (c) Sample preparation and mounting on the mechanical test stage.
   (d) Major challenges or drawbacks of the testing technique (if any).
   (e) Your own critical assessment if any modifications should be made for improving the test.

5. Please see the in-situ snapshots/load–displacement curves in the following four test scenarios (A-D). Which of these in-situ tests can be effectively conducted and observed in real time using an optical camera/microscope?

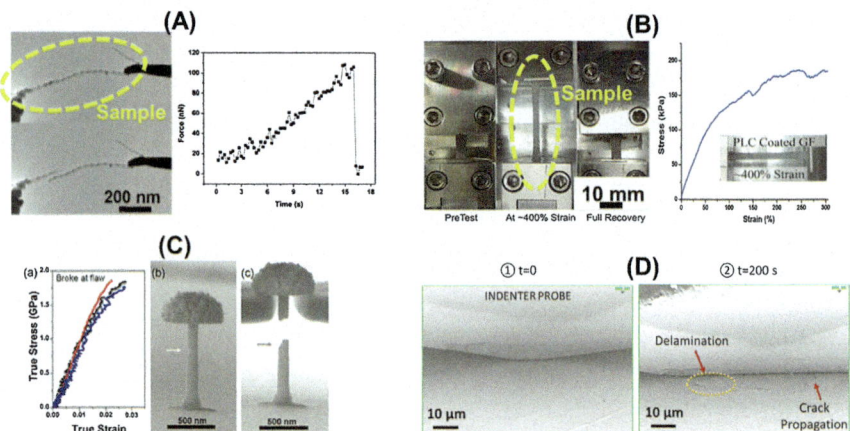

Images are reproduced/adapted with permissions from (Tang et al. 2011), (Nieto et al. 2015b), (Gu et al. 2013), (Nautiyal et al. 2019a)

6. The plot in Fig. 1.3b in Sect. 1.2.1 shows that microhardness of Al crystal is changing with the load. The hardness seems to be shooting up at ultralow loads. What could be the possible reasons for the variation of hardness (as a function of load) for the same material?

7. Figure 1.5a in Sect. 1.2.3 shows the design of a custom-built straining stage for tensile testing of metals inside TEM. Identify and read one relevant article on

the design and construction of an in-situ TEM tensile stage. Describe the **working principle** of the stage developed in the study for capturing the load-displacement response during the tensile test.

8. Why small-scale (or small-volume) testing of materials is important? List 5 reasons and elaborate with examples.

    Note: Please identify examples from real-world applications based on your daily observations.

9. *Case study question*: A team of materials and biomedical researchers embedded a titanium implant inside the femoral bone of a rat for accelerated formation of new bone. The bone was retrieved after 1 month (the picture shown below). The team now wants to measure the mechanical properties of the new grown bone (around the implantation site) using in-situ indentation inside SEM to observe the deformation behavior in real time. What could be the challenges associated with in-situ testing of the sample shown below "inside SEM"? What are your recommendations to mitigate those challenges? Suggest the sample preparation steps necessary for in-situ SEM imaging of this sample.

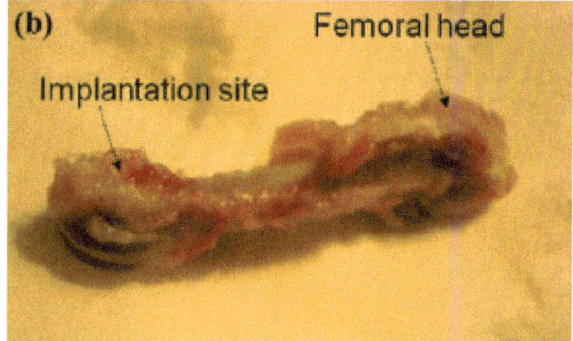

Reproduced with permission from (Facca et al. 2011)

10. *Application Question*: Structural integrity considerations are important during electrode material selection for lithium-ion batteries. Lithiation-delithiation processes cause changes in elastic modulus, failure strain and strength of the material. In a study published in Nano Energy (Song et al., Vol 53, p. 277), the researchers examined the tensile properties of SnO_2 nanowires (anode material) by in-situ SEM tests. The testing was performed on pristine, lithiated and delithiated SnO_2 samples. Go through the in-situ SEM videos for these three sample conditions and identify the plastic deformation/failure mechanisms. Compare the mechanisms for the three cases. The videos can be accessed through Nano Energy (Elsevier) website.

Link: https://www.sciencedirect.com/science/article/pii/S2211285518306189

References

Agrawal R, Nieto A, Chen H et al (2013) Nanoscale damping characteristics of boron nitride nanotubes and carbon nanotubes reinforced polymer composites. ACS Appl Mater Interfaces 5(22):12052–12057

Antillon M, Nautiyal P, Loganathan A et al (2018) Strengthening in boron nitride nanotube reinforced aluminum composites prepared by roll bonding. Adv Eng Mater 20:1–8. https://doi.org/10.1002/adem.201800122

Bakshi SR, Lahiri D, Patel RR, Agarwal A (2010) Nanoscratch behavior of carbon nanotube reinforced aluminum coatings. Thin Solid Films 518:1703–1711. https://doi.org/10.1016/j.tsf.2009.11.079

Balani K, Brito FC, Kos L, Agarwal A (2009a) Melanocyte pigmentation stiffens murine cardiac tricuspid valve leaflet. J R Soc Interface 6:1097–1102. https://doi.org/10.1098/rsif.2009.0174

Balani K, Lahiri D, Keshri AK et al (2009b) The nano-scratch behavior of biocompatible hydroxyapatite reinforced with aluminum oxide and carbon nanotubes. JOM 61:63–66

Bangert H, Wagendristel A (1985) Ultralow-load hardness tester for use in a scanning electron microscope. Rev Sci Instrum 56:1568–1572. https://doi.org/10.1063/1.1138154

Boesl B, Lahiri D, Behdad S, Agarwal A (2014) Direct observation of carbon nanotube induced strengthening in aluminum composite via in situ tensile tests. Carbon N Y 69:79–85. https://doi.org/10.1016/j.carbon.2013.11.061

Bustillos J, Montero D, Nautiyal P et al (2018) Integration of graphene in poly(lactic) acid by 3D printing to develop creep and wear-resistant hierarchical nanocomposites. Polym Compos 39:3877–3888. https://doi.org/10.1002/pc.24422

Bykowski M, Hudgins A, Deacon RM, Marder AR (2006) Failure analysis of the space shuttle Columbia RCC leading edge. J Fail Anal Prev 6:39–45. https://doi.org/10.1361/154770206X86509

Chandrasekaran M, Loh NL (2001) Effect of counterface on the tribology of UHMWPE in the presence of proteins. Wear 250:237–241. https://doi.org/10.1016/S0043-1648(01)00646-9

Chang S (2018) In-situ nanomechanical testing in electron microscopes. In: Hsueh C-H, Schmauder S, Chen C-S et al (eds) Handbook of mechanics of materials. Springer, Singapore, pp 1–47

Chen Y, Bakshi SR, Agarwal A (2009) Intersplat friction force and splat sliding in a plasma-sprayed aluminum alloy coating. ACS Appl Mater Interfaces 1(2):235–238. https://doi.org/10.1021/am800114h

Chen Q, Mangadlao JD, Wallat J et al (2017) 3D printing biocompatible polyurethane/poly(lactic acid)/graphene oxide nanocomposites: anisotropic properties. ACS Appl Mater Interfaces 9(4):4015–4023. https://doi.org/10.1021/acsami.6b11793

Choi SR, Bansal NP, Garg A (2007) Mechanical and microstructural characterization of boron nitride nanotubes-reinforced SOFC seal glass composite. Mater Sci Eng A 461:509–515. https://doi.org/10.1016/j.msea.2007.01.084

Deneen Nowak J, Mook WM, Minor AM et al (2007) Fracturing a nanoparticle. Philos Mag 87:29–37. https://doi.org/10.1080/14786430600876585

Dubnikova IL, Oshmyan VG, Gorenberg AY (1997) Mechanisms of participate filled polypropylene finite plastic deformation and fracture. J Mater Sci 32:1613–1622. https://doi.org/10.1023/A:1018547226983

Embrey L, Nautiyal P, Loganathan A et al (2017) Three-dimensional graphene foam induces multifunctionality in epoxy nanocomposites by simultaneous improvement in mechanical, thermal, and electrical properties. ACS Appl Mater Interfaces 9:39717–39727. https://doi.org/10.1021/acsami.7b14078

Espinosa HD, Bernal RA, Filleter T (2012) In situ TEM electromechanical testing of nanowires and nanotubes. Small 8:3233–3252. https://doi.org/10.1002/smll.201200342

Facca S, Lahiri D, Fioretti F et al (2011) In vivo osseointegration of nano-designed composite coatings on titanium implants. ACS Nano 5:4790–4799. https://doi.org/10.1021/nn200768c

Felkins K, Leigh HP, Jankovic A (1998) The royal mail ship Titanic: did a metallurgical failure cause a night to remember? JOM 50:12–18. https://doi.org/10.1007/s11837-998-0062-7

Fischer-Cripps AC (2004) A simple phenomenological approach to nanoindentation creep. Mater Sci Eng A 385:74–82. https://doi.org/10.1016/j.msea.2004.04.070

Fisher-Cripps A (2005) Nanoindentation, 3rd edn. Springer, New York

Foecke TJ (1998) Metallurgy of the RMS Titanic. NIST Interagency/Internal Rep - 6118. https://doi.org/10.6028/NIST.IR.6118

Fontoura L, Nautiyal P, Loganathan A et al (2018) Nacre-inspired graphene/metal hybrid by in situ cementation reaction and joule heating. Adv Eng Mater 20:1–8. https://doi.org/10.1002/adem.201800518

Gane N, Bowden FP (1968) Microdeformation of solids. J Appl Phys 39:1432–1435. https://doi.org/10.1063/1.1656376

Gao L, Song J, Jiao Z et al (2018) High-entropy alloy (HEA) -coated nanolattice structures and their mechanical properties. Adv Eng Mater 1700625:1–8. https://doi.org/10.1002/adem.201700625

Greer JR, Th J, De Hosson M (2011) Progress in materials science plasticity in small-sized metallic systems: intrinsic versus extrinsic size effect. Prog Mater Sci 56:654–724. https://doi.org/10.1016/j.pmatsci.2011.01.005

Gu XW, Wu Z, Zhang Y-W et al (2013) Microstructure versus flaw: mechanisms of failure and strength in nanostructures. Nano Lett 13:5703–5709. https://doi.org/10.1021/nl403453h

Hayes M, Edwards DSA (2015) Fractography in failure analysis of polymers. Elsevier, Amsterdam

Hirsch PB (1986) Direct observations of moving dislocations: reflections on the 30th anniversary of the 1st recorded observations of moving dislocations by transmission electron-microscopy. Mater Sci Eng 84:1–10

Hirsch PB, Horne RW, Whelan MJ (1956) LXVIII. Direct observations of the arrangement and motion of dislocations in aluminium. Philos Mag 1:677–684

Höche H-R, Schreiber J (1984) Anisotropic deformation behaviour of GaAs. Phys Status Solidi 86:229–236. https://doi.org/10.1002/pssa.2210860124

Huang S-C, Hall EL (1991) Plastic deformation and fracture of binary TiAl-base alloys. Metall Trans A 22:427–439. https://doi.org/10.1007/BF02656810

Hull D (1999) Fractography: observing, measuring and interpreting fracture surface topography. Cambridge University Press, Cambridge

Idowu A, Nautiyal P, Fontoura L et al (2018) Multi-scale damping of graphene foam-based polyurethane composites synthesized by electrostatic spraying. Polym Compos 40:E1862–E1870. https://doi.org/10.1002/pc.25178

Issa I, Amodeo J, Réthoré J et al (2015) In situ investigation of MgO nanocube deformation at room temperature. Acta Mater 86:295–304. https://doi.org/10.1016/j.actamat.2014.12.001

Jang D, Meza LR, Greer F, Greer JR (2013) Fabrication and deformation of three-dimensional hollow ceramic nanostructures. Nat Mater 12:893–898. https://doi.org/10.1038/nmat3738

Jiang C, Lu H, Zhang H et al (2017) Recent advances on in situ SEM mechanical and electrical characterization of low-dimensional nanomaterials. Scanning 2017:1985149. https://doi.org/10.1155/2017/1985149

Klimanek P, Pötzsch A (2002) Microstructure evolution under compressive plastic deformation of magnesium at different temperatures and strain rates. Mater Sci Eng A 324:145–150. https://doi.org/10.1016/S0921-5093(01)01297-7

Könönen MHO, Lavonius ET, Kivilahti JK (1995) SEM observations on stress corrosion cracking of commercially pure titanium in a topical fluoride solution. Dent Mater 11:269–272. https://doi.org/10.1016/0109-5641(95)80061-1

Lahiri D, Chen Y (2012) Insight into reactions and interface between boron nitride nanotube and aluminum. J Mater Res 27(21):2760–2770. https://doi.org/10.1557/jmr.2012.294

Lahiri D, Rouzaud F, Richard T et al (2010) Boron nitride nanotube reinforced polylactide – polycaprolactone copolymer composite: mechanical properties and cytocompatibility with osteoblasts and macrophages in vitro. Acta Biomater 6:3524–3533. https://doi.org/10.1016/j.actbio.2010.02.044

Lahiri D, Benaduce AP, Kos L, Agarwal A (2011) Quantification of carbon nanotube induced adhesion of osteoblast on hydroxyapatite using nano-scratch technique. Nanotechnology 22:355703. https://doi.org/10.1088/0957-4484/22/35/355703

Lahiri D, Hadjikhani A, Zhang C et al (2013) Boron nitride nanotubes reinforced aluminum composites prepared by spark plasma sintering: microstructure, mechanical properties and deformation behavior. Mater Sci Eng A 574:149–156. https://doi.org/10.1016/j.msea.2013.03.022

Lahiri D, Karp J, Keshri AK et al (2015) Scratch induced deformation behavior of hafnium based bulk metallic glass at multiple load scales. J Non Cryst Solids 410:118–126. https://doi.org/10.1016/j.jnoncrysol.2014.12.010

Lee DH, Seok MY, Zhao Y et al (2016) Spherical nanoindentation creep behavior of nanocrystalline and coarse-grained CoCrFeMnNi high-entropy alloys. Acta Mater 109:314–322. https://doi.org/10.1016/j.actamat.2016.02.049

Legros M (2014) In situ mechanical TEM: seeing and measuring under stress with electrons. Comptes Rendus Phys 15:224–240. https://doi.org/10.1016/j.crhy.2014.02.002

Liu F, Tang D-M, Gan H et al (2013) Individual boron nanowire has ultra-high specific young's modulus and fracture strength as revealed by in situ transmission electron microscopy. ACS Nano 7:10112–10120. https://doi.org/10.1021/nn404316a

Loganathan A, Sharma A, Rudolf C et al (2017) In-situ deformation mechanism and orientation effects in sintered 2D boron nitride nanosheets. Mater Sci Eng A 708:440–450. https://doi.org/10.1016/j.msea.2017.10.019

Louchet F, Kubin LP, Vesely D (1979) In situ deformation of b.c.c. crystals at low temperatures in a high-voltage electron microscope dislocation mechanisms and strain-rate equation. Philos Mag A 39:433–454. https://doi.org/10.1080/01418617908239283

Mindess S, Diamond S (1980) A preliminary SEM study of crack propagation in mortar. Cem Concr Res 10:509–519

Mulvany MJ, Halpern W (1976) Mechanical properties of vascular smooth muscle cells in situ. Nature 260:617–619

Nautiyal P, Jain J, Agarwal A (2015) A comparative study of indentation induced creep in pure magnesium and AZ61 alloy. Mater Sci Eng A 630:131–138. https://doi.org/10.1016/j.msea.2015.01.075

Nautiyal P, Jain J, Agarwal A (2016a) Influence of microstructure on scratch-induced deformation mechanisms in AZ80 magnesium alloy. Tribol Lett 61:1–7. https://doi.org/10.1007/s11249-016-0649-z

Nautiyal P, Jain J, Agarwal A (2016b) Influence of loading path and precipitates on indentation creep behavior of wrought Mg-6wt% Al-1wt% Zn magnesium alloy. Mater Sci Eng A 650:183–189. https://doi.org/10.1016/j.msea.2015.10.040

Nautiyal P, Rudolf C, Loganathan A et al (2016c) Directionally aligned ultra-long boron nitride nanotube induced strengthening of aluminum-based Sandwich composite. Adv Eng Mater 18:1747–1754. https://doi.org/10.1002/adem.201600212

Nautiyal P, Boesl B, Agarwal A (2017a) Harnessing three dimensional anatomy of graphene foam to induce superior damping in hierarchical polyimide nanostructures. Small 13:1–8. https://doi.org/10.1002/smll.201603473

Nautiyal P, Embrey L, Boesl B, Agarwal A (2017b) Multi-scale mechanics and electrical transport in a free-standing 3D architecture of graphene and carbon nanotubes fabricated by pressure assisted welding. Carbon N Y 122:298–306. https://doi.org/10.1016/j.carbon.2017.06.081

Nautiyal P, Gupta A, Seal S et al (2017c) Reactive wetting and filling of boron nitride nanotubes by molten aluminum during equilibrium solidification. Acta Mater 126:124–131. https://doi.org/10.1016/j.actamat.2016.12.034

Nautiyal P, Boesl B, Agarwal A (2018a) The mechanics of energy dissipation in a three-dimensional graphene foam with macroporous architecture. Carbon N Y 132:59–64. https://doi.org/10.1016/j.carbon.2018.02.028

References

Nautiyal P, Mujawar M, Boesl B, Agarwal A (2018b) In-situ mechanics of 3D graphene foam based ultra-stiff and flexible metallic metamaterial. Carbon N Y 137:502–510. https://doi.org/10.1016/j.carbon.2018.05.063

Nautiyal P, Zhang C, Boesl B, Agarwal A (2018c) Non-equilibrium wetting and capture of boron nitride nanotubes in molten aluminum during plasma spray. Scr Mater 151:71–75. https://doi.org/10.1016/j.scriptamat.2018.03.037

Nautiyal P, Zhang C, Champagne VK et al (2018d) In-situ mechanical investigation of the deformation of splat interfaces in cold-sprayed aluminum alloy. Mater Sci Eng A 737:297–309. https://doi.org/10.1016/j.msea.2018.09.065

Nautiyal P, Zhang C, Champagne V et al (2019a) In-situ creep deformation of cold-sprayed aluminum splats at elevated temperatures. Surf Coat Technol 372:353–360. https://doi.org/10.1016/j.surfcoat.2019.05.045

Nautiyal P, Zhang C, Loganathan A et al (2019b) High-temperature mechanics of boron nitride nanotube "Buckypaper" for engineering advanced structural materials. ACS Appl Nano Mater 2:4402–4416. https://doi.org/10.1021/acsanm.9b00817

Nieto A, Boesl B, Agarwal A (2015a) Multi-scale intrinsic deformation mechanisms of 3D graphene foam. Carbon N Y 85:299–308. https://doi.org/10.1016/j.carbon.2015.01.003

Nieto A, Dua R, Zhang C et al (2015b) Three dimensional graphene foam/polymer hybrid as a high strength biocompatible scaffold. Adv Funct Mater 25:3916–3924. https://doi.org/10.1002/adfm.201500876

Nili H, Kalantar-zadeh K, Bhaskaran M, Sriram S (2013) In situ nanoindentation: probing nanoscale multifunctionality. Prog Mater Sci 58:1–29. https://doi.org/10.1016/j.pmatsci.2012.08.001

Ohr SM (1985) An electron microscope study of crack tip deformation and its impact on the dislocation theory of fracture. Mater Sci Eng 72:1–35

Oyen ML (2007) Sensitivity of polymer nanoindentation creep measurements to experimental variables. Acta Mater 55:3633–3639. https://doi.org/10.1016/j.actamat.2006.12.031

Pan D, Wang C, Wang T, Yao Y (2017) Graphene foam: uniaxial tension behavior and fracture mode based on a mesoscopic model. ACS Nano 11:8988–8997. https://doi.org/10.1021/acsnano.7b03474

Pitchuka SB, Lahiri D, Sundararajan G, Agarwal A (2014) Scratch-induced deformation behavior of cold-sprayed aluminum amorphous/nanocrystalline coatings at multiple load scales. J Ther Spray Technol 23:502–513. https://doi.org/10.1007/s11666-013-0021-x

Rabe R, Breguet J-M, Schwaller P et al (2004) Observation of fracture and plastic deformation during indentation and scratching inside the scanning electron microscope. Thin Solid Films 469–470:206–213. https://doi.org/10.1016/j.tsf.2004.08.096

Rudolf C, Boesl B, Agarwal A (2015) In situ indentation behavior of bulk multi-layer graphene flakes with respect to orientation. Carbon N Y 94:872–878. https://doi.org/10.1016/j.carbon.2015.07.070

Rudolf C, Boesl B, Agarwal A (2016a) In situ mechanical testing techniques for real-time materials deformation characterization. JOM 68:136–142. https://doi.org/10.1007/s11837-015-1629-8

Rudolf CC, Agarwal A, Boesl B (2016b) TaC – NbC formed by spark plasma sintering with the addition of sintering additives. J Ceram Soc Japan 124(4):381–387

Rudolf CC, Eranezhuth B, Boesl B, Agarwal A (2017) Journal of the European Ceramic Society (Ta, Nb) C composites formed with graphene nanoplatelets by spark plasma sintering. J Eur Ceram Soc 37:3781–3790. https://doi.org/10.1016/j.jeurceramsoc.2016.11.017

Sernicola G, Giovannini T, Patel P et al (2017) In situ stable crack growth at the micron scale. Nat Commun 8:108. https://doi.org/10.1038/s41467-017-00139-w

Straffelini G (2015) Friction and wear methodologies for design and control. Springer International Publishing, Basel

Sumino K, Sato M (1979) In-situ HVEM observations of dislocation processes during high temperature deformation of silicon crystals. Krist und Tech 14:1343–1350. https://doi.org/10.1002/crat.19790141111

Syed Asif SA, Pethica JB (1997) Nanoindentation creep of single-crystal tungsten and gallium arsenide. Philos Mag A 76:1105–1118. https://doi.org/10.1080/01418619708214217

Takeuchi T (1973) Load-elongation curves of pure body-centred cubic metals at low temperatures. J Physical Soc Japan 35:1

Tang D-M, Ren C-L, Wei X et al (2011) Mechanical properties of bamboo-like boron nitride nanotubes by in situ TEM and MD simulations: Strengthening effect of interlocked joint interfaces. ACS Nano 5:7362–7368. https://doi.org/10.1021/nn202283a

Thomas T, Zhang C, Sahu A et al (2018) Effect of graphene reinforcement on the mechanical properties of Ti2AlC ceramic fabricated by spark plasma sintering. Mater Sci Eng A 728:45–53. https://doi.org/10.1016/j.msea.2018.05.006

Thomas T, Zhang C, Nautiyal P et al (2019) 3D graphene foam reinforced low-temperature ceramic with multifunctional mechanical, electrical, and thermal properties. Adv Eng Mater 1900085:1–9. https://doi.org/10.1002/adem.201900085

Wang CL, Lai YH, Huang JC, Nieh TG (2010) Creep of nanocrystalline nickel: a direct comparison between uniaxial and nanoindentation creep. Scr Mater 62:175–178. https://doi.org/10.1016/j.scriptamat.2009.10.021

Wheeler JM, Michler J (2013) Elevated temperature, nano-mechanical testing in situ in the scanning electron microscope. Rev Sci Instrum 84:045103. https://doi.org/10.1063/1.4795829

Wheeler JM, Armstrong DEJ, Heinz W, Schwaiger R (2015) High temperature nanoindentation: the state of the art and future challenges. Curr Opin Solid State Mater Sci 19:354–366. https://doi.org/10.1016/j.cossms.2015.02.002

Wu W, Qi D, Liao H et al (2018) Deformation mechanism of innovative 3D chiral metamaterials. Sci Rep 8:1–11. https://doi.org/10.1038/s41598-018-30737-7

Yue C, Liu W, Zhang L, Chen Y (2013) Fracture toughness and toughening mechanisms in a (ZrB2 – SiC) composite reinforced with boron nitride nanotubes and boron nitride nanoplatelets. Scr Mater 68:579–582. https://doi.org/10.1016/j.scriptamat.2012.12.005

Zhang P, Ma L, Fan F et al (2014) Fracture toughness of graphene. Nat Commun 5:3782. https://doi.org/10.1038/ncomms4782

Zhang C, Boesl B, Silvestroni L et al (2016) Deformation mechanism in graphene nanoplatelet reinforced tantalum carbide using high load in situ indentation. Mater Sci Eng A 674:270–275. https://doi.org/10.1016/j.msea.2016.07.110

Zhang C, Gupta A, Seal S et al (2017a) Solid solution synthesis of tantalum carbide-hafnium carbide by spark plasma sintering. J Am Ceram Soc 100:1853. https://doi.org/10.1111/jace.14778

Zhang Q, Lin D, Deng B et al (2017b) Flyweight, superelastic, electrically conductive, and flame-retardant 3D multi-nanolayer graphene/ceramic metamaterial. Adv Mater 29:1605506. https://doi.org/10.1002/adma.201605506

Chapter 2
In-Situ Mechanics: Experimental Tools and Techniques

This chapter introduces the tools and techniques useful for in-situ mechanical characterization of materials at multiple length scales. The chapter begins with an introduction to focused ion beam machining for sample preparation. Steps and strategies to fabricate miniature samples using ion beam milling are presented, along with some precautions and limitations that need to be taken into consideration. Thereafter, imaging tools for in-situ observation at multiples length scales are discussed, such as optical, scanning electron, and transmission electron microscopes. Some specific techniques under each class are also presented, such as high-speed camera, optical tweezer, electron backscatter diffraction, and dynamic transmission electron microscope. The final segment of the chapter covers mechanical instrumentation for studying deformation at different length scales, such as atomic force microscope inside electron microscopes, nanoindenters, microelectromechanical systems, micro-meso-mechanical stages/testers, and macroscale bulk sample testers. These instrumentation cover a wide range of loads and displacements, specimen sizes, and nature of loading. In addition to the introduction and application of different techniques, the chapter also discusses useful fundamental principles to provide the users a deeper understanding of how these techniques work, and what are some key scientific/technological considerations one needs to be aware of. The applicability of the tools and techniques is elucidated by relevant case studies. A schematic summarizing the content of this chapter is shown in Fig. 2.1. For clear demarcation and logical flow, the sections of the chapter are divided into three parts: Sect. 2.1 on sample preparation, Sect. 2.2 on in-situ imaging tools, and Sect 2.3 on in-situ mechanical instrumentation. The division of the chapter into three parts is necessitated by a wide diversity of tools required for the in-situ mechanical characterization of materials, and we believe this categorization will provide clarity to the readers.

Electronic supplementary material The online version of this chapter (https://doi.org/10.1007/978-3-030-43320-8_2) contains supplementary material, which is available to authorized users.

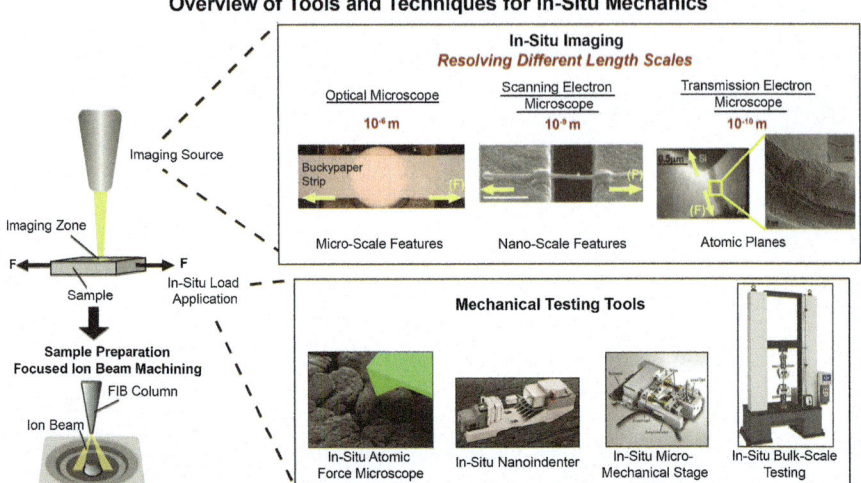

Fig. 2.1 Schematic summarizing the key tools and techniques used for the in-situ characterization of materials. (Some of the images are reproduced/adapted with permissions from Espinosa et al. (2012), Golberg et al. (2007), Nautiyal et al. (2019), Agarwal Group (FIU). Web sources: Bruker (Hysitron Nanomechanical Test Instruments), Qualitest (Universal Testing Machine—Tensile Tester), UL testing facilities, Germany (High-Speed Tensile Testing))

2.1 Miniature Sample Preparation

2.1.1 Focused Ion Beam for Machining Specimens

Focused ion beam machining (FIB) is a powerful approach to mill, image, and deposit materials inside a vacuum chamber. In FIB, the sample/region of interest is bombarded with an ion beam. The nature of interactions between the ions and the sample material determines the action during FIB. In the event the surface particles in the sample receive sufficient energy to overcome the binding energy, they are ejected from the surface (milling process). This phenomenon is called sputtering. FIB milling is a commonly used process for preparing specimens for in-situ testing. These could be standalone specimens, such as electron-transparent films for in-situ testing inside TEM. Another class of specimens could be specific geometries for mechanical testing in SEM. For instance, micro-/nanopillars (compression test), microbeams (bending or flexural test), or dogbone-shaped structure (miniature tensile test). The examples of different sample types fabricated by FIB for in-situ testing are shown in Fig. 2.2 (Uchic and Dimiduk 2005; Kiener et al. 2008; Zeng et al. 2017; Rafael Velayarce et al. 2018).

There are some FIB-milling artifacts that should be taken into consideration during sample preparation. The sputtered material (due to ion bombardment) can deposit on the surfaces in near vicinity causing damage or unintended shape transformations. For instance, it is a common observation that milling of a hole by FIB

2.1 Miniature Sample Preparation

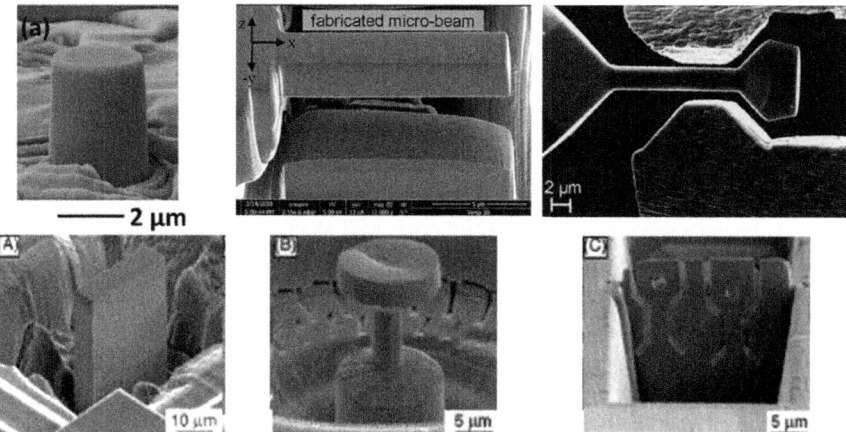

Fig. 2.2 Miniature samples with different geometries fabricated by FIB. (Reproduced/adapted with permissions from Zeng et al. (2017), Rafael Velayarce et al. (2018), Kiener et al. (2008), Uchic and Dimiduk (2005))

creates tapered or V-shapes due to *redeposition of sputtered material*. This phenomenon becomes prominent for high aspect ratio structures. This is illustrated in Fig. 2.3a, where three successive trenches with increasing aspect ratio show an increasing trend in the degree of taper. This is because the escape of sputtered material is restricted during deep subsurface milling. Therefore, as the sputtering yield of the material is increased, the likelihood of redeposition becomes higher. It is for this reason that redeposition artifacts due to milling increase with high beam currents. Beam current is a measure of the rate of delivery of ions to the sample surface and should be carefully controlled during sample preparation. It has also been studied that sputtering yield is dependent on the angle of ion incidence. This is shown in Fig. 2.3b for Si and Cu (Giannuzzi et al. 2005). The maxima (for sputtering yield) varies for different materials. Typically, angles of incidence around ~75–80° result in high sputtering yields and are worst from the standpoint of redeposition. Thereafter, the yield drops to very low values with a further increase in angle. Hence, glancing angles (close to ~90°) are preferred for sample preparation by FIB. The issue of redeposition is crucial for preparing mechanical testing specimens, such as micropillars for compression testing. The high aspect ratio pillars exhibit tapering, which affects stress distribution/concentration during mechanical loading. It should be noted from Fig. 2.3b that sputtering yield is material-dependent (Giannuzzi et al. 2005). For instance, the yield for Cu (dotted lines) is consistently higher than Si for all the angles. Therefore, sample preparation for some materials is more challenging and requires additional precautions than others. Another concern with FIB milling is damage caused due to ion irradiation. The implantation of ions bombarded during milling can produce *surface amorphous phase* in an otherwise crystalline sample. Materials have varying degrees of sensitivity to amorphization. For instance, crystalline metals and alloys generally resist amorphization.

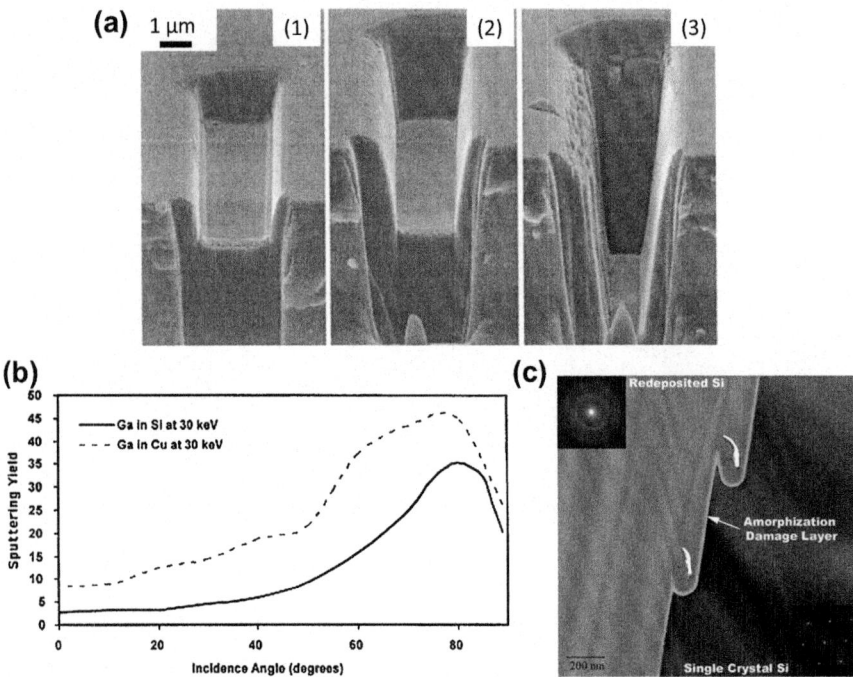

Fig. 2.3 (**a**) Tapering phenomena during FIB milling of high aspect ratio geometries due to redeposition, (**b**) dependence of sputtering yield on ion incidence angle and material type, and (**c**) TEM microanalysis showing amorphization of crystalline Si due to ion beam bombardment. (Reproduced/adapted with permissions from Giannuzzi et al. (2005), Matteson et al. (2002))

However, Si is known to experience amorphization during ion beam milling (shown in Fig. 2.3c) (Matteson et al. 2002).

FIB is an effective sample preparation technique for nano- and microscale testing of materials, since bulk-scale machining techniques cannot be used for preparing the test specimens. For fabricating pillar specimens, concentric annular patterns are milled in the area of interest (Fig. 2.4a). The large cavity allows imaging of pillar sidewalls during deformation and reduces the probability of redeposition of sputtered material during FIB machining. Typically, high beam currents (>1 nA) are used, with sample oriented normal to the FIB column for this step. Subsequently, lower beam currents (several hundreds of pA) are used for fine machining of the pillar to obtain final desired dimensions and uniform cross section. This is done at relatively lower tilt angles between the FIB column and the plane of the sample, such that the milling is performed almost parallel to the sample surface. Figure 2.4b shows the milling of "sub-micrometer" wide dogbone-shaped tension samples from single crystals (Kim and Greer 2009). A two-step process is employed for making tension samples: (a) a thin vertical lamella is defined by ion beam patterning, followed by milling of the two sides of the lamella, and (b) shoulders are milled from the top of the lamella for the purpose of gripping during the tensile test. It has been

2.1 Miniature Sample Preparation

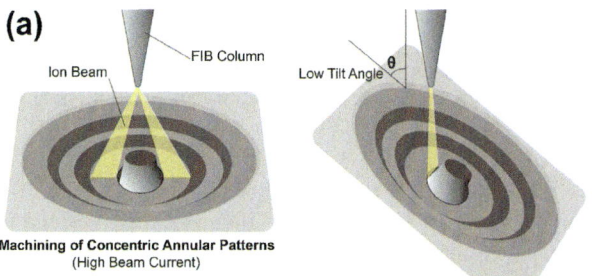

Machining of Pillar

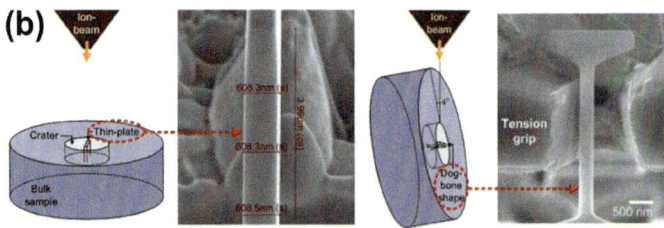

Machining of Dogbone

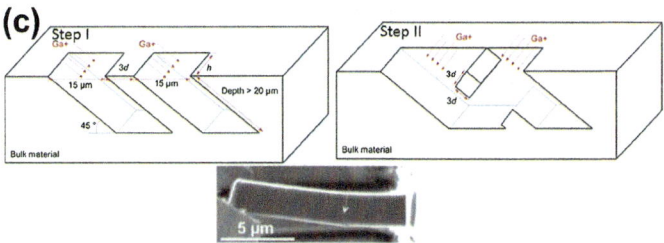

Machining of Cantilever

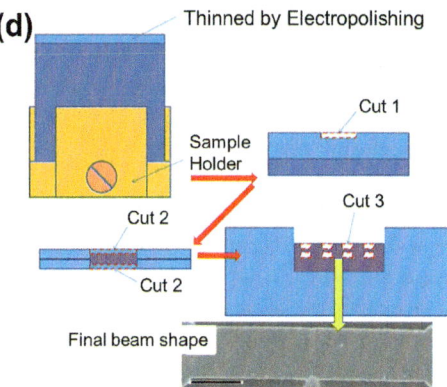

Machining of Beam for 3-Point Bending

(continued)

Fig. 2.4 (**a**) Milling of concentric annular patterns for fabricating micro- and nanopillar specimens, (**b**) machining steps for fabricating dogbone tensile samples, (**c**) schematic representation of milling of trenches to obtain cantilever samples, and (**d**) milling steps to fabricate ultrathin beam specimen for the 3-point bending test. (Reproduced/adapted with permissions from Kim and Greer (2009), Liu and Flewitt (2017), Hintsala et al. (2015))

demonstrated that tilting the sample stage during etching either sides of the lamella controls tapering in the samples. For instance, Kim and Greer used the tilt angle of ±0.6° to minimize taper in the lamella. The widths marked in Fig. 2.4b show this technique is effective to obtain reasonably uniform cross sections. FIB is also used for fabricating miniature cantilever beam specimens. The process involves milling of trenches at a 45° angle to the sample surface. Two trenches are milled in two steps, such that the beam is made incident on the sample at +45° and −45°. The process is illustrated in Fig. 2.4c (Liu and Flewitt 2017). During milling, the section that separates the two trenches is usually kept larger than the final cantilever dimension. The final dimension of the cantilever is then achieved by using much smaller beam currents to minimize Ga^+ ion-induced damage. FIB milling can be employed for preparing ultrathin films or beams for 3-point bending tests. The sample preparation strategy involves milling to produce thin lamellae. The final beam shape is then obtained by removing the material above and below the beam region. The milling steps and final product are shown in Fig. 2.4d (Hintsala et al. 2015).

FIB is also used to mill pre-notches and pre-cracks in mechanical testing specimens. Due to excellent accuracy and control in ion milling (when using low beam currents), structural flaws with desired dimensions can be introduced at specific locations in the samples. Figure 2.5a illustrates the in-situ bending of a pre-notched cantilever beam inside SEM (Cao et al. 2016). A pre-crack notch, which is ~40% of the beam thickness (1.2 μm deep) was introduced by FIB milling. The ability to machine notches with precise depth is useful for fracture toughness calculations. FIB has also been used to introduce pre-cracks in 2D graphene for fracture toughness calculations by in-situ tensile testing (Fig. 2.5b). Initial crack lengths in graphene were varied from ~33 nm to over 1 μm, and fracture toughness evaluated using the Griffith theory of brittle fracture (Zhang et al. 2014).

In addition to sample preparation, FIB is also a useful technique to machine grips in desired shape and size for different kinds of mechanical tests. For instance, a tension/compression grip fabricated by FIB is shown in Fig. 2.6a (Kim and Greer 2009). The grip consists of a through-hole for grasping the dogbone samples from the collar. The grip shown also has a flat bottom for compression testing. The grip shown in Fig. 2.6a is made of diamond, one of the common materials used for fabricating mechanical testing tips and probes. FIB can also be used to machine desired grips from polycrystalline tungsten needles, a typical material used in micromanipulators. Grips with custom sizes can be fabricated by ion milling for testing a wide range of sample sizes (sub-micrometer range to several hundreds of micrometers). The process of alignment and gripping of the sample is illustrated through the

2.1 Miniature Sample Preparation 31

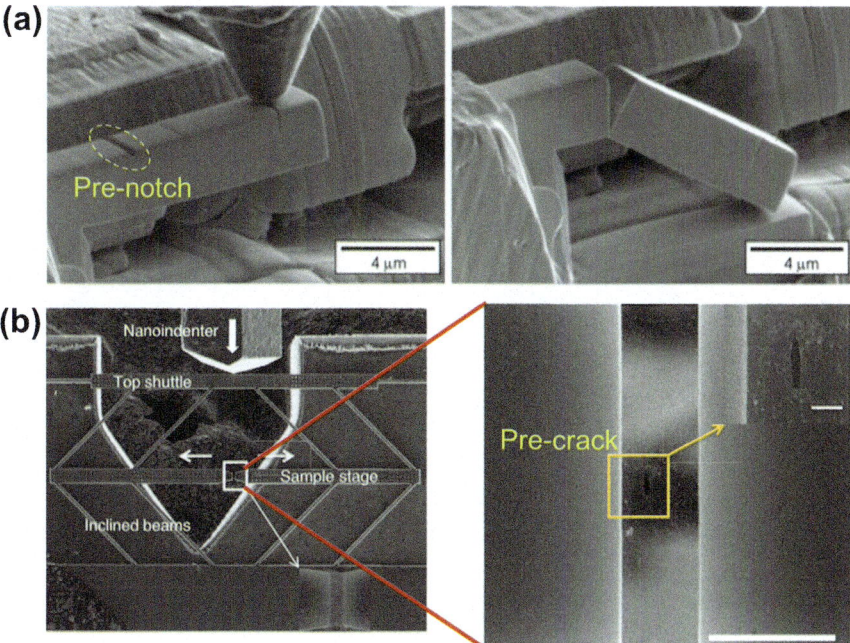

Fig. 2.5 Introduction of structural flaws in the samples by ion milling for in-situ testing: (**a**) pre-notch in a cantilever beam, and (**b**) pre-crack in a graphene sheet under tensile loading. (Reproduced/adapted with permissions from Cao et al. (2016), Zhang et al. (2014))

SEM images in Fig. 2.6b (Kiener et al. 2008). Precise alignment is critical for mechanical testing. Ability to machine fine dimensions using FIB is useful for fabricating custom-dimension and custom-design grips to ensure proper alignment for tests.

The fundamental concepts and case studies discussed in this section highlight FIB is an extremely useful tool for in-situ mechanical characterization of materials. The technique is used for preparing a wide variety of samples for in-situ mechanical testing inside electron microscopes. The application of FIB can be as simple as milling local flaws and features, or as complex as fabricating intricate geometries/structures and ultrathin electron-transparent samples. Irradiation damage and redeposition are key concerns during ion milling, which require careful process strategies to overcome or limit these issues.

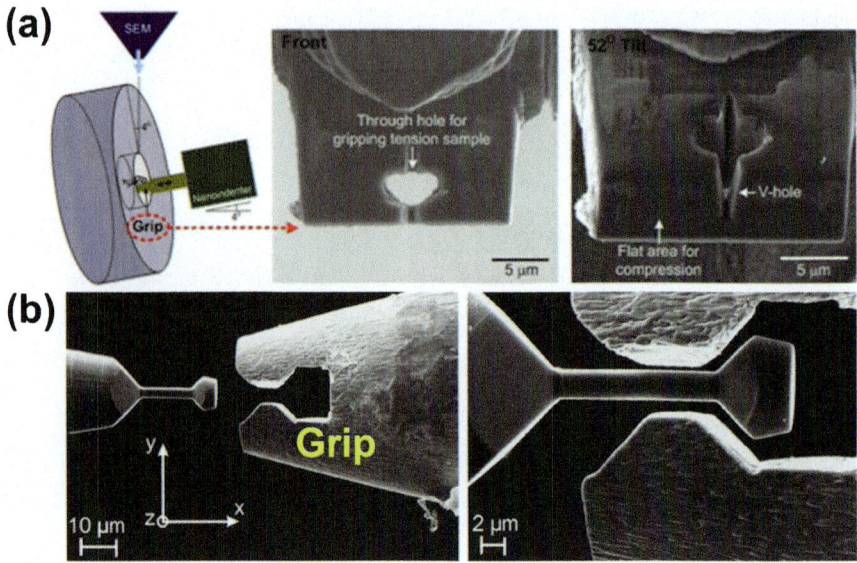

Fig. 2.6 (**a**) Machining of a tension–compression grip for in-situ testing, and (**b**) SEM micrographs showing the alignment of the sample with the grip for mechanical loading. (Reproduced/adapted with permissions from Kim and Greer (2009), Kiener et al. (2008))

2.2 Imaging Tools and Techniques for In-Situ Observation

2.2.1 Optical Techniques for In-Situ Imaging

In-situ optical microscope-based mechanical tests are conducted under white light for real-time observation of the deformation. The best achievable resolution by optical microscopes can be up to ~0.2 μm. Therefore, in-situ tests can possibly reveal phenomena at sub-micron length scales at higher magnifications. However, it should be noted that the field of view for high magnification optical imaging is limited. Field of view or field size is related to magnification by the formula:

$$\text{Field size} = \frac{\text{Eyepiece Field Number}}{\text{Objective Magnification}} \quad (2.1)$$

where eyepiece field number (F.N.) is the diameter of the view field at the intermediate image plane. From this equation, it can be seen that increasing the objective lens magnification from 10× to 100× will result in a one-tenth field of view. Therefore, the ability to resolve fine features comes at the expense of field of view. This tradeoff is of importance during in-situ testing. Resolving ultrafine features might not be feasible during the in-situ testing of large samples.

One of the applications of in-situ optical imaging is strain computation during a tensile test. The strain gage is not effective for measuring strains in slender, small-sized, fragile, or soft specimens. For such samples, tensile strains can be calculated by image correlation technique (Sutton et al. 2009; Jerabek et al. 2010; Nautiyal et al. 2016). The video of tensile deformation is captured, and strain evolution is determined by comparing the real-time snapshots with respect to the starting image of the sample (prior to loading). Figure 2.7 shows the progression of tensile deformation of a polymer sample captured under an optical camera (Snapshots 1 through 4 in the increasing order of tensile strain). During tensile tests, the samples often undergo slippage at the grips. This can result in overestimation of strains as the physical movement of crosshead is not the true indicator of how much the sample truly deforms. Image correlation is based on tracking of the specific pixel(s) as the sample deforms, allowing precise calculation of strain. The detailed application of the image correlation technique is discussed in Chap. 6.

Apart from quantitative measurements, in-situ optical microscope based tests can be useful to decipher the deformation and failure mechanisms in novel materials. Figure 2.8 illustrates the deformation behavior of a polymer composite, reinforced with a cellular graphene foam (Embrey et al. 2017). Failure mechanisms can be correlated with microstructure features (resolvable by optical microscope). In

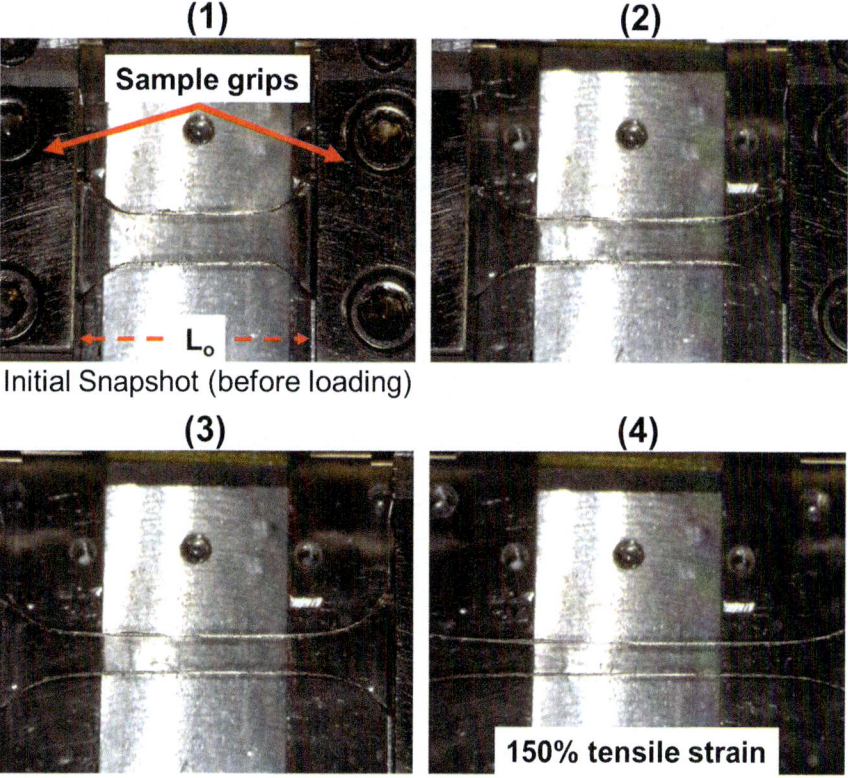

Fig. 2.7 Snapshots from the in-situ optical camera during the tensile test of highly stretchable PDMS polymer. (Courtesy: Agarwal group (FIU))

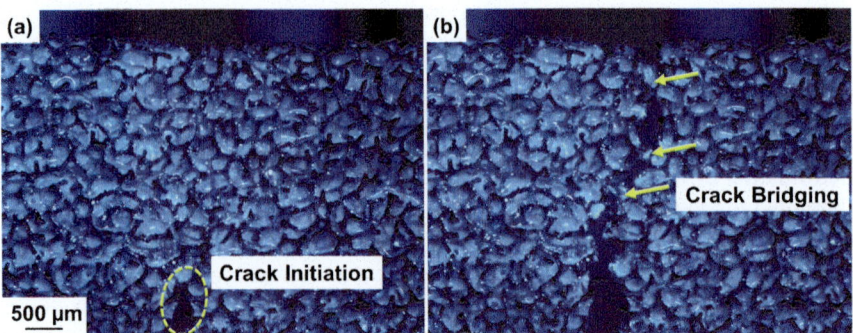

Fig. 2.8 In-situ optical microscopy to study deformation mechanisms. Real-time snapshots showing: (**a**) crack initiation, and (**b**) crack propagation, deflection, and bridging mechanisms. (Reproduced/adapted with permission from Embrey et al. (2017))

Fig. 2.8, it can be seen that graphene foam cells provide structural strength and resist deformation. Once a crack initiates (Fig. 2.8a), the graphene branches tend to arrest, deflect, and suppress its propagation. This can be seen from Fig. 2.8b, where graphene branches act as crack bridges. Therefore, in-situ imaging prior to and during failure allows the determination of the key mechanisms and microstructure elements that play the most important role in determining the strength of materials.

Conventional optical cameras are not capable of capturing rapid deformation phenomena happening in the millisecond timescale. For instance, capturing crack propagation phenomena requires high-speed imaging. The capability of a high-speed camera is quantified in terms of the number of stills (frames) it can capture in one second, denoted by the unit frames per second (fps). Conventional optical cameras typically record deformation at rates less than 50 fps. High-speed cameras, on the other hand, acquire real-time deformation snapshots at rates exceeding 1000 fps. Figure 2.9a shows a sequence of deformation mechanisms (1 through 6) observed by the in-situ high-speed camera during the tensile stretching of a nanotube buckypaper (Nautiyal et al. 2019). Crack initiation, deflection, and bridging mechanisms were seen to be activated during tensile loading. These mechanisms could be resolved because the sample deformation video was captured at 5000 fps. The readers are referred to Supplementary Video, Video 2.1, showing in-situ high-speed camera imaging of deformation/failure of buckypaper. Another application of high-speed cameras is to study the "impact" behavior of materials. Impact tests involve very high strain rates (10^2–10^3 s^{-1}). Figure 2.9b shows the impact-deformation of poly(L-lactide)/poly(ε-caprolactone) polymer blend by split Hopkinson pressure bar method (Gustafsson et al. 2015). These snapshots are captured 600 μs after impact, highlighting the capability of high-speed camera for resolving events in sub-millisecond timescales. The captured images are used for measuring specimen cross sections for calculating true stress values. This helps in establishing true stress–strain relationship during impact-deformation, as shown in Fig. 2.9b.

2.2 Imaging Tools and Techniques for In-Situ Observation

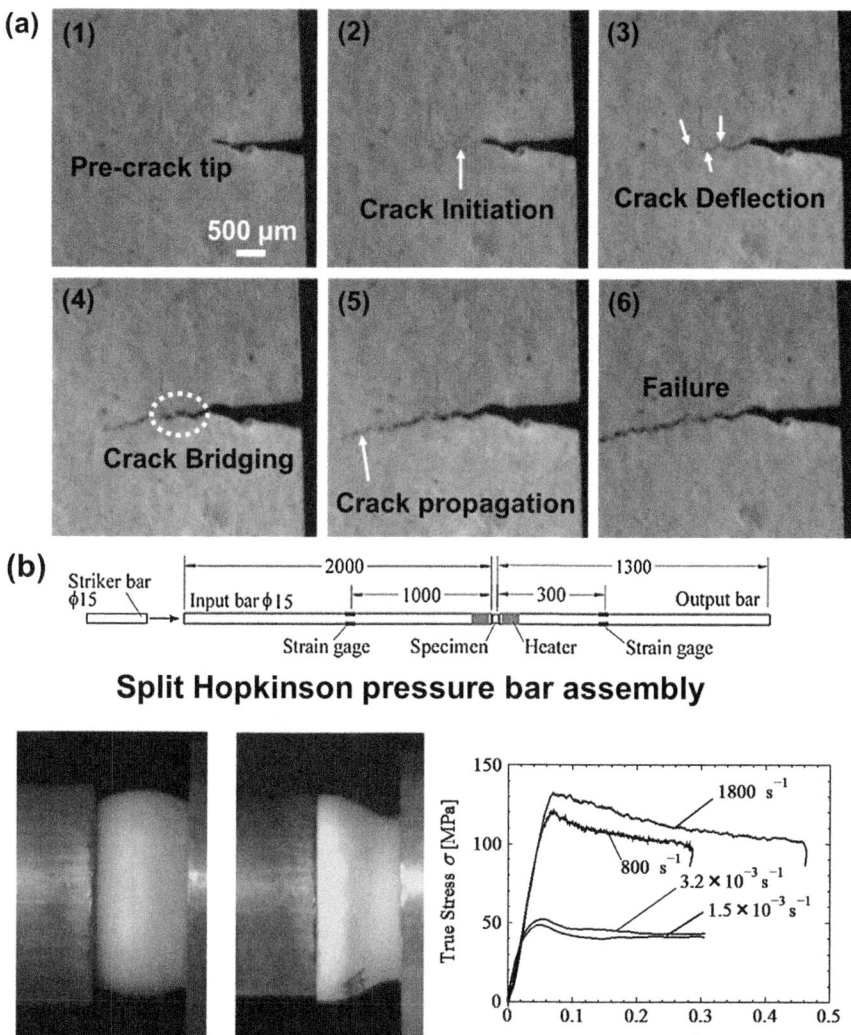

Fig. 2.9 (a) A sequence of deformation events recorded by the high-speed camera at 5000 frames per second during tensile testing of boron nitride nanotube buckypaper, and (b) high-speed camera images captured for impact testing of poly(L-lactide)/poly(ε-caprolactone) polymer blend by split Hopkinson pressure bar method. The high-speed camera snapshots for the two conditions correspond to lubricated (#1) and un-lubricated (#2) contacts between the striker and the specimen. True stress–strain behavior is determined and plotted from the high-speed camera videos. (Reproduced/adapted with permissions from Nautiyal et al. (2019), Gustafsson et al. (2015))

Observing high-speed impact phenomena requires extremely fast exposure times and small inter-frame duration. Ultrafast stroboscopic microscopy is useful to capture the dynamics of impact phenomena. In stroboscopic microscopy, ultrafast white light pulses enable the capturing of multiple exposed optical snapshots. This is shown in Fig. 2.10, where aluminum, nylon, Kevlar, and carbon nanotube fibers were impacted by glass microsphere with supersonic velocities (Xie et al. 2019). Real-time images were captured by a femtosecond laser, electro-optic modulators, and a digital camera. These images were captured with exposure time <1 ps and inter-frame time of 25.14 ns. The ultrafast imaging is vital to determine dynamic parameters, such as transverse wave speed (c_T) and deflection angle (γ), as shown in Fig. 2.10. The transverse wave speed, c_T is a measure of ultrahigh strain-rate energy dissipation ability of fibers. It was noticed that the specific energy dissipation power (SEDP) for the fibers was linearly related to c_T (Xie et al. 2019):

$$\text{SEDP} = \left(7.5 \times 10^{10}\,\text{kg}^{-1}\right) c_T \tag{2.2}$$

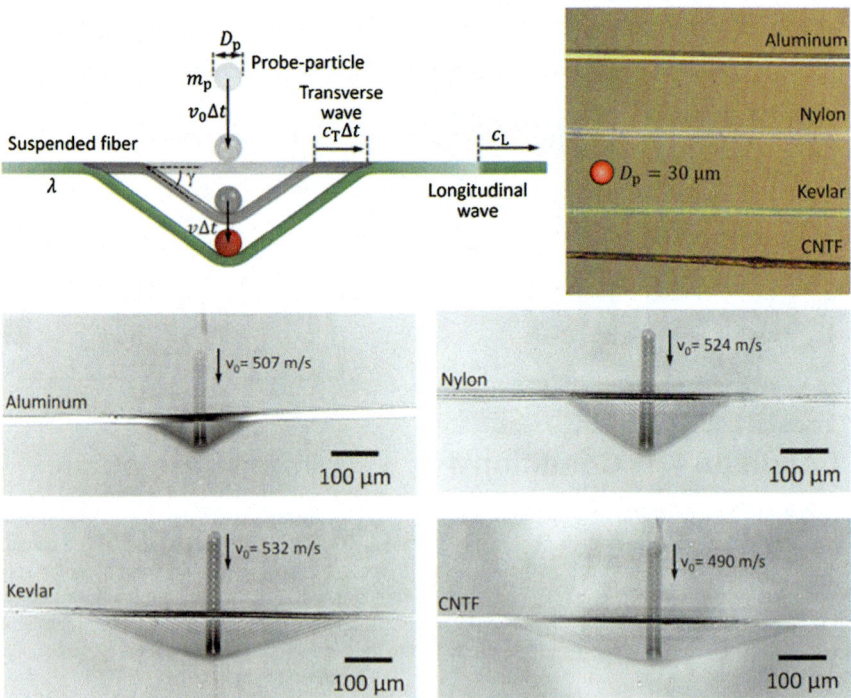

Fig. 2.10 In-situ stroboscopic microscopy for imaging supersonic speed impact of a probe-particle on aluminum, nylon, Kevlar and carbon nanotube fibers. (Reproduced with permission from Xie et al. (2019))

2.2 Imaging Tools and Techniques for In-Situ Observation

Therefore, ultrafast in-situ imaging is highly informative to unravel the dynamic mechanics of materials. Real-time imaging is useful for qualitative insights into impact-response, as well as quantitative determination of dynamic parameters.

For probing the mechanics of single particles, superior resolution of force and displacement is desirable. Optical tweezer, also known as optical trap, is a promising in-situ optical technique for applying forces on and monitoring associated displacements of single particles (Grier 2003; Moffitt et al. 2008; Neuman and Nagy 2008; Maragò et al. 2013; Jones et al. 2015; Koch and Shaevitz 2017; Nautiyal et al. 2018a). A laser is focused with a high numerical aperture microscope objective, creating two opposing optical forces: (a) scattering forces, which push the particles, and (b) gradient forces, which pull the particles. An optical trap is created when the gradient forces exceed the scattering forces. Optical tweezers can effectively manipulate dielectric particles since they can be polarized by the optical field. Therefore, dielectric particles (such as polystyrene or silica beads) serve as handles and probes for manipulating the specimens of interest and detecting the forces and displacements. For small displacements (typically <150 nm), the optical trap behaves like a linear spring such that the force and displacement are linearly proportional. The displacements of the trapped particle are determined using a position detector, and the corresponding forces can be calculated using Hooke's law:

$$F = -kx \tag{2.3}$$

where k is the stiffness of the optical trap. Stiffness (k) of the beads is calibrated for measuring the forces. High-resolution optical tweezers can measure forces and distances as low as ~0.02 pN and 0.2 nm, respectively (Zhang et al. 2013). Figure 2.11a illustrates the application of optical tweezers for the mechanical deformation of red blood cells (RBCs) (Li and Liu 2008). The cells were attached to silica beads and stretched by optical tweezers. The associated strains in the transverse (D) and longitudinal (L) directions can be quantified from the digital images captured in real time. Optical tweezers are useful to study adhesion phenomena at mesoscale. Their application has been demonstrated for determining the probability of virus–cell adhesion, and to compute binding strength between fibrinogen–integrin pairs. The optical trap-based scheme shown in Fig. 2.11b is employed to make contact, followed by bond rupture between the two species to compute binding forces (Litvinov et al. 2002). The manipulation of the species is visualized in real time by using a video camera. Excellent force resolution of optical tweezers has been exploited to quantify folding-unfolding phenomena in proteins. Figure 2.11c depicts the methodology where the titin molecule attached to a bead is stretched and the forces are continuously monitored. It was observed that the molecules unfold when the stretching forces exerted on titin exceeds ~20 pN. Upon unloading, the refolding of the molecules occurs, albeit at much lower forces (<5 pN). The unfolding-folding behavior of molecules can be loading-rate dependent. Coincident stretching-relaxation curves were obtained for an RNA molecule when the loading rates were low (1 pN/s or lower), as opposed to nonequilibrium response for 10 pN/s loading rate for the same molecule (Fig. 2.11c) (Liphardt et al. 2001). Therefore, optical trap

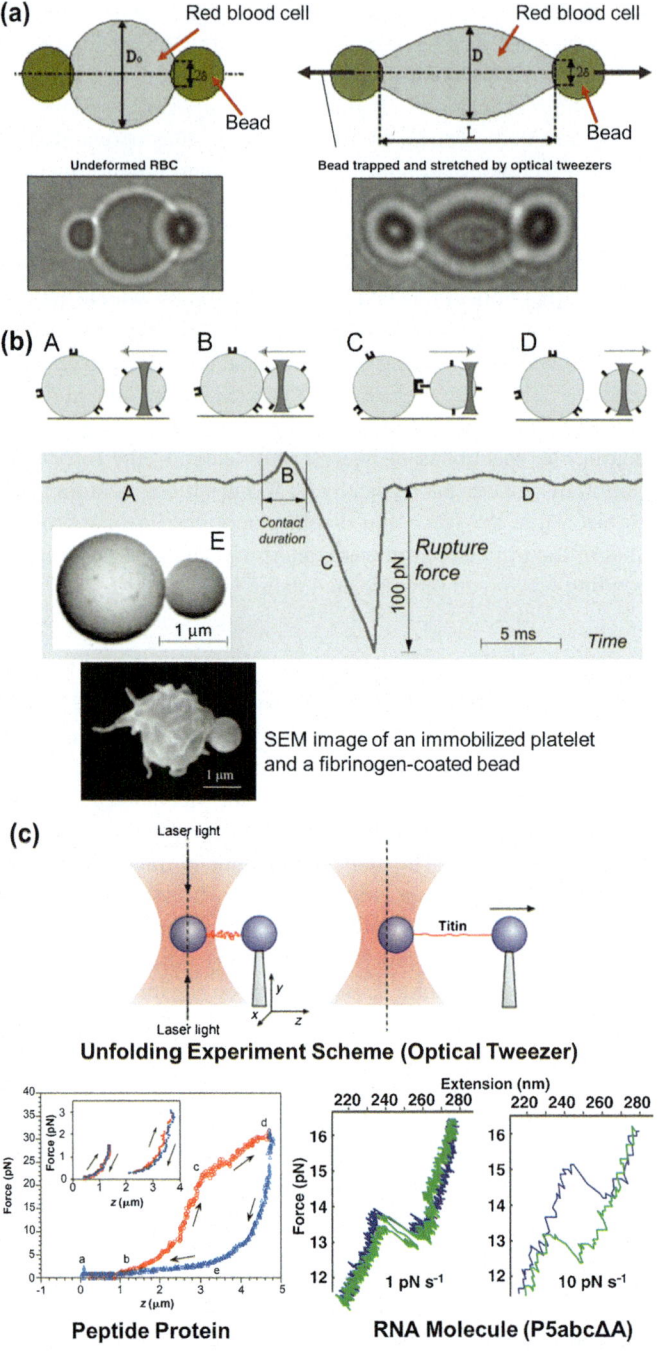

Fig. 2.11 (**a**) The optical tweezer technique for stretching red blood cells and measuring strains using real-time digital images, (**b**) adhesion strength measurement using optical trap scheme, and

is a highly informative technique for deformation and mechanical phenomena at micro- and meso-length scales, with amazing force and displacement resolutions.

2.2.2 Electron Microscopy for High-Resolution Real-Time Imaging

Optical techniques are not sufficient to resolve dimensions and phenomena happening in nanometer length scales. For instance, deformation mechanisms associated with nanomaterials, nanostructured materials, nano-dimension specimens, defects in the materials or fine microstructure features/constituents cannot be effectively resolved by white light. Additionally, optical microscopes have a limited depth of focus, limiting their applicability for visualizing deformation mechanisms in complex 3D architectures. In-situ testing inside electron microscopes provides superior resolution and magnification capabilities. Based on the desired mechanical information and nature of the samples, in-situ tests can be conducted in scanning and transmission electron microscopes. The subsequent subsections discuss the in-situ mechanical characterization of materials in SEM and TEM, respectively.

In-Situ SEM Mechanical Characterization

A scanning electron microscope uses an electron beam to image and analyze specimens. Interactions between the electrons and the specimen can produce different kinds of signals (schematically shown in Fig. 2.12). Of these modes, secondary electrons (SE) and backscattered electrons (BSE) based imaging are the most relevant modes from the standpoint of in-situ mechanical testing. Secondary electrons are the excited electrons emitted from the specimen surface due to the energy absorbed from the incident beam. This mode is useful for identifying, probing and observing the deformation of fine surface features in the sample. Contrary to this, BSE images are obtained when the incident electrons (in the beam) are backscattered from the sample. The extent of scattering is dependent on the atomic number of the sample. Larger atoms can strongly scatter electrons compared to lighter atoms. Therefore, BSE imaging is useful to distinguish phases in the microstructure and evaluate the deformation behavior associated with these individual phases.

Fig. 2.11 (continued) (**c**) mechanics of unfolding of biomolecules using optical tweezers. Associated force-extension curves for protein and RNA molecules are shown and enable the quantification of mechanical forces. (Reproduced/adapted with permissions from Li and Liu (2008), Litvinov et al. (2002) Copyright (2002) National Academy of Sciences, U.S.A. The images from Kellermayer et al. (1997), Liphardt et al. (2001) are purchased for reproduction from The American Association for the Advancement of Science)

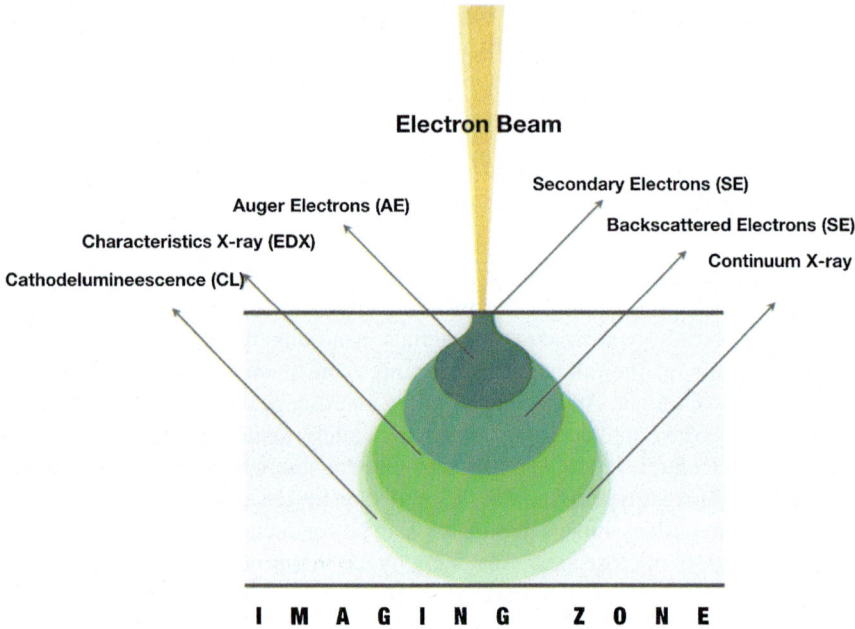

Fig. 2.12 Schematic representation of electron-specimen interaction during scanning electron microscopy

However, the resolution for BSE imaging is inferior to SE imaging. The imaging modes should be selected based on the feature size, microstructure makeup, and the information desired from in-situ characterization.

Imaging in SEM has two benefits: first, it facilitates testing of nano-sized samples to study size effects, and second, it allows visualization of nanoscale deformation mechanisms in the materials (which can be micro- to bulk-scale specimens). This is illustrated in Fig. 2.13. Mechanical characteristics, such as yield stress and ultimate tensile strength were historically believed to be intensive material properties that are independent of sample size. However, mechanical properties start showing size dependence in micro- to sub-micrometer length scales (Greer et al. 2011). It has been observed that the strength of sub-micrometer sized specimens exceeds their bulk "macro-sized" counterparts. This necessitates a systematic investigation of the deformation behavior of materials as a function of the sample dimension. In-situ characterization in SEM facilitates resolving and testing nano-sized samples and observe deformation mechanisms in real time. Figure 2.13a shows mechanisms activated upon compression of nanopillars made from single metal crystals. In-situ investigations reveal different mechanisms activated in a host of metals at different stresses, such as the formation of slip lines, bending/buckling, and viscous flow of material (Kim and Greer 2009; Lee et al. 2011). The extent of plastic deformation and the critical stresses for activation of different mechanisms demonstrate sample size dependence. Yield stress vs. pillar dimension plot for a

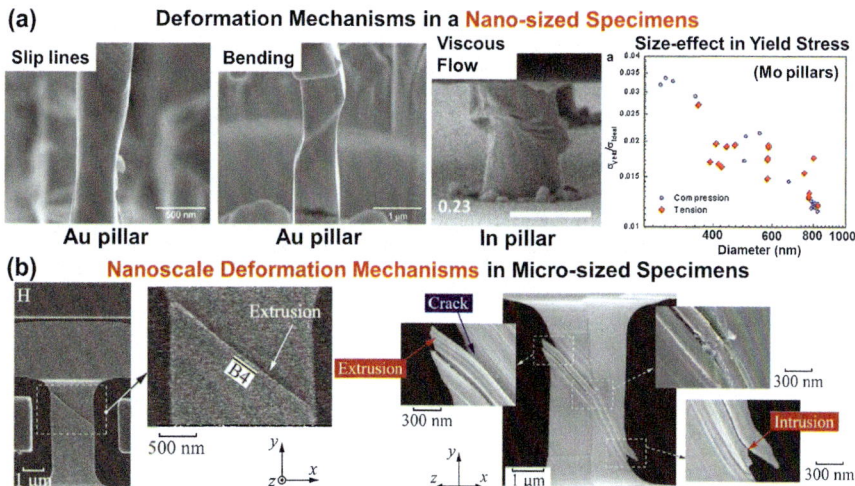

Fig. 2.13 (a) In-situ investigation of deformation mechanisms in nano-sized samples to evaluate size effect in mechanical properties, and (b) nanoscale deformation mechanisms captured during in-situ mechanical testing in SEM. (Reproduced/adapted with permissions from Kim and Greer (2009), Lee et al. (2011), Sumigawa et al. (2018))

Molybdenum single crystal is shown in Fig. 2.13a. In this specific case study, it is observed that the experimental yield stress is a relatively small fraction (0.015) of the theoretical stress when the pillar diameter is approaching 1 μm. However, for very fine-diameter nanopillars (with diameters less than 400 nm), there is more than three-fold enhancement in yield stress and the experimental yield stress attains a significant fraction (>0.03) of the theoretical yield stress. This is seen to hold true for different loading scenarios. For instance, the yield stress vs. diameter plot in Figure 2.13a compares the values for tension as well as compression-induced plastic deformation. There is a definite trend of increasing deformation-resistance for small diameters. The second important advantage of in-situ SEM characterization is to observe and study "nanoscale deformation mechanisms" in relatively larger samples (micro- to bulk-sizes). These effects are not due to extrinsic sample size, but due to intrinsic factors associated with sample microstructure. Figure 2.13b shows nanoscale extrusions and intrusions captured during in-situ SEM testing of a single crystal Cu (Sumigawa et al. 2018). These intrusions/extrusions arose because of locally concentrated slip bands during fatigue loading (tension–compression cycles). In-situ SEM observations indicate that crystallographic slip is localized to the slip band, which gives rise to nanoscale extrusion/intrusion at the sample surface.

In addition to plasticity mechanisms, in-situ SEM characterization is also used to identify crack initiation and propagation events during mechanical loading of materials. Figure 2.14a shows indentation deformation mechanisms in a TiN/SiN$_x$ thin film corresponding to different segments of the load–displacement curve (Rabe et al. 2004). Real-time snapshots show initial plastic deformation/pileup of material

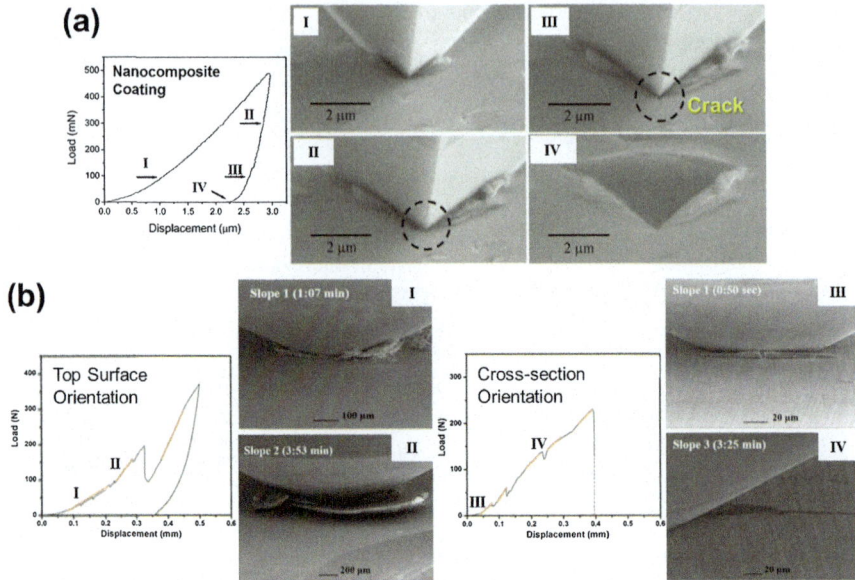

Fig. 2.14 Load–displacement curves and real-time SEM images showing deformation mechanisms during: (**a**) in-situ nanoindentation of TiN/SiNx thin film, and (**b**) a bulk pellet composed of boron nitride platelets. (Reproduced/adapted with permissions from Rabe et al. (2004), Loganathan et al. (2017))

(I). During unloading, there is the formation and propagation of crack. These mechanisms are shown in images II and III (encircled region). This crack formation mechanism is not obvious by just looking at the final, post-indentation image (IV) because of a large amount of simultaneous plastic deformation. This highlights the importance of in-situ mechanical testing to identify critical mechanisms that can be activated when a material experiences mechanical stresses. For bulk-scale samples, cracks and fracture can be probed by high-load indentation, where the applied loads are several hundreds of Newton and the penetration depths are in the order of several hundreds of microns to enhance the deformation zone and extract the properties and mechanisms that are truly representative of the sample. The load–displacement curves and deformation mechanisms due to high-load indentation of a sintered boron nitride pellet are illustrated in Fig. 2.14b (Loganathan et al. 2017). BN platelet is a 2D material and hence, anisotropy is expected in its mechanical response. Figure 2.14b highlights how in-situ imaging can be effective in capturing mechanisms associated with different orientations (along and perpendicular to basal plane in this example). Top surface indentation exhibits material sliding, pileup, bending of BN sheets, and delamination mechanisms (Images I and II). Contrary to this, the indentation on the cross section is dominated by crack initiation and propagation (Images III and IV). These two case studies highlighted in Fig. 2.14a, b convey the importance of in-situ SEM tests to understand failure initiation mechanisms in materials at different microstructural length scales.

2.2 Imaging Tools and Techniques for In-Situ Observation

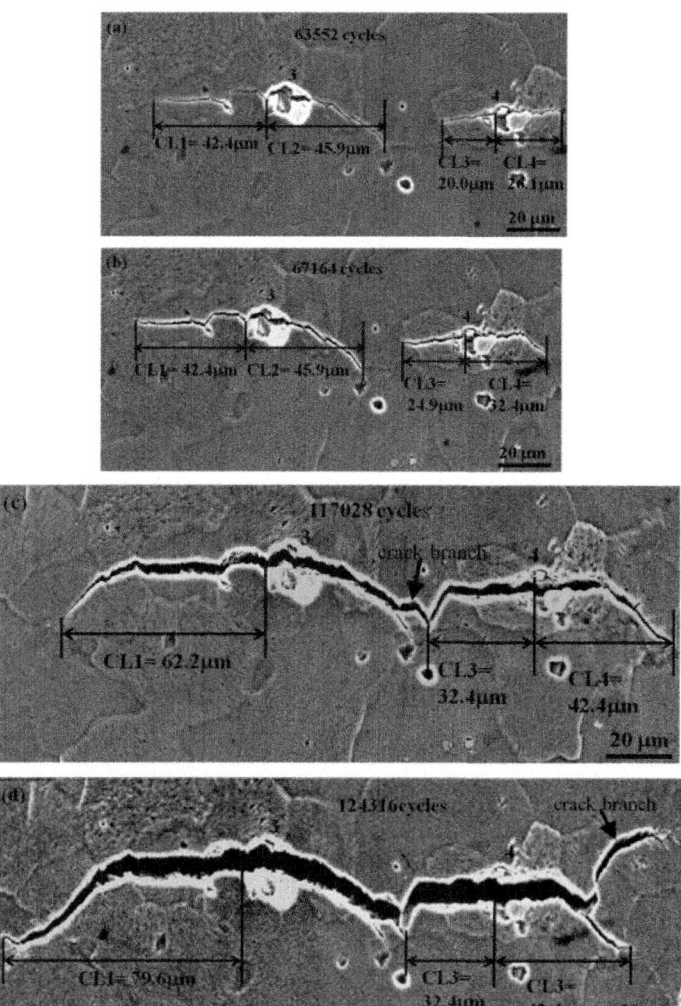

Fig. 2.15 In-situ SEM observation of crack initiation, propagation, coalescence, branching, and opening mechanisms due to fatigue loading of aluminum alloy. (Reproduced with permission from Yan and Fan (2016))

In-situ SEM investigations can also provide useful insights into fatigue or time-dependent deformation of materials. Real-time imaging enables the observation of crack expansion by monitoring the crack length over many loading cycles (Fig. 2.15a–d). The case study in Fig. 2.15 is for 2524 aluminum alloy (Yan and Fan 2016). In-situ investigations show that inclusion particle size influences the possibility of crack initiation. High-resolution imaging also provides insights into grain boundary-crack interaction during fatigue loading: for instance, there is no crack growth for CL3 and CL4 as the loading cycles are increased from 117,028

(Fig. 2.15c) to 124,316 (Fig. 2.15d) because the cracks encounter grain boundary. The interactions can lead to crack branching, which is captured in Fig. 2.15c. The crack opening is seen after 124,316 cycles due to the coalescence of multiple cracks (Fig. 2.15d). It is difficult to obtain these mechanistic insights into fatigue failure of engineering materials without high-resolution in-situ SEM investigation.

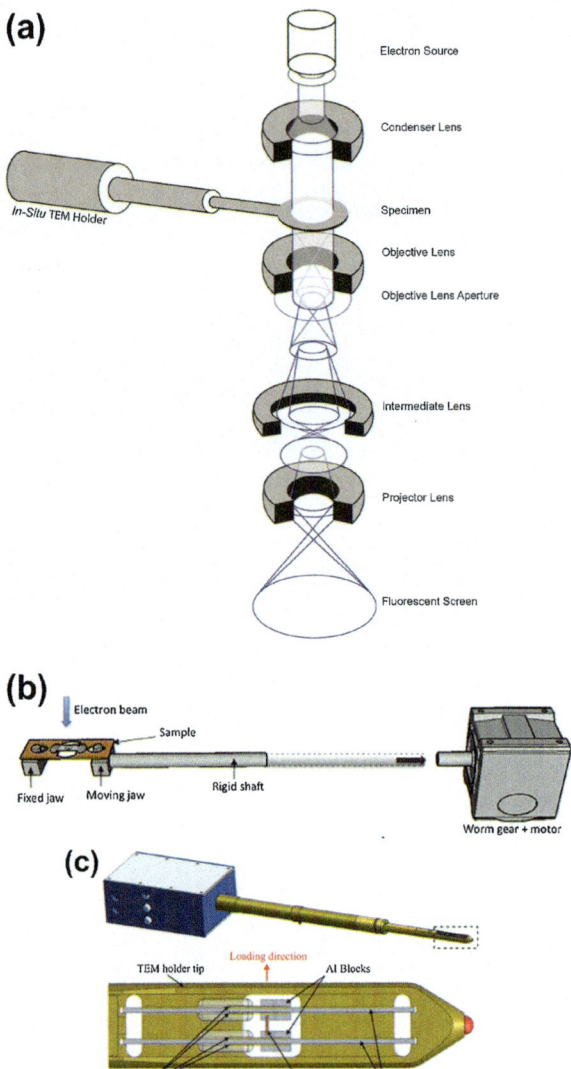

Fig. 2.16 (**a**) Working principle of TEM imaging, (**b**) in-situ TEM holder (straining stage), and (**c**) special TEM holder for high frame rate mechanical loading in a dynamic TEM (DTEM). (Reproduced with permissions from Legros (2014), Voisin et al. (2016))

In-Situ TEM Mechanical Characterization

In-situ testing inside a TEM is performed to decipher deformation mechanisms at a more fundamental atomic length scale. TEM imaging requires the sample to be electron transparent, as illustrated in Fig. 2.16a. Typically, the sample thickness should be about or less than 100 nm for TEM imaging. The incident electron beam is scattered by the sample, which is focused by an objective lens to form the primary image. Thereafter, additional lenses are used to magnify the primary image. Mechanical testing devices can be fit in the column for in-situ characterization. Figure 2.16b shows an in-situ TEM holder to apply mechanical strains. The holders can be equipped with heating/cooling options for mechanical testing at desired temperatures. High strain-rate mechanical loading (as high as 10^3 s^{-1}), where deformation events happen at much smaller time scales (~nanoseconds), cannot be effectively captured using conventional in-situ TEM testing techniques. The maximum frame rate for conventional TEMs is typically restricted to 30 frames per second, limiting their ability to image high frame rate deformation. A dynamic TEM (DTEM) has time resolution in the order of tens of nanosecond, which can record deformation phenomena during rapid mechanical loading (Voisin et al. 2016). DTEM uses a laser-driven Ta photocathode instead of a conventional electron gun. Bombarding the photocathode with a high-intensity ultraviolet laser produces electron exposure in a short period of time (~30 ns), resulting in excellent time resolution. A high strain-rate TEM holder is used to apply strain rates as high as 4×10^3 s^{-1}. The holder achieves high strain rates because of two piezoelectric actuators, which bend upon voltage application. The holder design is shown in Fig. 2.16c.

Nanoindentation holders are also used for indenting specimens inside TEM. Flat punch indenter tips are used for compression testing of nanoparticles. Figure 2.17a demonstrates real-time TEM imaging during the application of load on alumina nanoparticle (Calvié et al. 2012). The failure of the nanoparticle was observed and simultaneously recorded in the load–displacement curve. Nanoindentation holders can also be employed for nanopillar compression experiments. Because of excellent imaging resolution, in-situ TEM pillar compression can provide mechanistic information regarding deformation at nanometer length scales. Figure 2.17b illustrates the in-situ compression of Cu nanopillars in TEM for a range of pillar diameters, from over 1 μm to less than 100 nm (Kiener and Minor 2011). In addition to enhanced yield stress for smaller pillars, the deformation pattern is also visibly different. There is a transition from multiple slips (for larger pillars) to a single slip mechanism (finer pillars).

Real-time TEM imaging during mechanical loading is a useful means to interrogate dislocation dynamics in materials. In-situ tests can decode the underlying plasticity and failure mechanisms. Figure 2.18a highlights dislocation nucleation mechanisms active in a single crystal aluminum upon tensile loading (Oh et al. 2009). It is seen that loop formation and operation of single-armed sources are responsible for tensile deformation. In addition to nucleation, there can also be a loss or emission of dislocations as the material is mechanically loaded. This is shown in Fig. 2.18b. The balance between nucleation and loss plays a decisive role

Fig. 2.17 Application of in-situ TEM holders for: (**a**) nanoparticle compression, and (**b**) nanopillar compression experiments. (Reproduced/adapted with permissions from Calvié et al. (2012), Kiener and Minor (2011))

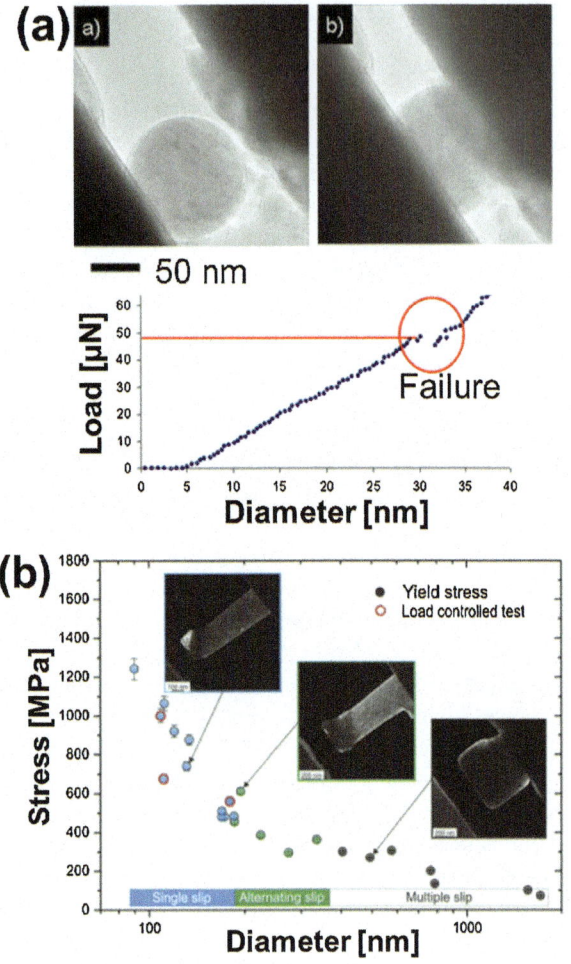

in deformation behavior. In-situ TEM investigations can be used to determine the net dislocation densities in the sample during deformation. For instance, dislocation density during the tensile loading of the single crystal aluminum case shown in Fig. 2.18b is found to be strain-rate sensitive. When the single crystal Al was strained at 10^{-4} s^{-1}, the dislocation density remained statistically constant. However, raising the strain rate to 10^{-3} s^{-1} caused an enhanced nucleation rate, resulting in higher dislocation density in the sample. In-situ TEM tests are also used to gain insights into fracture behavior. Real-time TEM images demonstrating the progression of tensile deformation of a high-entropy alloy coating are shown in Fig. 2.18c (Cai et al. 2017). It is seen that dislocation motion is responsible for crack formation prior to failure. In-situ TEM imaging shows dislocation accumulation (encircled region). Fracture of the material takes place along the zone of dislocation accumulation.

2.2 Imaging Tools and Techniques for In-Situ Observation

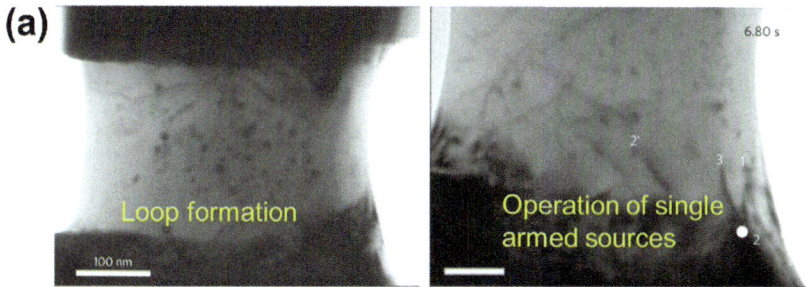

Dislocation nucleation mechanisms

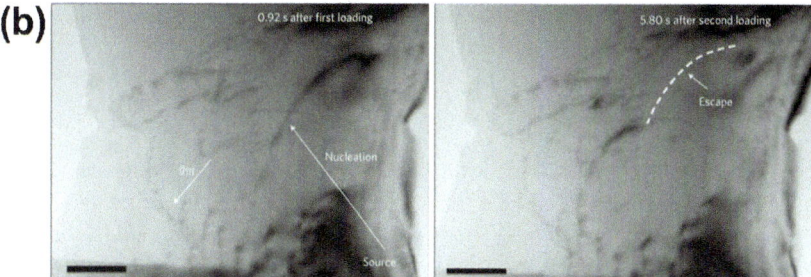

Dislocation escape process

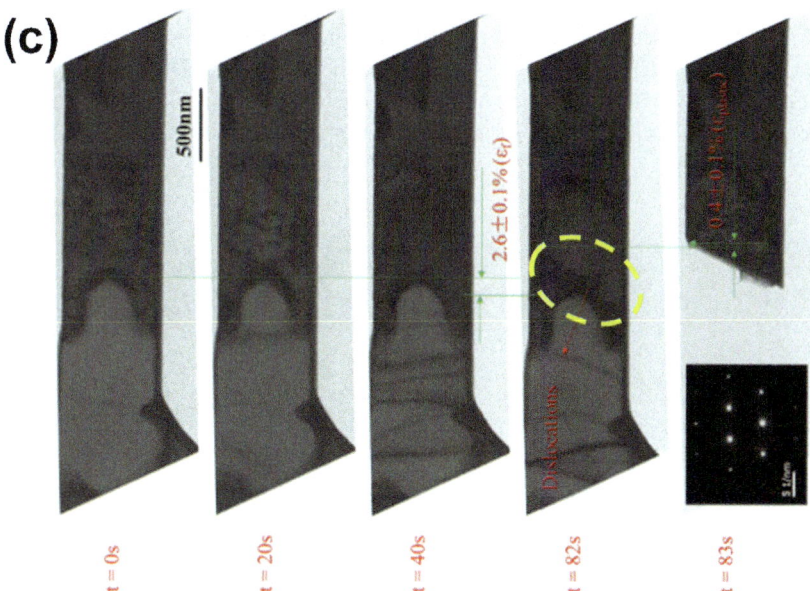

Dislocation accumulation

Fig. 2.18 In-situ TEM testing to understand dislocation dynamics during deformation: (**a**) dislocation nucleation mechanisms, (**b**) dislocation emission/loss event, and (**c**) dislocation motion and accumulation leading to material fracture. (Reproduced and adapted with permissions from Oh et al. (2009), Cai et al. (2017))

2.2.3 Electron Backscatter Diffraction During In-Situ Mechanical Testing

Electron backscatter diffraction (EBSD) is a technique used in conjunction with SEM to interrogate the crystallographic structure of materials. During EBSD characterization, an electron beam is focused on a highly tilted sample, with tilt angles ~70° (Fig. 2.19a). The diffraction/interaction zone scatters the electrons in multiple directions. The scattered electrons interact with crystallographic planes in the sample and are coherently diffracted out by the planes. These diffracted electrons produce EBSD patterns, which provide information about the crystal lattice structure. An example of the EBSD pattern for nickel is shown in Fig. 2.19b. The thickness of the bands provide information about d-spacing based on Bragg's law:

$$\lambda = 2d_{hkl} \sin\theta \qquad (2.4)$$

where λ is the wavelength of electrons, d_{hkl} is the interplanar spacing, and θ is the angle between the incident and the scattered wave. The width of the bands in EBSD patterns is proportional to sin θ. Therefore, thicker bands correspond to the planes with smaller d-spacing. Figure 2.19b illustrates the application of the EBSD pattern to identify the interplanar spacings for different regions in the microstructure. The d-spacing values, in turn, provide information about the index (hkl) of the plane.

The application of EBSD for in-situ mechanical characterization is demonstrated in Fig. 2.20a (Cepeda-Jimenez et al. 2016). The figure shows EBSD mapping in the region of interest (from SEM) before and after the tensile test. The maps shown are inverse pole figure (IPF) maps, which differentiate different orientations using different colors. IPF maps enable the correlation between grain orientations and differ-

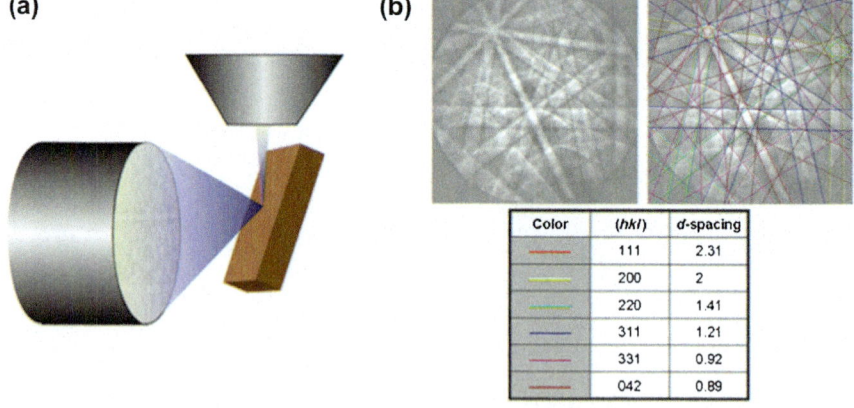

Fig. 2.19 (a) Schematic illustrating electron beam-sample interaction for capturing EBSD pattern, and (b) example showing the use of EBSD pattern to determine interplanar spacing (for nickel). (Reproduced/adapted with permission from Adams et al. (2013))

2.2 Imaging Tools and Techniques for In-Situ Observation 49

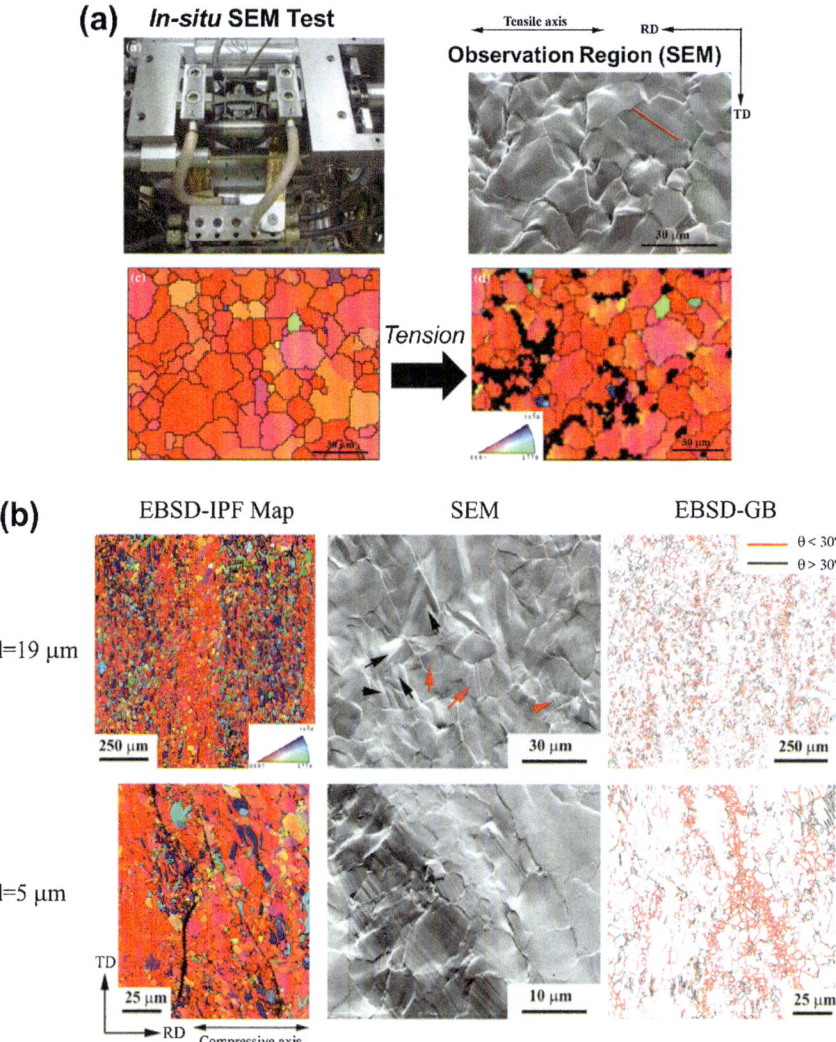

Fig. 2.20 (**a**) EBSD analysis coupled with in-situ SEM tensile test to identify orientation changes in a region of interest, and (**b**) EBSD-IPF and GB maps at 10% strain to compare deformation mechanisms (slip systems/twinning) in fine- and coarse-grained polycrystals. (Adapted with permission from Cepeda-Jimenez et al. (2016))

ent deformation mechanisms. The example in Fig. 2.20a identifies the orientation-slip trace relationship. Figure 2.20b shows EBSD-IPF maps in Mg polycrystals after a strain of 10%. The figure compares IPF maps for the deformation of fine-grained (5 μm) and coarse-grained (19 μm) polycrystals. EBSD characterization reveals that 60% of the grains were twinned for coarse-grained polycrystal. Contrary to this, only 14% of the grains were twinned for the fine-grained sample. EBSD analysis is also used for identifying slip systems associated with the slip traces. The fine-grained polycrystal in Fig. 2.20b comprises of 73% basal, 13% are pyramidal, and 14% prismatic slip traces (after 10% strain). The figure also shows EBSD-GB maps to determine grain boundary misorientation. The GB misorientation maps show two cases: $\theta < 30°$ and $\theta > 30°$. A higher fraction of boundaries with $\theta < 30°$ denote better grain connectivity. A comparison of the GB maps reveals superior connectivity for fine-grained microstructure, thereby explaining the prevalence of basal slip (as opposed to twinning for coarse-grained microstructure).

EBSD mapping during in-situ mechanical testing can also be useful to understand lattice rotation during deformation. Figure 2.21a illustrates micropillar compression in SEM, while simultaneously acquiring EBSD map (Di Gioacchino and Clegg 2014). Positive values of the angle ϑ_3 (about the normal axis) in the map signifies that the rotation is predominantly taking place in the counter-clockwise direction. The map shows pronounced rotation along the slip direction (encircled by white-colored dash lines in the EBSD map): it can be seen that the rotation increases from ~0 radians at the lower right end to 0.07 radian (~4°) at the upper right end of the encircled region. Contrary to this, ϑ_3 remains constant in the direction perpendicular to slip (shown by a straight, black dashed line). Kernel Average Misorientation (KAM) map based on EBSD characterization is useful for interrogating the evolution of subgrain structures during in-situ mechanical loading. KAM represents local misorientation with respect to the surrounding pixels. Figure 2.21b shows the evolution of KAM maps during the in-situ loading of the sample in SEM (from 0% to 2% deformation). These high resolution-EBSD scans focus on a triple junction in Al 6061 (Yoo et al. 2019). The mapping reveals the formation of rotation boundary at 2% strain, attested by misorientation angle values reaching 20°. A plot of misorientation angles is shown for a region (red-colored line) in the lower right corner of the KAM map. KAM maps can also provide estimates about the dislocation density:

$$\rho^{GND} = \frac{2KAM}{Lb} \quad (2.5)$$

where KAM is the first nearest neighbor kernel average misorientation, b is the Burgers vector, and L is the step size at which the data is collected.

The EBSD examples presented so far provide surface-level information of a sample undergoing mechanical loading. In order to holistically understand mechanisms associated with the whole volume of the material, a three-dimensional high-resolution EBSD (3D HR-EBSD) is employed. This is accomplished with the aid of FIB, to perform serial sectioning of the specimen under deformation. The sectioning reveals a new surface that is then mapped by EBSD. The principle of 3D EBSD is shown in Fig. 2.22a (Kalácska et al. 2020). The slice-by-slice EBSD measurement allows the construction of 3D models (using 2D slices). A case study on 3D HR-EBSD of Cu micropillars is shown in Fig. 2.22b. 3D-EBSD was employed to examine dislocation distribution in the single crystal specimen. The evolution of geometrically necessary dislocations (GNDs) was mapped by stopping the compression test at different strains. It was observed that a dislocation cell structure was

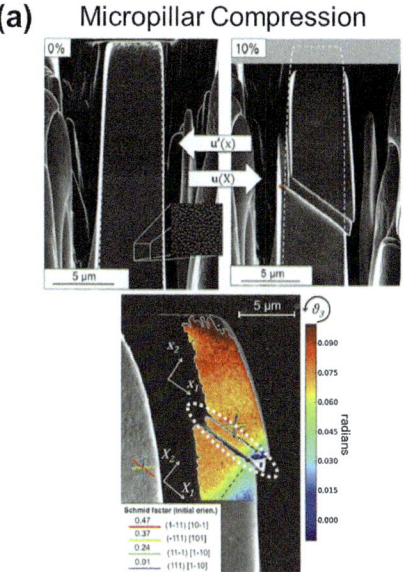
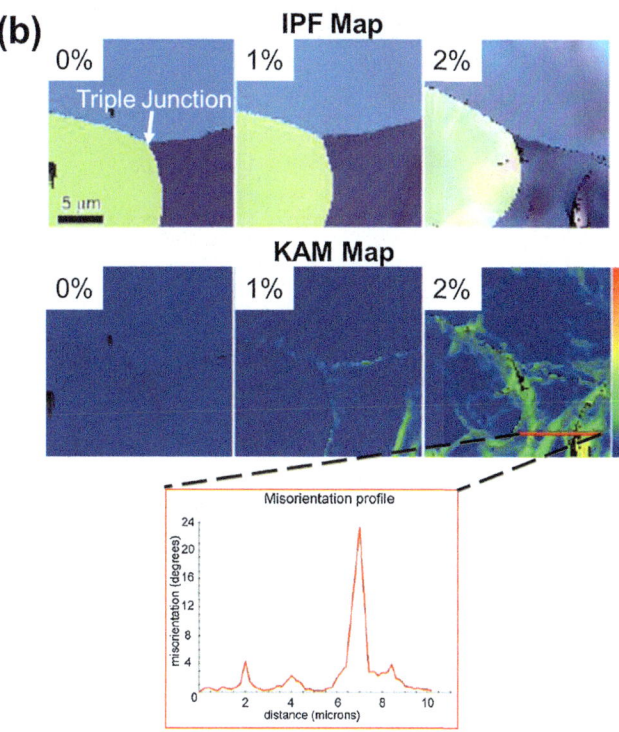

Fig. 2.21 (**a**) EBSD mapping during micropillar compression to extract lattice rotation in Cu, and (**b**) evolution of IPF and KAM maps at a selected triple junction in the microstructure during deformation of Al 6061 up to 2% strain. The misorientation profile plot shows the formation of a new rotation boundary due to deformation. (Adapted with permissions from Di Gioacchino and Clegg (2014), Yoo et al. (2019))

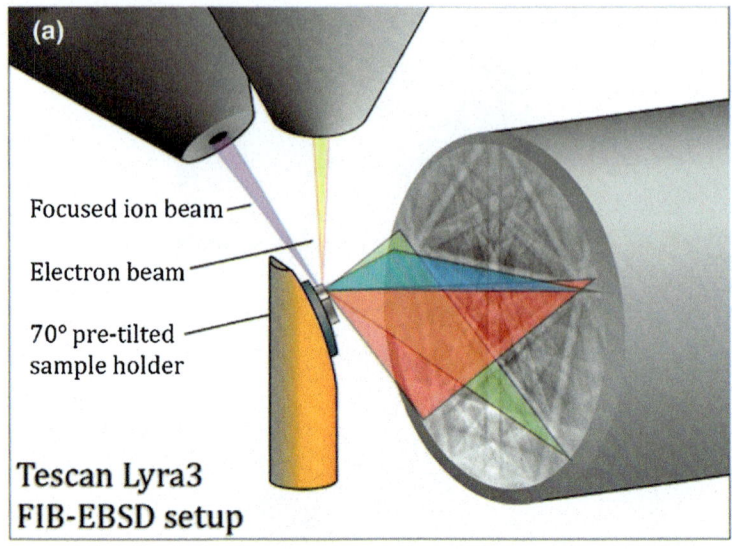

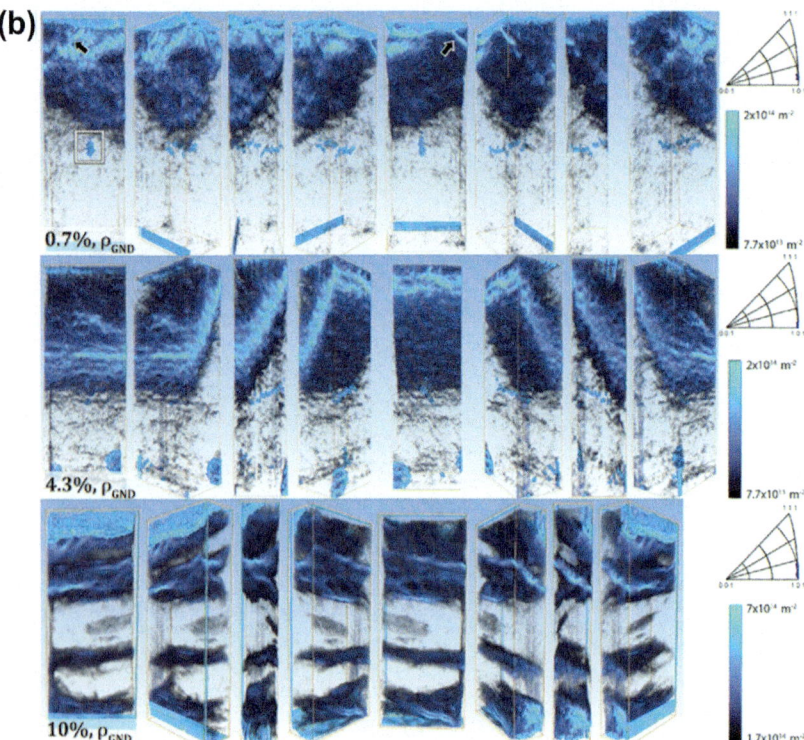

Fig. 2.22 (**a**) Schematic representation of 3D high-resolution EBSD mapping by FIB-assisted serial slicing, and (**b**) 3D models of GND density values for Cu micropillars under mechanical loading (at three different compressive strains). (Reproduced with permission from Kalácska et al. (2020))

built up during compression of the micron-sized pillar, although the GND density was much lower than the bulk-sized sample.

These examples illustrate the significance of the EBSD technique in conjunction with SEM for in-situ mechanical characterization. EBSD mapping can be useful to decipher microstructure deformation mechanisms at multiple length scales: low-magnification scanning provides an overall picture of microstructure response for different grains and grain boundaries, and high-resolution EBSD mapping is useful to understand sub-grain mechanisms such as twinning or different slip systems.

2.2.4 In-Situ Tomography for 3D Internal Imaging

The optical and electron microscopy techniques discussed so far capture 2D surface-level deformation events. It is vital to image the internal structure for evaluating bulk deformation mechanisms. FIB machining is one of the strategies employed to

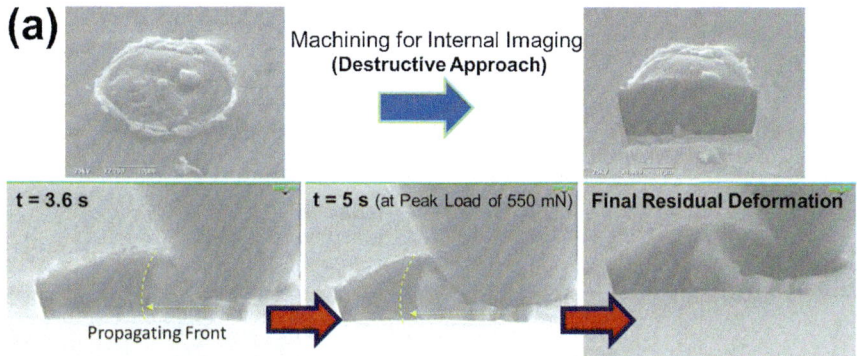

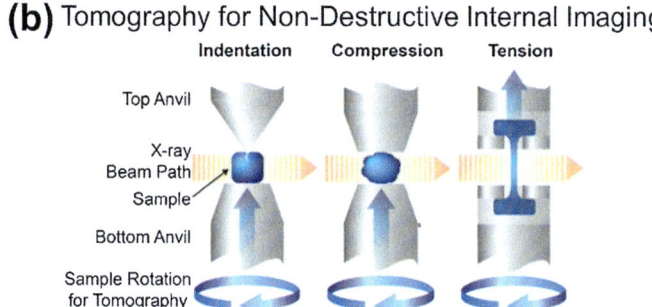

Fig. 2.23 (**a**) FIB machining to reveal the internal structure (cross section) of a cold-sprayed Al splat for in-situ imaging during indentation loading (destructive approach) [Unpublished work], and (**b**) schematic representation of X-ray assisted in-situ testing for imaging internal structure (nondestructive approach). (Reproduced/adapted with permissions from Agarwal group (FIU), Patterson et al. (2016))

reveal the internal structure. This is shown in Fig. 2.23a for a cold-sprayed metallic splat, where a trench is milled to expose the particle cross section prior to mechanical loading. The real-time images showcase shear band propagation in the particle due to indentation loading. However, FIB milling is a destructive approach for internal imaging. The use of X-rays allows for nondestructive sample imaging without any need for material removal. 2D X-ray radiographic imaging is based on the attenuation or phase change of X-ray photons when they pass through the material. In order to obtain a 3D image, hundreds-to-thousands of angular projection images are collected and reconstruction algorithms convert the radiographs into volumetric image (Ying et al. 2011; Patterson et al. 2016). Figure 2.23b shows the in-situ imag-

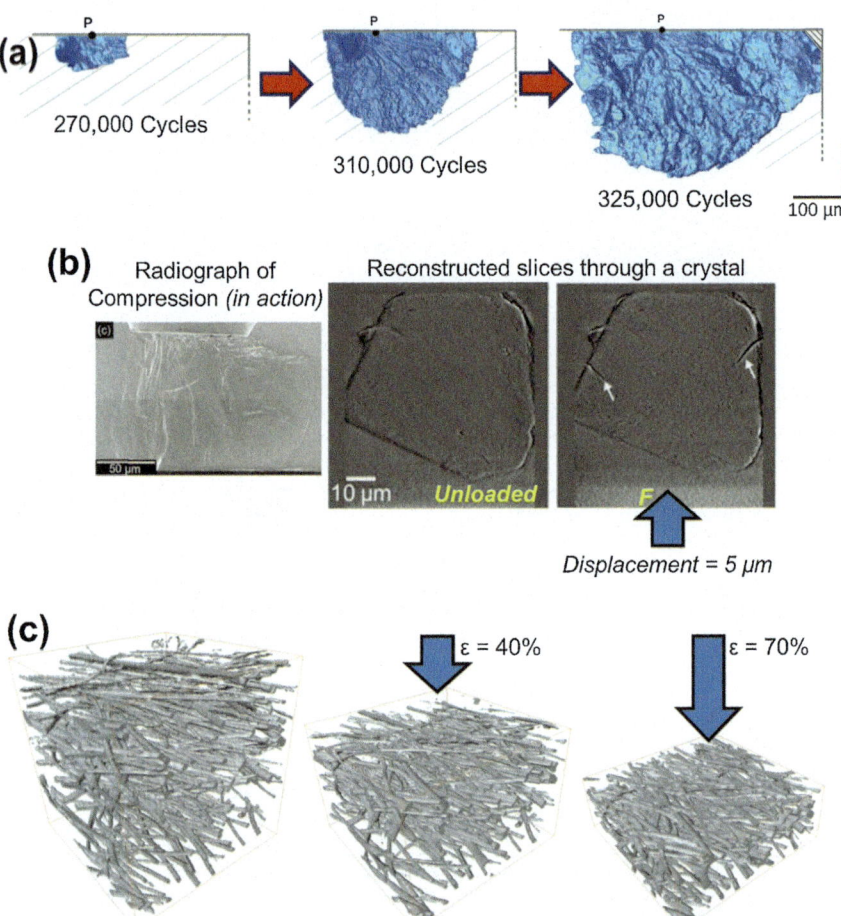

Fig. 2.24 (**a**) Fatigue crack growth in an Al-Si alloy, (**b**) development of micro-cracks in an explosive single crystal prior to its catastrophic failure, and (**c**) compression of a porous, fibrous sample composed of steel fibers. (Reproduced/adapted with permissions from Buffiere et al. (2010); Patterson et al. (2016))

ing via tomography while the mechanical tests are conducted (Patterson et al. 2016). Transmission X-ray microscopy (TXRM) can provide resolutions as fine as 30–50 nm, making this technique highly effective for probing nano-micro-scale deformation mechanisms.

Figure 2.24a shows the application of volumetric imaging (using synchrotron radiation) to image crack growth during fatigue loading of cast Al-Si alloy (Buffiere et al. 2010). The crack initiated at point P, and the crack front was seen to be irregular in shape. The tomographic scans were captured as a function of loading cycles (up to 325,000 cycles), as shown in the figure. X-ray tomography can be useful to probe the mechanisms responsible for the catastrophic failure of materials. Figure 2.24b illustrates the compression of a high-explosive single crystal (Patterson et al. 2016). The real-time imaging showed the formation of micro-cracks prior to the shattering of the crystal. It was also observed that crack propagation followed the hexagonal crystal structure. Another example of 3D rendition during compression is shown in Fig. 2.24c, where a laboratory-tomography is used to visualize the change in relative density, contact points and fiber-orientation for a porous sample composed of entangled steel fiber (Buffiere et al. 2010). These examples show the importance of tomography-assisted volumetric imaging to decipher the bulk deformation mechanisms in different classes of materials.

2.3 Mechanical Instrumentation for Multi-Scale Characterization

2.3.1 *In-Situ Atomic Force Microscope Inside Electron Microscopes*

Atomic force microscopes (AFM) are conventionally used for surface imaging and nanomechanical characterization of materials. Mechanical measurements in AFM are based on the deflection of the cantilever, since deflection is proportional to force (Hooke's law). AFM provides impressive force resolution (up to 10 nN), making it a suitable tool for small-scale mechanical testing. Therefore, AFM can be used inside an electron microscope for sensing ultra-small forces while simultaneously imaging the material deformation in real time. Figure 2.25a–c illustrate the application of AFM for tensile, compressive and bending tests, respectively (Liu et al. 2013). This can be done by clamping one side of the sample with the cantilever while the other side is clamped to the tungsten tip. The schematics in Fig. 2.25 pertain to mechanical testing of 1-D nanowire in TEM. The piezo-driven W tip deforms the sample by applying force in the desired direction: axial loading for tension/compression tests and vertical to nanowire axis for bending. It is important to ensure the samples are clamped strongly enough to avoid slippage. This is done by electron beam-induced carbon deposition (EBID) technique. Testing in TEM allows precise determination of cantilever deflection as well as sample deformation (elongation/

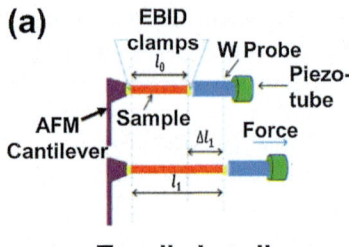

Tensile Loading

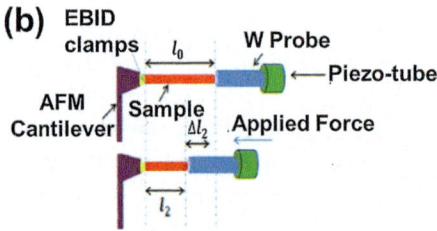

Compression Loading

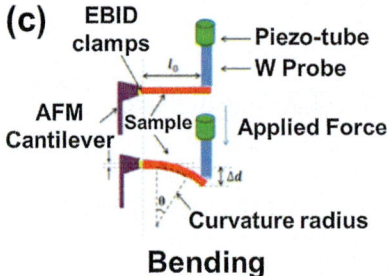

Bending

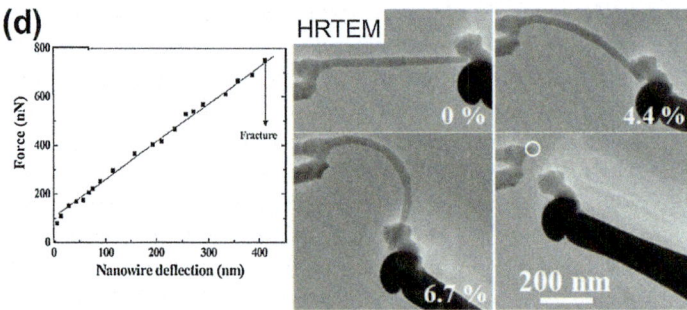

Fig. 2.25 Schematics showing the application of AFM cantilever for in-situ TEM mechanical testing under: (**a**) tensile, (**b**) compressive, and (**c**) bending loads. (**d**) Example of the force–deflection curve and corresponding real-time HRTEM snapshots for bending deformation of a nanowire. (Reproduced/adapted with permission from Liu et al. (2013))

compression/bending deformation). A typical force vs. deflection plot for AFM-based characterization is shown in Fig. 2.25d. Additionally, real-time imaging in HRTEM allows the observation of sample deformation. This is illustrated by showing TEM images of bending of nanowire for different strains, up to nanowire failure (~10%). These images also allow the calculation of bending angles when the sample is strained. For the example shown in Fig. 2.25d, the bending angle was observed to reach ~129.6° prior to brittle failure. This clearly highlights the importance of coupling AFM and TEM for in-situ characterization to extract detailed mechanistic information about sample failure, otherwise difficult to understand from ex-situ testing and postmortem imaging.

AFM cantilever can be used for in-situ measurements inside SEM. Figure 2.26 depicts tensile testing inside SEM by using AFM cantilever as the load sensor (Zhu et al. 2009). The sample is strained using a nanomanipulator, and the deformation is observed in real time. Figure 2.26a, b show real-time snapshots of a nanowire in tension and post-fracture, respectively. In-situ SEM imaging is used to measure sample elongation (deformation). The ability to measure small strains depends on the resolution of the SEM. High magnification images provide superior strain resolution for ultrafine samples, like nanowires shown in Fig. 2.26. Stresses can be

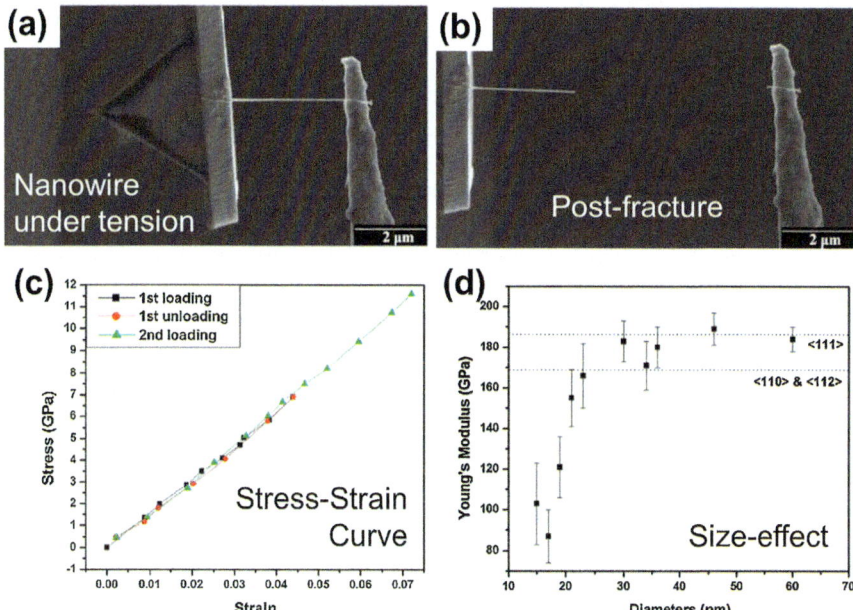

Fig. 2.26 In-situ tensile testing of silver nanowire using AFM cantilever and a nanomanipulator inside SEM: (**a**) nanowire under tension, (**b**) fracture/failure of a nanowire, (**c**) corresponding stress–strain curve for the in-situ tensile test, and (**d**) size effect in mechanical properties examined by correlating modulus with nanowire diameters. (Reproduced and adapted from Zhu et al. (2009))

determined from the loads sensed by the AFM cantilever. A representative stress-strain curve obtained from the in-situ AFM-SEM tensile test is shown in Fig. 2.26c. The plot also demonstrates stress–strain characteristics for multiple loading–unloading cycles. Performing the tests inside SEM allows precise measurement of sample dimensions, which can be useful to obtain further mechanistic insights. Figure 2.26d illustrates the correlation between the nanowire diameter and Young's modulus. A clear size effect in mechanical response is observed: there is a softening trend for diameters less than 30 nm for the case study shown in Fig. 2.26.

2.3.2 Instrumented Nanoindenter Inside Electron Microscopes

Instrumented indentation testing involves penetration of a sample surface by a probe or tip, while recording the load-depth profile. The load–displacement curves can be used for quantifying properties of the material, such as elastic modulus and nanohardness. For real-time visualization, the indenter can be installed inside the SEM or TEM. Figure 2.27a shows an in-situ nanoindenter being installed on an SEM stage. The indenter consists of a load transducer, which applies and senses forces during the test. A tip holder is connected to the transducer, where indenter probes of desired sizes and geometries can be installed for performing the tests. Sample(s) to be tested are mounted on a sample stage. The sample stage can translate in X, Y, and Z axes. Figure 2.27b shows the zoomed view of the indenter probe and the specimen to be tested installed on the sample stage. The sample translation stage allows to move and position the sample/feature of interest directly under the tip prior to the test (Fig. 2.27c). The indentation progression is captured by the incident electron beam in a vacuum chamber. The SEM stage is typically tilted during the test, to be able to resolve and observe the sample surface during indentation. For instance, the sample image in Fig. 2.23c is captured at an 18° tilt angle. Figure 2.27d shows a side-entry instrument for performing indentation testing inside the TEM. This TEM sample holder comprises of a tip, sample mount, and a transducer (Chang 2018). The in-situ nanoindenters can operate in load control and displacement control modes, with achievable force and displacement resolutions in sub-μN and sub-nm regime, respectively. In addition to the depth-sensing tests, nanoindenters can also be employed for scratch testing by incorporating lateral load sensing ability in the transducer (Hintsala et al. 2017). In-situ nanoscratch studies can be useful to examine nanoscale tribological phenomena. The principles, applications, and case studies for in-situ nanoindentation in electron microscopes are discussed in detail in the next chapter.

2.3 Mechanical Instrumentation for Multi-Scale Characterization

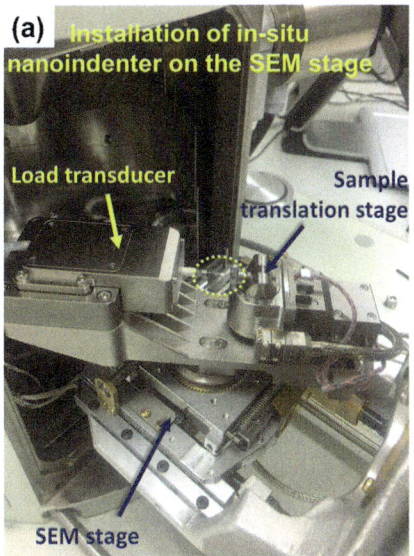

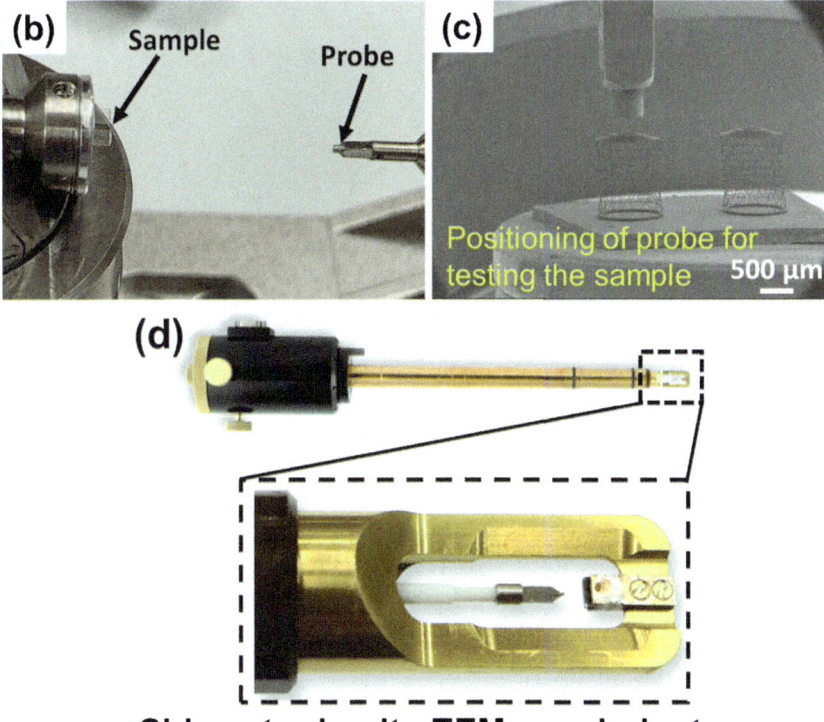

Fig. 2.27 (**a**) Installation of an in-situ nanoindenter on a SEM stage, (**b**) zoomed image of probe and sample (encircled region in (**a**)), (**c**) application of XYZ-sample translation stage for precise positioning of the sample/probe prior to the test, and (**d**) a side-entry instrument for in-situ indentation inside TEM. (Reproduced/adapted with permission from the Agarwal group (FIU), Chang (2018))

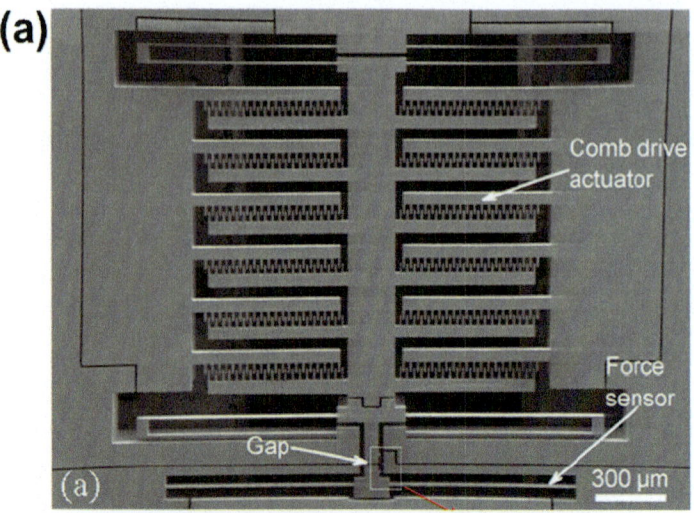
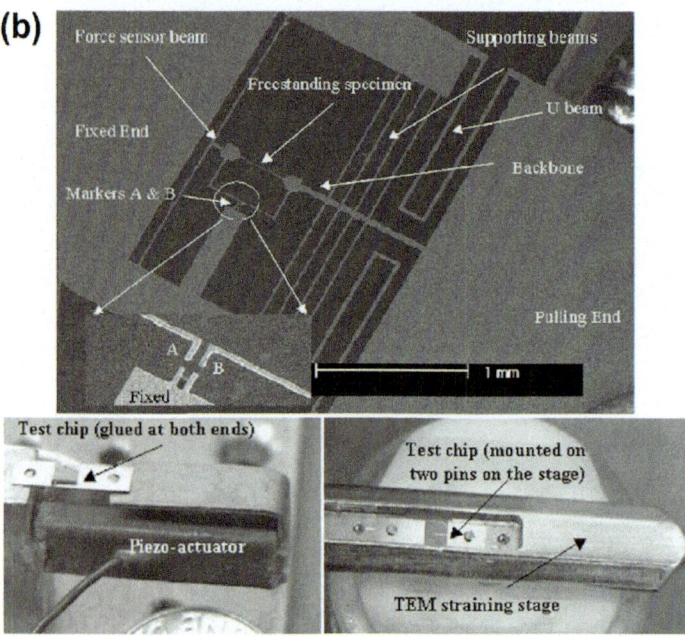

Fig. 2.28 MEMS-based mechanical testing platforms with integrated actuator and load sensors in a single chip (**a, b**), and with external load application by nanoindenter using a push-to-pull micromechanical device (**c, d**). (Reproduced/adapted with permissions from Zhang et al. (2009), Haque and Saif (2002), Lu et al. (2010), Liao et al. (2017) CC BY 4.0)

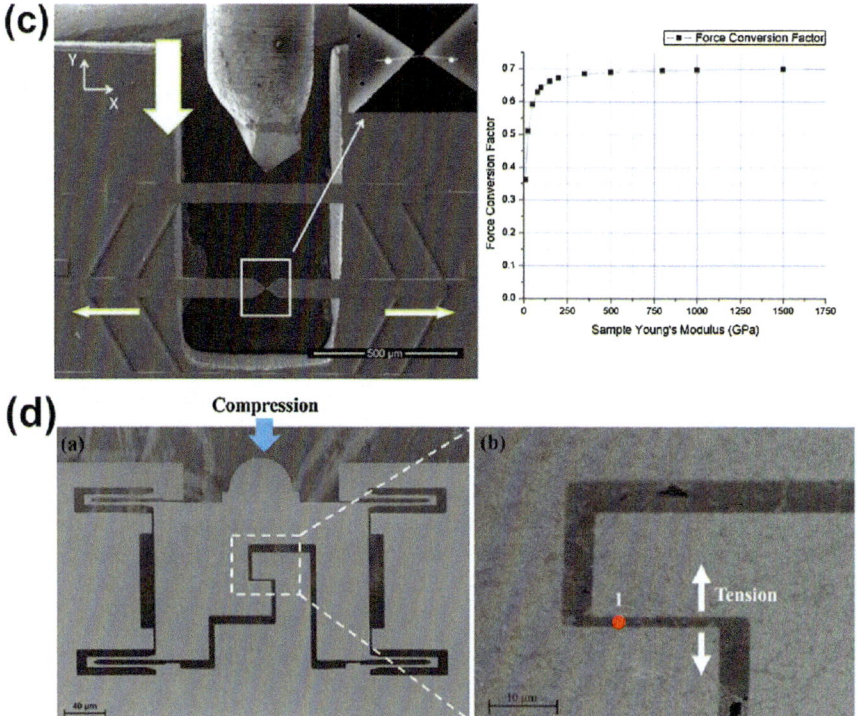

Fig. .28 (continued)

2.3.3 Microelectromechanical Systems for Mechanical Testing

Microelectromechanical systems (MEMS) consist of actuator-sensor-electronics assembly, used to probe, manipulate, and sense a wide diversity of responses from the specimens, such as mechanical, electrical, thermal, chemical, optical, etc. MEMS have found application for in-situ mechanical testing of materials because of their small size (easy to fit inside microscope chambers), compatibility with vacuum (important for testing in electron microscope), excellent displacement/force resolutions (essential for small-scale or nanomaterial testing) and integration of temperature sensors/heating or cooling pads is possible (for environmental mechanical testing) (Haque and Saif 2002).

MEMS-based stages with different designs and operating principles have been developed for in-situ mechanical testing (shown in Fig. 2.28). The MEMS tensile stage shown in Fig. 2.28a comprises of a beam force sensor, which has a gap to suspend the sample to be tested. The sample is deformed by applying DC voltage, which generates electrostatic force between the fingers of the comb drive actuator shown in the figure (Zhang et al. 2009). The force sensor works on the principle of bending of a beam with known spring constant. During the in-situ test, the sample elongation and beam deflection are measured from the frames captured using SEM. The MEMS device design shown in Fig. 2.28b consists of a silicon beam-based force sensor (Haque and Saif 2002). One end of the sample rests on the sensor, while the other end is supported by a backbone structure. A piezoelectric actuator is used to pull the backbone on which one end of the sample rests, thereby applying tensile strains. The force resolution of the device can be varied by changing the beam design. The device chip can be used for mechanical testing in a TEM as well as SEM (the test-setups are shown in Fig. 2.28b). The platforms shown in Fig. 2.28a, b comprise of "in-built" load sensors along with actuators. There are some testing platforms that are dependent on external source for load application. Figure 2.28c shows one such micromechanical device, which is used along with a nanoindenter for in-situ mechanical testing (Lu et al. 2010). The device works on the "push-to-pull" principle: the compression force (push) exerted by indenter is translated to the tensile loading (pull). The indenter applies force in the vertical direction and the four inclined beams (shown in the figure) translate this into horizontal force, which is applied on the sample for tensile testing. The inset of Fig. 2.28c shows the magnified view of the sample under tension using the push-to-pull device. It is noteworthy that the accurate determination of mechanical readings can be difficult for low modulus materials. The problem can be alleviated by running finite element simulations to obtain the relationship between system stiffness and Young's modulus of the samples. This exercise provides load conversion factors for different modulus materials. A plot of conversion factors vs. Young's modulus is shown in

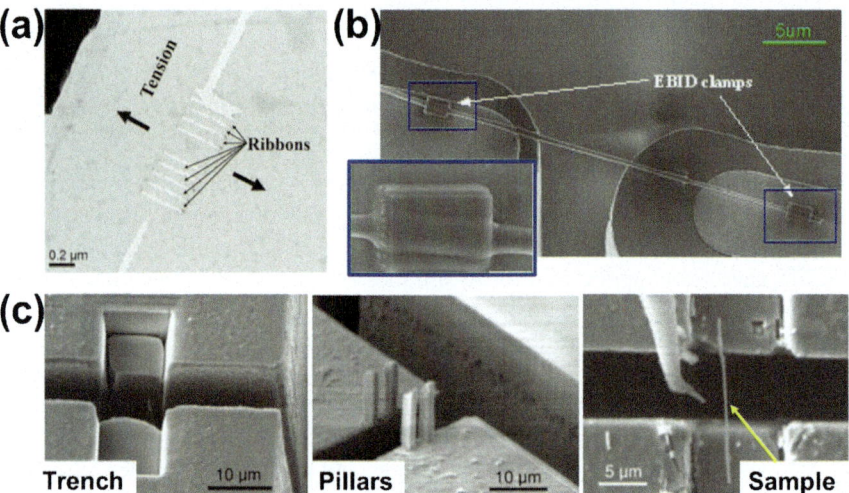

Fig. 2.29 Sample preparation strategies for in-situ mechanical characterization using MEMS devices: (**a**) focused electron beam to pattern the samples suspended on MEMS platform, (**b**) EBID-assisted clamping of sample on MEMS device to avoid slippage, and (**c**) modification of MEMS platform surface topography for effective sample placement/alignment on the device. (Reproduced/adapted with permissions from Liao et al. (2017) CC BY 4.0, Lu et al. (2006), Zhang et al. (2009))

Fig. 2.28c. The suitable conversion factor can then be used for accurately determining mechanical data from the in-situ tests. The device can be coupled with a nanoindenter for in-situ SEM as well as TEM testing. A variation of push-to-pull device design is shown in Fig. 2.28d (Liao et al. 2017).

The samples do not have to be in their final form to load them on the MEMS devices. It is possible to follow different sample preparation protocols after their integration with the platform. As an example, the magnified view of the push-to-pull device in Fig. 2.28d shows a successfully transferred graphene on to the device. The transfer process involved heating graphene-PMMA film in oven followed by the dissolution of PMMA in alcohol in a critical point dryer. The device is robust enough to experience oven heating or mild chemical exposure. Prior to in-situ experiments, the electron beam is also used for final sample preparation steps with the sample installed in the MEMS device. Patterning of graphene to obtain graphene nanoribbons is demonstrated in Fig. 2.29a (Liao et al. 2017). The patterning is performed using a focused electron beam, and the obtained ribbons are then tested using the device. EBID is useful to clamp the sample with the MEMS device to avoid slippage during mechanical loading. Sample clamping for an extremely fine 1D-nanotubular sample is demonstrated in Fig. 2.29b (Lu et al. 2006). Placing extremely fine/low-dimension samples on MEMS devices can be challenging as there are several factors to be considered: avoiding the crash of nanomanipulator with the device, making sure the sample is well-aligned along the loading axis and the sample is properly clamped to avoid slippage. One potential approach is to modify the surface topography of the MEMS testing platform for effectively holding the sample. Figure 2.29c illustrates fabrication of trench or pillar on the device surfaces (Zhang et al. 2009). The sample can then be dropped along the sidewalls of the fabricated structures. The fabricated structures are well-aligned on both the ends where the sample is supposed to be clamped. This helps in avoiding misorientation during sample placement. An example demonstrating the placement of a fine nanowire on the MEMS platform (with a trench on the surface) is shown in Fig. 2.29c.

These case studies show the importance of MEMS platforms for testing the mechanical properties of materials inside high-resolution electron microscopes. The devices are excellent for "miniature" sample testing and sense extremely small loads. This allows the correlation of quantitative and qualitative mechanical characteristics of a wide diversity of materials.

2.3.4 *Micromechanical Stage for In-Situ Testing*

Micro-meso load frames for in-situ testing are useful to bridge the gap between nanoscale and macroscale mechanical characterization. Figure 2.30a shows the line diagram of a micro-load frame, consisting of a load cell (to sense forces), actuator (for deforming the samples), extensometer (to determine displacements/strains), and fixtures for different sample types and sizes (Rudolf et al. 2016). The compact

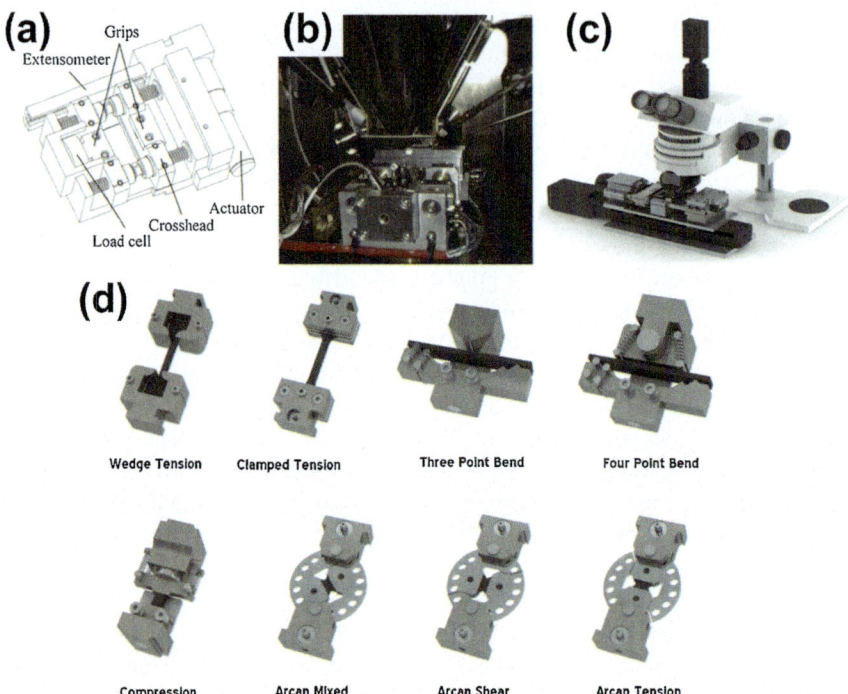

Fig. 2.30 (a) A line diagram providing an overview of the design and components of a micromechanical tester. Use of the micromechanical tester for in-situ imaging is shown in: (**b**) scanning electron microscope, and (**c**) optical microscope. (**d**) Examples of different fixtures for a micromechanical stage to perform mechanical testing under different loading conditions. (Some of the images are reproduced/adapted with permission from Rudolf et al. (2016). Sources: MTI Instruments (Tensile Stage), Psylotech (Micro Test System), Boesl group (FIU))

design of the stage allows installation inside an electron microscope chamber (Fig. 2.30b) or under an optical microscope (Fig. 2.30c) for real-time imaging. Micromechanical stages provide sub-micrometer displacement resolutions. The force range is dependent on the load cell installed in the tester, with achievable resolutions down to a few milli-Newtons and peak load sensing ability can be as high as several kilo-Newtons. The stages are capable of displacement- and load-controlled tests, as well as precise control over crosshead speed to investigate the effect of varying strain rates on the deformation of materials. Custom designed fixtures can be used to perform different kinds of mechanical tests, such as tension, compression, three- and four-point bending, indentation, double cantilever, and shear (Fig. 2.30d). The stages can also be customized to include heating elements to probe the high-temperature mechanical properties of materials.

There are some disadvantages or limitations of micromechanical stages. Due to space constraint, the total deformation span is often limited. This makes the testing of ultra-stretchable samples up to failure difficult (Nieto et al. 2015). This is illus-

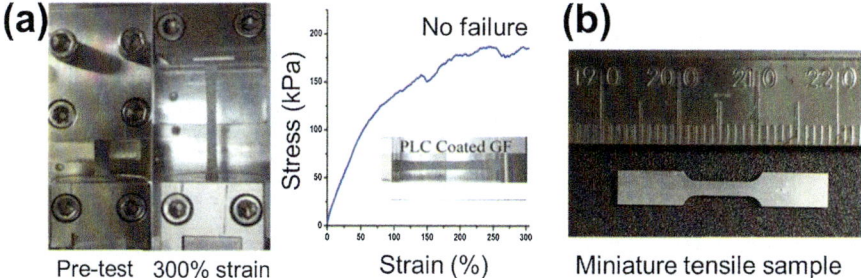

Fig. 2.31 (a) Illustration of limited test space in a micromechanical stage insufficient for achieving failure of a stretchable polymer, and (b) a miniature dogbone shaped tensile specimen (~2.5 cm in length, 0.2 cm in thickness and 0.2 cm in gage width) of a high-strength metal matrix composite is fabricated and used for testing to be able to achieve failure stresses. (Reproduced/adapted with permissions from Nieto et al. (2015), Nautiyal et al. (2016), Agarwal group (FIU))

trated in Fig. 2.31a, where the elastomer does not fail even after 300% strain (which is the upper limit of the available displacement span in the specific micromechanical tester used). Another limitation is the peak load capacity practically achievable in the micromechanical testers (few kN). These loads might not be sufficient for testing strong structural materials. One possible strategy to overcome this limitation is to prepare miniature specimens (illustrated in Fig. 2.31b), so that the failure stresses can be achieved with relatively lower loads (Nautiyal et al. 2016). However, one might start observing size effects in mechanical properties as the specimen size is significantly lowered. The effect of multiple grains/grain boundaries, different microstructure phases, and structural flaws might not be effectively captured in miniature specimens. Although such investigations provide fundamental insights into multiscale stress-transfer mechanisms, the overall properties and response can deviate significantly from the bulk material response. This necessitates macroscale in-situ testing, where the limitations of load and displacement spans can be overcome.

2.3.5 Macroscale In-Situ Characterization

A universal testing machine (UTM) is typically used for macroscale mechanical testing of bulk samples. A UTM comprises of the load-bearing column(s) and a moveable crosshead for straining the samples during the test. These testers are employed for different kinds of deformation modes, such as tensile, compressive, and flexural loading. The range and resolution of forces during the test depends on the load cell being used. The load frames with load capacities as high as several Mega-Newtons can be used for testing of high-strength materials. A high-load capacity UTM is shown in Fig. 2.32a. On the other side of the spectrum, low-load frames are used for testing softer samples. These low-load frames differ from the

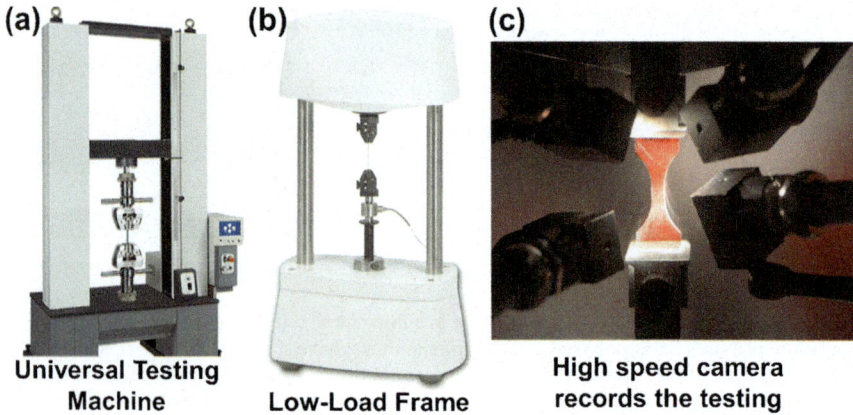

Fig. 2.32 Tools and techniques for macroscale testing of bulk materials: (**a**) a high-load capacity universal testing machine, (**b**) a low-load frame for testing soft and slender samples, and (**c**) illustration of in-situ imaging using a high-speed camera to capture the progression of deformation during bulk testing. (Website Sources: Qualitest (Universal Testing Machine—Tensile Tester), Electroforce (Mechanical Test Instruments), UL testing facilities—Germany (High-Speed Tensile Testing))

micromechanical stages in terms of test space, as the available displacement span is much higher (>150 mm). Contrary to this, micromechanical stages allow rather limited span (less than 50 mm). Therefore, the low-load frames overcome the limitation associated with micromechanical testers and allow testing of macro-sized specimens with excellent force resolution. For instance, testing of soft tissues or tissue constructs, long and slender fibers and soft elastomers requires large test space and low-load resolutions. An example of a low-load frame is shown in Fig. 2.32b. To use a macroscale tester for in-situ characterization, an optical camera can be installed and used for real-time imaging while the mechanical test is conducted. This allows observation of failure zones and failure mechanisms (Embrey et al. 2017). The in-situ imaging is useful for determining strains by digital image correlation (DIC) analysis of the real-time snapshots. Additionally, strain evolution in the entire sample can be examined by DIC analysis, providing insights into strain redistribution in the microstructure during mechanical loading (Nautiyal et al. 2018b, c). The application of the DIC technique for analyzing in-situ videos is discussed in great detail in Chap. 6. For high strain-rate testing, an ultrahigh-speed camera can be employed for monitoring the progression of deformation and failure of materials (Fig. 2.32c).

2.3.6 Summary

There are three key categories of tools essential for in-situ mechanical characterization of materials: (a) sample preparation tools, (b) microscopes for in-situ imaging, and (c) mechanical testing equipment/instrumentation. FIB is the most commonly used technique to fabricate miniature samples for in-situ characterization. It is important to take into consideration ion implantation, redeposition, and amorphization phenomena associated with FIB milling to prepare samples with predetermined shapes/dimensions while avoiding undesired artifacts. The optical microscope, SEM and TEM are three broad classes of microscopes used for in-situ imaging as the samples are subjected to mechanical loads. The selection of microscope depends on the sample size, features of interest, and the nature of deformation mechanisms that are to be investigated. Other aspects, such as vacuum compatibility, sample conductivity, rate of deformation and sensitivity to imaging source are also important for in-situ imaging of sample deformation. Additional tools are used for special imaging or analysis requirements. A high-speed camera is useful to capture deformation videos with very high frame rates (for phenomena happening at very small timescales). Optical tweezer is a specialized optics-based technique for simultaneously probing and manipulating samples to capture their mechanical response. EBSD can be coupled with SEM for in-situ mechanical testing to determine crystallographic deformation mechanisms. The samples can be subjected to localized or global deformation by using different mechanical testers/platforms. Atomic force microscope, nanoindenter, MEMS device, micromechanical stage, and universal testing machine are useful for mechanical testing of materials at different length and load scales. The selection of mechanical testing equipment influences the achievable force/stress resolution. Strain/displacement resolutions can depend either on the mechanical equipment or the microscope depending on the nature of the test and instrumentation involved. The case studies presented in the chapter elucidate the application of these tools and techniques to understand the mechanics of materials.

Questions and Assignments

1. Which one of the following in-situ imaging technique(s) is/are suitable for identifying slip mechanisms associated with the indentation deformation of an Al-Mg alloy specimen?

    (a) In-situ optical microscopy coupled with a high-speed camera.
    (b) In-situ scanning electron microscopy coupled with electron backscatter diffraction.
    (c) In-situ transmission electron microscopy coupled with selected area electron diffraction.

2. While using an atomic force microscope inside an electron microscope for in-situ tensile testing, which of the following statements hold true?

   (a) Force resolution is dependent on the resolution capability of the electron microscope.
   (b) Stress resolution (in stress–strain plot for sample deformation) is dependent on the AFM cantilever used.
   (c) Strain resolution (in stress–strain plot for sample deformation) is dependent on the resolution capability of the electron microscope.
   (d) AFM cantilever deflection cannot be measured by electron imaging in SEM/TEM.

3. Determination of stress–strain characteristics is one of the key objectives of mechanical testing. Tensile testing is often influenced by the slippage of the sample at the grip. This can lead to incorrect estimation of strains and elastic modulus. How can in-situ imaging be used for accurate determination of strains during deformation?

4. Donald's research project is focused on the fabrication and mechanical properties of ultrahigh-temperature ceramics. He is considering using an in-situ SEM-based mechanical characterization technique to study the deformation mechanisms in real time. What could be some of the challenges he might experience during the in-situ imaging of ceramic failure?

5. Jenniffer applies in-situ SEM-based mechanical characterization techniques to study the deformation of a multiphase composite system. The microstructure of the material consists of BN and Fe phases. She wants to use in-situ imaging to identify the differences in the deformation behavior/mechanisms for the two phases. What mode of SEM imaging should she use for in-situ imaging to be able to distinguish the two phases clearly?

6. Arrange the following mechanical instrumentation in the increasing order of the best achievable load resolutions for mechanical testing:

   (a) Micromechanical stage.
   (b) Universal testing machine.
   (c) Nanoindenter.
   (d) Atomic force microscope.

7. The figure below shows indents made on ferritic stainless steel sample by the in-situ SEM indentation technique. The tests were done to obtain hardness values (as shown by a superimposed color-coded map below). The tests are conducted to understand the role of hydrogen charging on nanomechanical properties. The number under each indent represents the hydrogen charging time in minutes. By looking at this map, what inference would you make on the effect of hydrogen charging on the mechanics of this material? Is there any obvious correlation?

References

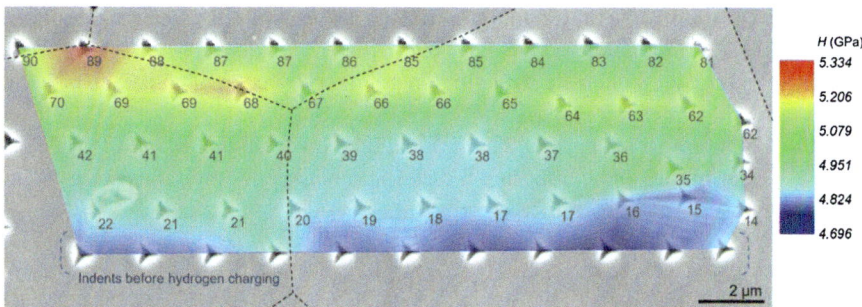

Source: Kim et al. International Journal of Hydrogen Energy 44 (2019) 6333–6343. Reproduced with permission (Kim and Tasan 2019)

8. Micropillar compression study was conducted on an Mg6Zn alloy and Mg2Zn containing a dense dispersion of SiC nanoparticles. The stress–strain curves and the SEM images of the pillars post-compression are shown below. Comment on difference in the magnitude and nature of the stress–strain curves for the two specimens. Correlate these differences with the post-compression SEM micrographs showing the deformation mechanisms.

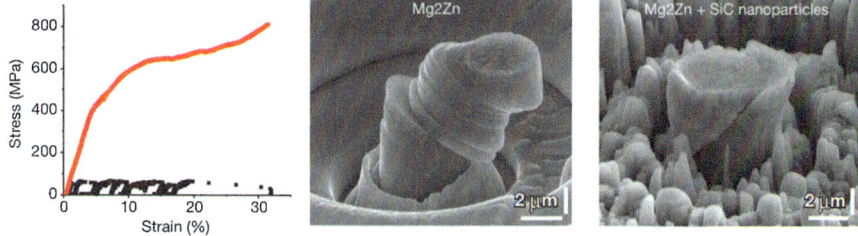

Stress–strain curve: black curve corresponds to Mg_2Zn and red curve corresponds to Mg_2Zn + SiC. Images are reproduced with permission from (Chen et al. 2015)

9. List some of the important considerations/challenges for installing an in-situ mechanical tester inside a scanning electron microscope for real-time testing, and what could be some of the measures one needs to take to address these. Take the following points into account for framing your answer:

    (a) Overall space inside the vacuum chamber.
    (b) Communication between the in-situ tester inside the chamber and the controller/electronics stationed outside.
    (c) Time to evacuate the vacuum chamber after installing the in-situ tester.
    (d) Quality of imaging: Fast SEM scan vs. slow SEM scan for capturing real-time videos.
    (e) Imaging at high magnifications to resolve fine features vs. having large enough field of view to identify the region of failure initiation.

10. Name the mechanical instrumentation(s) suitable for each of the following characterizations (answer separately for each part):

(a) Determining the elastic modulus of a nanofiber, which is *10 nm in diameter and 1 μm in length.*
(b) Finding out the tensile strength of a polymeric sheet, which is *25 mm long, 5 mm wide and 2 mm thick.*
(c) Compressive strength of a cylindrical billet of titanium, *30 cm in radius.*
(d) Determining and comparing the elastic modulus of "individual phases" in a material with two different phases.

References

Adams BL, Kalidindi SR, Fullwood DT (2013) Electron backscatter diffraction microscopy and basic stereology. In: Microstructure sensitive design for performance optimization. Elsevier Inc., pp 341–371
Buffiere J-Y, Maire E, Adrien J et al (2010) In situ experiments with X ray tomography: an attractive tool for experimental mechanics. Exp Mech 50:289–305. https://doi.org/10.1007/s11340-010-9333-7
Cai Z, Cui X, Liu E et al (2017) Fracture behavior of high-entropy alloy coating by in-situ TEM tensile testing. J Alloys Compd 729:897–902. https://doi.org/10.1016/j.jallcom.2017.09.233
Calvié E, Joly-pottuz L, Esnouf C et al (2012) Real time TEM observation of alumina ceramic nano-particles during compression. J Eur Ceram Soc 32:2067–2071. https://doi.org/10.1016/j.jeurceramsoc.2012.02.029
Cao W, Marvel C, Yin D et al (2016) Correlations between microstructure, fracture morphology, and fracture toughness of nanocrystalline Ni-W alloys. Scr Mater 113:84–88. https://doi.org/10.1016/j.scriptamat.2015.09.030
Cepeda-Jimenez CM, Molina-Aldareguia JM, Perez-Prado MT (2016) EBSD-assisted slip trace analysis during in situ SEM mechanical testing: application to unravel grain size effects on plasticity of pure mg polycrystals. JOM 68:116–126. https://doi.org/10.1007/s11837-015-1521-6
Chang S (2018) In-situ nanomechanical testing in electron microscopes. In: Hsueh C-H, Schmauder S, Chen C-S et al (eds) Handbook of mechanics of materials. Springer, Singapore, pp 1–47
Chen LY, Xu JQ, Choi H et al (2015) Processing and properties of magnesium containing a dense uniform dispersion of nanoparticles. Nature 528:539–543. https://doi.org/10.1038/nature16445
Di Gioacchino F, Clegg WJ (2014) Mapping deformation in small-scale testing. Acta Mater 78:103–113. https://doi.org/10.1016/j.actamat.2014.06.033
Embrey L, Nautiyal P, Loganathan A et al (2017) Three-dimensional graphene foam induces multifunctionality in epoxy nanocomposites by simultaneous improvement in mechanical, thermal, and electrical properties. ACS Appl Mater Interfaces 9:39717–39727. https://doi.org/10.1021/acsami.7b14078
Espinosa HD, Bernal RA, Filleter T (2012) In situ TEM electromechanical testing of nanowires and nanotubes. Small 8:3233–3252. https://doi.org/10.1002/smll.201200342
Giannuzzi LA, Prenitzer BI, Kempshall BW (2005) Ion - solid interactions. In: Giannuzzi LA, Stevie FA (eds) Introduction to focused ion beams: instrumentation, theory, techniques and practice. Springer, Boston, pp 13–52
Golberg D, Costa PMFJ, Lourie O et al (2007) Direct force measurements and kinking under elastic deformation of individual multiwalled boron nitride nanotubes. Nano Lett 7:2146–2151. https://doi.org/10.1021/nl070863r
Greer JR, Th J, De Hosson M (2011) Progress in materials science plasticity in small-sized metallic systems: intrinsic versus extrinsic size effect. Prog Mater Sci 56:654–724. https://doi.org/10.1016/j.pmatsci.2011.01.005
Grier DG (2003) A revolution in optical manipulation. Nature 424:810–816

Gustafsson G, Nishida M, Ito Y et al (2015) Mechanical characterization and modelling of the temperature-dependent impact behaviour of a biocompatible poly(L-lactide)/poly(ε-caprolactone) polymer blend. J Mech Behav Biomed Mater 51:279–290. https://doi.org/10.1016/j.jmbbm.2015.07.007

Haque MA, Saif MTA (2002) Application of MEMS force sensors for in situ mechanical characterization of nano-scale thin films in SEM and TEM. Sensors Actuators A Phys 97–98:239–245. https://doi.org/10.1016/S0924-4247(01)00861-5

Hintsala E, Kiener D, Jackson J, Gerberich WW (2015) In-situ measurements of free-standing, ultra-thin film cracking in bending. Exp Mech 55:1681–1690. https://doi.org/10.1007/s11340-015-0069-2

Hintsala ED, Stauffer DD, Oh Y, Asif SAS (2017) In situ TEM scratch testing of perpendicular magnetic recording multilayers with a novel MEMS tribometer. JOM 69:51–56. https://doi.org/10.1007/s11837-016-2154-0

Jerabek M, Major Z, Lang RW (2010) Strain determination of polymeric materials using digital image correlation. Polym Test 29:407–416. https://doi.org/10.1016/j.polymertesting.2010.01.005

Jones PH, Maragò OM, Volpe G (2015) Optical tweezers: principles and applications. Cambridge University Press, Cambridge

Kalácska S, Dankházi Z, Zilahi G et al (2020) Investigation of geometrically necessary dislocation structures in compressed Cu micropillars by 3-dimensional HR-EBSD. Mater Sci Eng A 770:138499. https://doi.org/10.1016/j.msea.2019.138499

Kellermayer MSZ, Smith SB, Granzier HL, Bustamante C (1997) Folding-unfolding transitions in single titin molecules characterized with laser tweezers. Science 276:1112–1116. https://doi.org/10.1126/science.276.5315.1112

Kiener D, Minor AM (2011) Source-controlled yield and hardening of Cu (1 0 0) studied by in situ transmission electron microscopy. Acta Mater 59:1328–1337. https://doi.org/10.1016/j.actamat.2010.10.065

Kiener D, Grosinger W, Dehm G, Pippan R (2008) A further step towards an understanding of size-dependent crystal plasticity: in situ tension experiments of miniaturized single-crystal copper samples. Acta Mater 56:580–592. https://doi.org/10.1016/j.actamat.2007.10.015

Kim JY, Greer JR (2009) Tensile and compressive behavior of gold and molybdenum single crystals at the nano-scale. Acta Mater 57:5245–5253. https://doi.org/10.1016/j.actamat.2009.07.027

Kim J, Tasan CC (2019) Microstructural and micro-mechanical characterization during hydrogen charging: an in situ scanning electron microscopy study. Int J Hydrog Energy 44:6333–6343. https://doi.org/10.1016/j.ijhydene.2018.10.128

Koch MD, Shaevitz JW (2017) Introduction to optical tweezers. In: Optical tweezers - methods and protocols. Springer, New York, pp 3–24

Lee G, Kim JY, Burek MJ et al (2011) Plastic deformation of indium nanostructures. Mater Sci Eng A 528:6112–6120. https://doi.org/10.1016/j.msea.2011.04.065

Legros M (2014) In situ mechanical TEM: seeing and measuring under stress with electrons. Comptes Rendus Phys 15:224–240. https://doi.org/10.1016/j.crhy.2014.02.002

Li C, Liu KK (2008) Nanomechanical characterization of red blood cells using optical tweezers. J Mater Sci Mater Med 19:1529–1535. https://doi.org/10.1007/s10856-008-3382-9

Liao Z, Sandonas LM, Zhang T et al (2017) In-situ stretching patterned graphene nanoribbons in the transmission electron microscope. Sci Rep 7:211. https://doi.org/10.1038/s41598-017-00227-3

Liphardt J, Onoa B, Smith SB et al (2001) Reversible unfolding of single RNA molecules by mechanical force. Science 292:733–737. https://doi.org/10.1126/science.1058498

Litvinov RI, Shuman H, Bennett JS, Weisel JW (2002) Binding strength and activation state of single fibrinogen-integrin pairs on living cells. Proc Natl Acad Sci U S A 99:7426–7431. https://doi.org/10.1073/pnas.112194999

Liu D, Flewitt PEJ (2017) Deformation and fracture of carbonaceous materials using in situ micro-mechanical testing. Carbon N Y 114:261–274. https://doi.org/10.1016/j.carbon.2016.11.084

Liu F, Tang D-M, Gan H et al (2013) Individual boron nanowire has ultra-high specific young's modulus and fracture strength as revealed by in situ transmission electron microscopy. ACS Nano 7:10112–10120. https://doi.org/10.1021/nn404316a

Loganathan A, Sharma A, Rudolf C et al (2017) In-situ deformation mechanism and orientation effects in sintered 2D boron nitride nanosheets. Mater Sci Eng A 708:440–450. https://doi.org/10.1016/j.msea.2017.10.019

Lu S, Guo Z, Ding W et al (2006) In situ mechanical testing of templated carbon nanotubes. Rev Sci Instrum 77:125101. https://doi.org/10.1063/1.2400212

Lu Y, Ganesan Y, Lou J (2010) A multi-step method for in situ mechanical characterization of 1-D nanostructures using a novel micromechanical device. Exp Mech 50:47–54. https://doi.org/10.1007/s11340-009-9222-0

Maragò OM, Jones PH, Gucciardi PG et al (2013) Optical trapping and manipulation of nanostructures. Nat Nanotechnol 8:807–819. https://doi.org/10.1038/nnano.2013.208

Matteson TL, Schwarz SW, Houge EC et al (2002) Electron backscattering diffraction investigation of focused ion beam surfaces. J Electron Mater 31:33–39

Moffitt JR, Chemla YR, Smith SB, Bustamante C (2008) Recent advances in optical tweezers. Annu Rev Biochem 77:205–228. https://doi.org/10.1146/annurev.biochem.77.043007.090225

Nautiyal P, Rudolf C, Loganathan A et al (2016) Directionally aligned ultra-long boron nitride nanotube induced strengthening of aluminum-based sandwich composite. Adv Eng Mater 18:1747–1754. https://doi.org/10.1002/adem.201600212

Nautiyal P, Alam F, Balani K, Agarwal A (2018a) The role of nanomechanics in healthcare. Adv Healthc Mater 7:1700793. https://doi.org/10.1002/adhm.201700793

Nautiyal P, Mujawar M, Boesl B, Agarwal A (2018b) In-situ mechanics of 3D graphene foam based ultra-stiff and flexible metallic metamaterial. Carbon N Y 137:502–510. https://doi.org/10.1016/j.carbon.2018.05.063

Nautiyal P, Zhang C, Champagne VK et al (2018c) In-situ mechanical investigation of the deformation of splat interfaces in cold-sprayed aluminum alloy. Mater Sci Eng A 737:297–309. https://doi.org/10.1016/j.msea.2018.09.065

Nautiyal P, Zhang C, Loganathan A et al (2019) High-temperature mechanics of boron nitride nanotube "Buckypaper" for engineering advanced structural materials. ACS Appl Nano Mater 2:4402–4416. https://doi.org/10.1021/acsanm.9b00817

Neuman KC, Nagy A (2008) Single-molecule force spectroscopy: optical tweezers, magnetic tweezers and atomic force microscopy. Nat Methods 5:491–505. https://doi.org/10.1038/nmeth.1218

Nieto A, Dua R, Zhang C et al (2015) Three dimensional graphene foam/polymer hybrid as a high strength biocompatible scaffold. Adv Funct Mater 25:3916–3924. https://doi.org/10.1002/adfm.201500876

Oh SH, Legros M, Kiener D, Dehm G (2009) In situ observation of dislocation nucleation and escape in a submicrometre aluminium single crystal. Nat Mater 8:95–100. https://doi.org/10.1038/nmat2370

Patterson BM, Cordes NL, Henderson K et al (2016) In situ laboratory-based transmission X-ray microscopy and tomography of material deformation at the nanoscale. Exp Mech 56:1585–1597. https://doi.org/10.1007/s11340-016-0197-3

Rabe R, Breguet J-M, Schwaller P et al (2004) Observation of fracture and plastic deformation during indentation and scratching inside the scanning electron microscope. Thin Solid Films 469–470:206–213. https://doi.org/10.1016/j.tsf.2004.08.096

Rafael Velayarce J, Zamanzade M, Torrents Abad O, Motz C (2018) Influence of single and multiple slip conditions and temperature on the size effect in micro bending. Acta Mater 154:325–333. https://doi.org/10.1016/j.actamat.2018.05.054

Rudolf C, Boesl B, Agarwal A (2016) In situ mechanical testing techniques for real-time materials deformation characterization. JOM 68:136–142. https://doi.org/10.1007/s11837-015-1629-8

References

Sumigawa T, Byungwoon K, Mizuno Y et al (2018) In situ observation on formation process of nanoscale cracking during tension-compression fatigue of single crystal copper micron-scale specimen. Acta Mater 153:270–278. https://doi.org/10.1016/j.actamat.2018.04.061

Sutton MA, Orteu J-J, Schreier H (2009) Image correlation for shape, motion and deformation measurements; basic concepts, theory and applications. Springer, Boston, MA

Uchic MD, Dimiduk DM (2005) A methodology to investigate size scale effects in crystalline plasticity using uniaxial compression testing. Mater Sci Eng A 400–401:268–278. https://doi.org/10.1016/j.msea.2005.03.082

Voisin T, Grapes MD, Zhang Y et al (2016) DTEM in situ mechanical testing: defects motion at high strain rates. In: Casem D, Lamberson L, Kimberley J (eds) Dynamic behavior of materials, vol 1. Springer, Cham, pp 209–213

Xie W, Zhang R, Headrick RJ et al (2019) Dynamic strengthening of carbon nanotube fibers under extreme mechanical impulses. Nano Lett 19:3519–3526. https://doi.org/10.1021/acs.nanolett.9b00350

Yan L, Fan J (2016) In-situ SEM study of fatigue crack initiation and propagation behavior in 2524 aluminum alloy. Mater Des 110:592–601. https://doi.org/10.1016/j.matdes.2016.08.004

Ying X, Barlow NJ, Feuston MH (2011) Micro-CT and volumetric imaging in developmental toxicology. In: Gupta RC (ed) Reproductive and developmental toxicology. Academic Press, pp 983–1000

Yoo J, Carroll J, Emery J, Kacher J (2019) Understanding heterogeneous deformation in polycrystalline Al 6061 by in situ SEM deformation and HREBSD characterization. Microsc Microanal 23:770–771. https://doi.org/10.1017/S1431927617004512

Zeng XM, Du Z, Tamura N et al (2017) In-situ studies on martensitic transformation and high-temperature shape memory in small volume zirconia. Acta Mater 134:257–266. https://doi.org/10.1016/j.actamat.2017.06.006

Zhang D, Breguet JM, Clavel R et al (2009) In situ tensile testing of individual co nanowires inside a scanning electron microscope. Nanotechnology 20:365706. https://doi.org/10.1088/0957-4484/20/36/365706

Zhang X, Ma L, Zhang Y (2013) High-resolution optical tweezers for single- molecule manipulation. Yale J Biol Med 86:367–383

Zhang P, Ma L, Fan F et al (2014) Fracture toughness of graphene. Nat Commun 5:3782. https://doi.org/10.1038/ncomms4782

Zhu Y, Xu F, Qin Q et al (2009) Mechanical properties of vapor–liquid–solid synthesized silicon nanowires. Nano Lett 9:3934–3939. https://doi.org/10.1021/nl902132w

Chapter 3
Test Methods for In-Situ Mechanical Characterization

The in-situ approach can be applied to understand the mechanical response of materials under different loading scenarios. This chapter introduces the mechanical test methods and how they can be combined with real-time imaging at multiple length scales. The chapter starts with the nanoindentation technique for selectively probing the response of local features in a material. Application of real-time imaging for tribological studies is discussed from the standpoint of wear, interfacial friction, debonding, rolling, and sliding. Nano- and micro-pillar compression method enables localized testing with a uniaxial stress-state, overcoming limitations associated with the nanoindentation method. Beam bending approach is introduced and its application for in-situ fatigue and fracture studies is elucidated. The tensile test method is discussed for uniaxial loading of materials and the utility of real-time imaging is highlighted for studying notch sensitivity, inhomogeneous deformation behavior, and failure mechanisms. In the end, the double cantilever beam testing method is introduced for inducing stable crack growth in otherwise brittle materials to study fracture under controlled condition. This chapter is an extension of the previous chapter as we discuss the applications of the tools and instrumentation introduced in Chap. 2 for performing different kinds of mechanical tests and to extract a wide range of mechanical properties of materials (schematically illustrated in Fig. 3.1).

3.1 Indentation for Localized Deformation Study

Indentation-based mechanical characterization involves penetration and retraction of a specimen by a hard tip or probe with preprogrammed load or displacement (Fig. 3.2a) (Oliver and Pharr 1992, 2010; Chudoba et al. 2004; Fisher-Cripps 2005;

Electronic supplementary material The online version of this chapter (https://doi.org/10.1007/978-3-030-43320-8_3) contains supplementary material, which is available to authorized users.

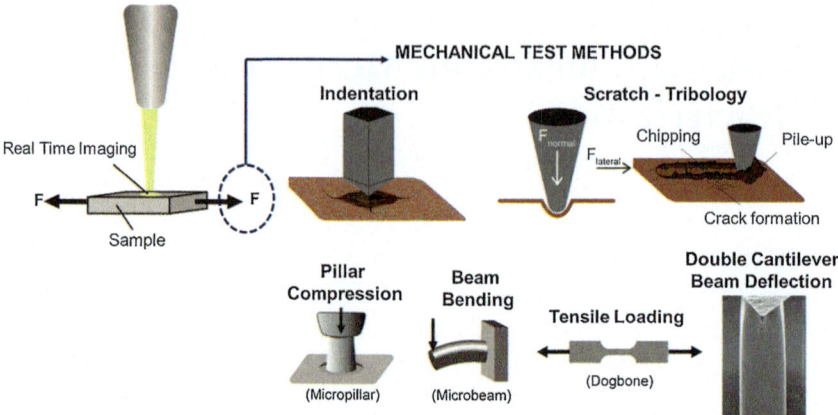

Fig. 3.1 Schematic summary of in-situ mechanical test methods discussed in this chapter (see next page). (One of the images used is adapted with permission from Sernicola et al. (2017))

Nautiyal et al. 2015). A hard material, such as diamond, is typically used for fabricating the tips to avoid tip damage or excess tip deformation during the test. The load–displacement (F–h) response is recorded during the test (Fig. 3.2b). Indentation load–displacement curves have elastic and plastic components. Material removal due to tip penetration is responsible for permanent impression/indent visible after the test. The F–h curves can be analyzed to determine the mechanical properties of the material being probed, such as elastic modulus and hardness. The projected area of tip-sample contact (A_c) is required to calculate these mechanical properties. This area can be computed by taking into consideration the tip geometry and the contact depth during indentation. The formulae for the A_c for different tip geometries as a function of contact depth, h_c are shown in Table 3.1. Once the contact area is known, the indentation modulus, also known as reduced modulus or combined modulus, can be determined using the equation (Fisher-Cripps 2005):

$$E_r = \frac{1}{2} \frac{dF}{dh} \frac{\sqrt{\pi}}{\sqrt{A_c}} \quad (3.1)$$

where dF/dh is called contact stiffness and is computed at the point of onset of unloading (shown in Fig. 3.2b). The elastic modulus of the sample (E_s) can then be calculated using the following equation:

$$\frac{1}{E_r} = \frac{(1-v_s^2)}{E_s} + \frac{(1-v_i^2)}{E_i} \quad (3.2)$$

where v is the Poisson's ratio, and the subscripts "s" and "i" stand for sample and indenter, respectively.

3.1 Indentation for Localized Deformation Study

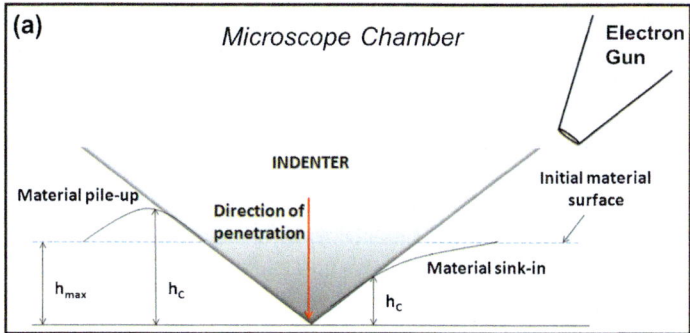

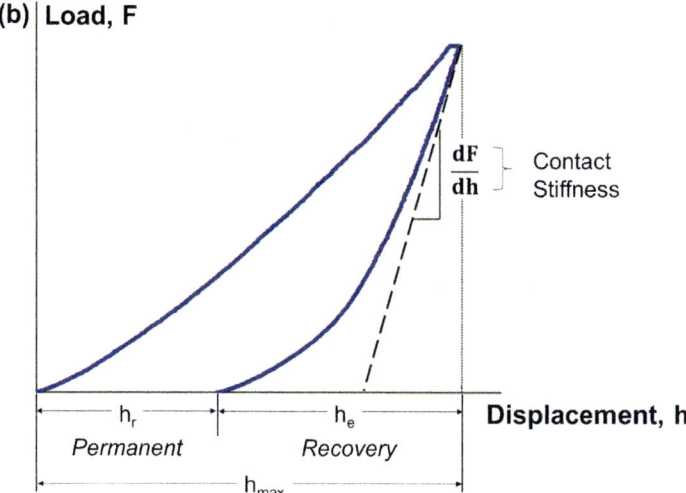

Fig. 3.2 (a) Schematic illustration of in-situ indentation testing of material, and (b) a load–displacement curve

Table 3.1 Area of contact for different nanoindenter tip geometries

Tip geometries	Area of contact, A_c
Berkovich	$3\sqrt{3}h_c^2 \tan^2\theta$; $\theta = 65.27°$
Cube-corner	$3\sqrt{3}h_c^2 \tan^2\theta$; $\theta = 35.26°$
Conospherical	$2\pi R h_c$
Knoop	$2h_c^2 \tan\theta_1 \tan\theta_2$; $\theta_1 = 86.25°, \theta_2 = 65°$
Vickers	$4h_c^2 \tan^2\theta$; $\theta = 68°$
Conical	$\pi h_c^2 \tan^2\alpha$; α is the semi cone angle

Unlike bulk-scale tests, such as tension, compression, torsion, and bending, indentation testing facilitates the evaluation of localized, site-specific elastic modulus. This allows to measure and compare the mechanical response of individual microstructure features or constituents. Some of the features can be extremely fine, not possible to resolve by optics. In-situ indentation inside an electron microscope enables inspection of these fine features. In-situ imaging allows observation of indentation mechanisms in real-time, making possible holistic understanding of qualitative and quantitative aspects related to mechanical deformation. Figure 3.3a shows an in-situ nanoindentation investigation of transition metal nitrides inside SEM (Rzepiejewska-Malyska et al. 2009). SEM imaging showed the prevalence of radial cracks for TiN, whereas pileup was the primary deformation mechanism for CrN film. Nanoindentation of multilayer TiN/CrN revealed a mixed deformation mode. A combination of radial cracks and pileup can be seen in the SEM micrographs for this multilayer film. Therefore, real-time imaging can inform the engineering of multicomponent structures and assemblies to achieve the desired mechanical response. Another case study where in-situ nanoindentation is employed to establish structure-mechanics correlation is shown in Fig. 3.3b (Heiroth et al. 2011). The figure compares the deformation mechanisms in thin films

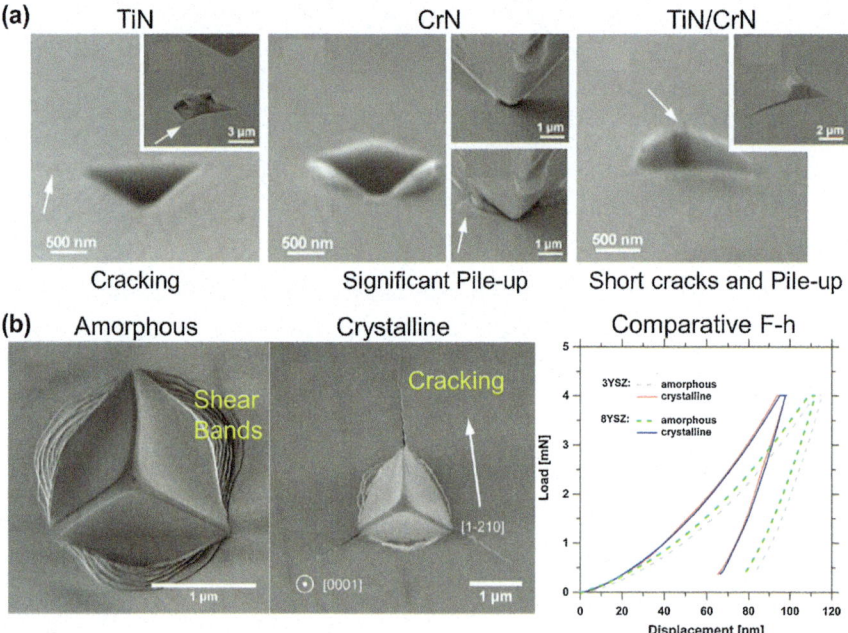

Fig. 3.3 (**a**) In-situ indentation of transition metal nitride thin films in SEM reveals different deformation mechanisms, and (**b**) qualitative (mechanisms) and quantitative ($F–h$ curves) comparison of nanoindentation response for amorphous and crystalline YSZ by in-situ SEM nanoindentation. (Reproduced/adapted with permissions from Rzepiejewska-Malyska et al. (2009), Heiroth et al. (2011))

($t \sim$ 600–700 nm) of crystalline and amorphous yttria-stabilized zirconia (YSZ). The micrographs showed altogether different mechanisms: while amorphous film deformed by the formation of shear bands, the crystalline films demonstrated cracking during indentation. The understanding and observation of these mechanisms by in-situ imaging are helpful to put into perspective the load–displacement curves obtained in ex-situ indentation testing (comparative F–h curves for crystalline and amorphous YSZ are shown in Fig. 3.3b).

These case studies discussed so far involve testing of nano- to micrometer thin films. Hence, low-load nanoindentation, with depths ranging from few nm to hundreds of nm and loads in the µN-mN range are typically employed. However, for characterizing bulk samples and to capture the response at much larger length scales, a high-load in-situ indentation technique is used. Figure 3.4a shows F–h response for high-load testing (Rudolf et al. 2015). The loads during high-load indentation can vary from a few N to hundreds of N, whereas the indentation depths typically range from a few µm to hundreds of µm. The higher load and displacement ranges allow assessment of mechanical response for large volumes of material.

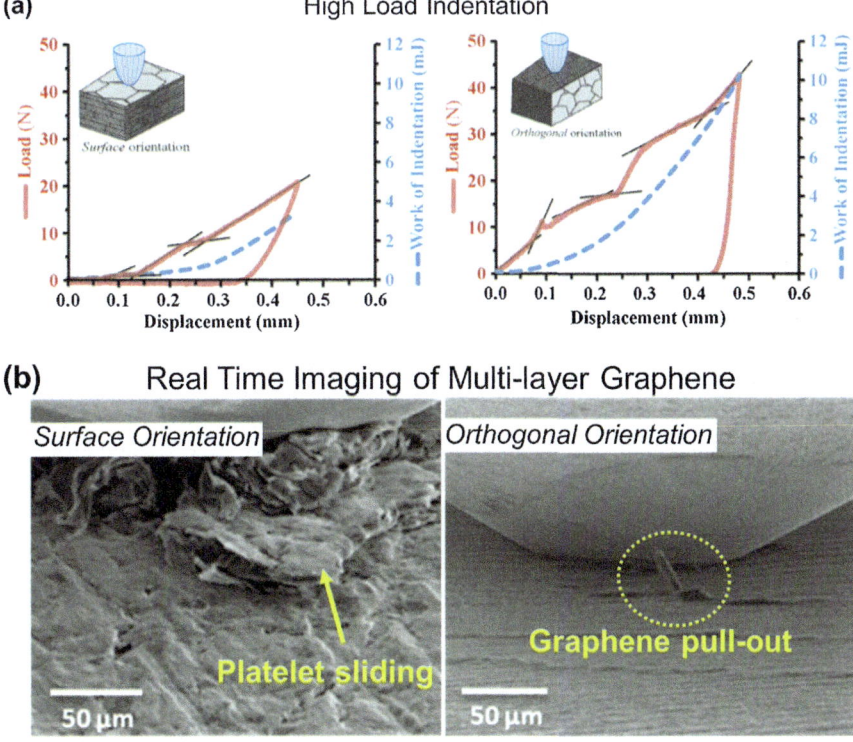

Fig. 3.4 High-load in-situ indentation measurements to study the effect of orientation on mechanical deformation: (**a**) comparison of F–h curves, and (**b**) real-time mechanisms observed in SEM. (Reproduced/adapted with permission from Rudolf et al. (2015))

Figure 3.4b shows real-time imaging of indentation-induced deformation in a multilayer graphene monolith. Activation of mechanisms as a function of "graphene orientation" was studied. It was observed that surface indentation results in sliding of graphene platelets, whereas indentation of the cross section or along the orthogonal orientation is dominated by platelet pop out/pullout. The corresponding $F–h$ curves during high-load indentation were also compared (which were shown in Fig. 3.4a), clearly distinguishing the effect of these different deformation mechanisms on the load-bearing capability. It is noteworthy that low-load nanoindentation is not a suitable technique to capture these mechanisms involving interactions between multiple platelets. Similarly, high-load indentation is not suitable for probing fine nano-to-micrometer sized features because of the lack of low-load resolution. Therefore, the selection of indentation loads and displacements should be based on the nature of the sample and the desired mechanical information. To observe the multi-scale nature of in-situ indentation, readers are referred to the real-time videos in Videos 3.1 and 3.2.

The case studies shown in Figs. 3.3 and 3.4 pertain to the in-situ indentation in SEM. To resolve and study deformation mechanisms at nm-µm length scales, in-situ indentation tests are performed in a TEM. In-situ TEM imaging requires specialized sample preparation, as the specimen should be thin enough to ensure electron transparency. Figure 3.5a illustrates the setup for in-situ TEM indentation: the indenter tip approaches the sample perpendicular to the electron beam (Stach et al. 2001). TEM imaging during indentation allows direct observation of dislocation nucleation and propagation, as shown in Fig. 3.5b for aluminum. These geometrically necessary dislocations (GNDs) accommodate shape change during nanoindentation. Interactions between multiple dislocations can also be observed due to indentation forces. For the example shown in Fig. 3.5b, the moving dislocations are seen to stop upon encountering other dislocations in the specimen (Stach et al. 2001). Real-time imaging can be useful to contrast and compare the dislocation activity in different materials. Figure 3.5c shows in-situ images captured during the indentation of a much harder TiC. Contrary to Al, indentation-induced dislocation motion is significantly restricted in the TiC sample. The TEM micrographs captured during the test reveal that the deformation is concentrated in the vicinity of the tip (within 500 nm) and the dislocation motion is not seen deep within the sample. In-situ nanoindentation of single crystal specimens can be useful to observe the motion of bend contours. The region beneath the tip in the TEM micrographs in Fig. 3.5d shows the bend contours in silicon. Since Si is a brittle material, crack nucleation and propagation is observed during in-situ TEM testing. Figure 3.5d clearly shows the crack in the specimen along a {111} type plane (Stach et al. 2001).

Real-time TEM images vis-à-vis indentation curves are quite useful for precise mapping of mechanical stresses with microstructural events/phenomena. An example of *indentation stress–depth response for multiple deformation cycles* in highly twinned aluminum is shown in Fig. 3.6a (Bufford et al. 2014). The curves capture plastic yielding and strain burst events during deformation. In-situ TEM imaging revealed dislocation nucleation under the indenter tip in Cycle 1, dislocation avalanche consisting of large dislocation loops in Cycle 2, pileup of dislocations at

3.1 Indentation for Localized Deformation Study

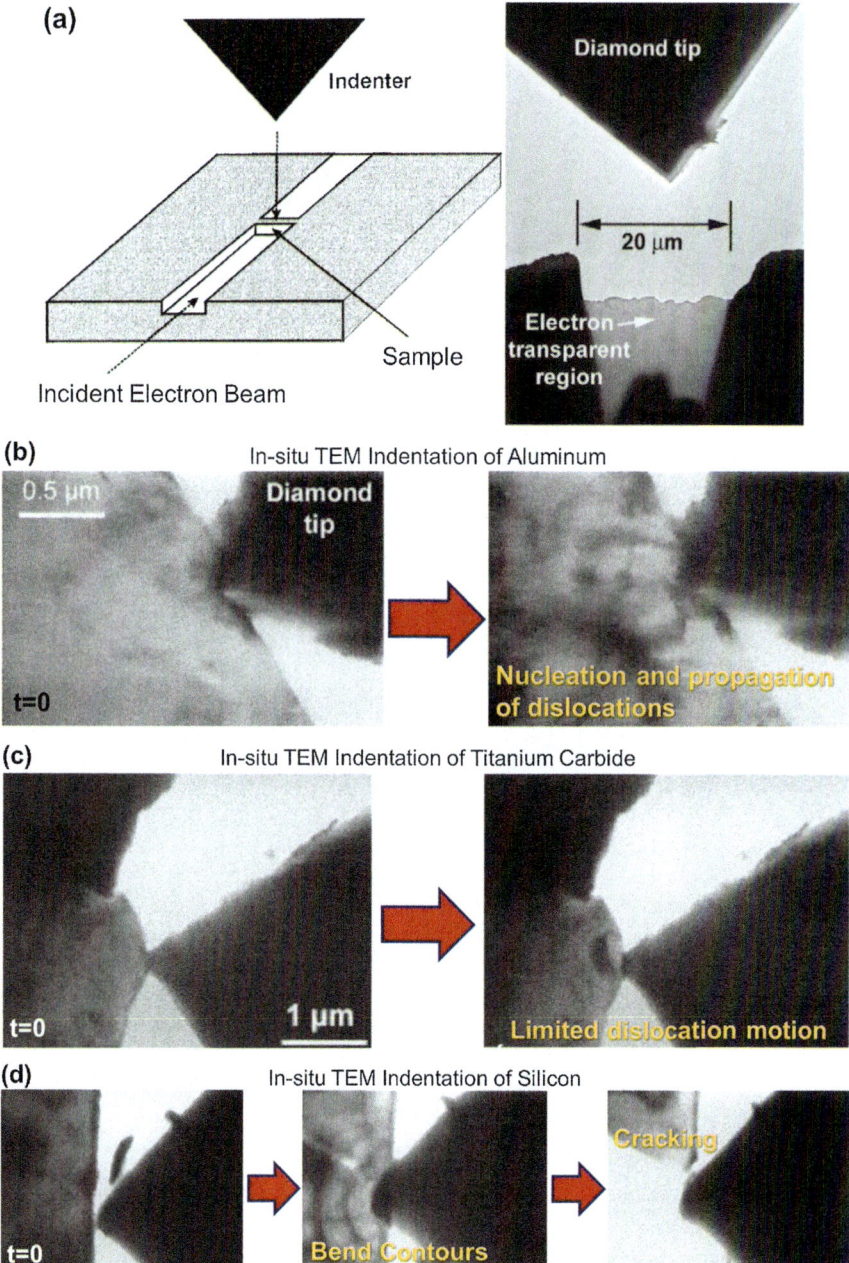

Fig. 3.5 (**a**) Illustration of in-situ nanoindentation and real-time imaging in a TEM. Case studies showing real-time mechanisms during in-situ TEM indentation for different kinds of materials: (**b**) propagation of dislocations in Al (metallic specimen), (**c**) localized dislocation motion in TiC (ceramic specimen), and (**d**) bend contours and cracking of Si (single crystal specimen). (Reproduced/adapted with permission from Stach et al. (2001))

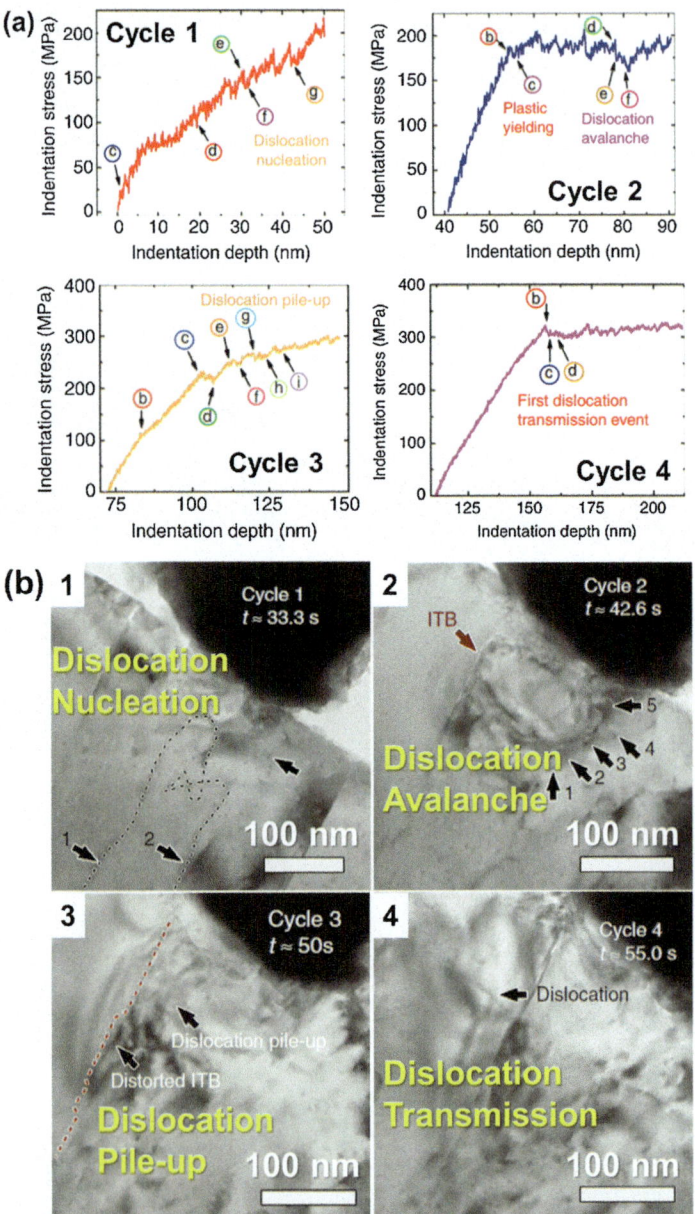

Fig. 3.6 Mapping of events/phenomena in F–h curves (**a**) with real-time deformation mechanisms (**b**) during in-situ TEM nanoindentation of highly twinned aluminum. Strain bursts/stress drops corresponding to dislocation nucleation, dislocation avalanche, dislocation pileup, and dislocation transmission are characterized. (Reproduced/adapted with permission from Bufford et al. (2014))

3.1 Indentation for Localized Deformation Study

higher indentation stresses against the twin boundaries (Cycle 3), and transmission of piled-up dislocations across these boundaries in Cycle 4 (Fig. 3.6b). Real-time observation coupled with high precision measurements enables the quantification of the critical stresses for these mechanisms. This is shown in Fig. 3.6a, where specific mechanisms at specific instants are marked in the indentation stress–depth curves. For the case study in Fig. 3.6, stress drops were determined to be in the range of ~10–20 MPa for individual dislocation nucleation phenomena. The stress drops for dislocation avalanches were observed to be much higher in magnitude (up to 50 MPa).

Cyclic in-situ indentation experiments can be used to study work hardening behavior in materials. Figure 3.7a demonstrates nanoindentation-induced work hardening in nanocrystalline nickel (Lee et al. 2013). The indentation force required for the onset of plastic yielding is seen to increase from ~4.9 µN in the first cycle to ~6.6 µN in the second cycle. This further increases to 7 µN in the third cycle. The associated increase in dislocation density can be computed from real-time imaging. During the in-situ TEM indentation of Ni, the density is found to increase from ~$7 \times 10^{15}/m^2$ in cycle 1 (Fig. 3.7b) to ~$1.2 \times 10^{16}/m^2$ (Fig. 3.7c). In-situ high-resolution characterization provides the ability to decipher these dynamic and time-dependent mechanisms which are otherwise difficult to examine by postmortem imaging.

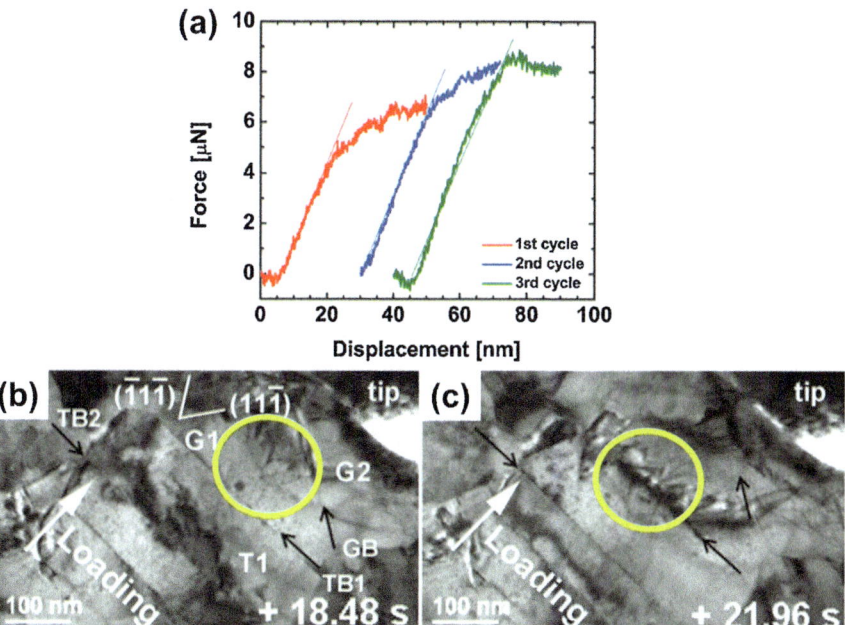

Fig. 3.7 (a) Investigation of work hardening in nanocrystalline nickel by in-situ "cyclic" nanoindentation. The corresponding increase in dislocation density is observed by real-time snapshots for cycle 1 (b) and cycle 2 (c). (Reproduced/adapted with permission from Lee et al. (2013))

3.2 In-Situ Tribology to Study Surface Interactions

Interfacial effects arising due to contact between two surfaces, such as sliding, wear, friction, or rolling, are critical for mechanical integrity, performance, and application of materials. Tribological studies at multiple length scales are performed to study these phenomena and the associated outcome, such as material removal, cracking, de-bonding, and plastic deformation. These surface interactions are complex and difficult to decipher by postmortem imaging. In-situ observation during tribological tests can provide insights into the contact effects.

Nanoscratch technique is often used for studying adhesion and delamination behavior of thin films deposited on substrates. A nanoindenter tip penetrates the sample surface and moves in the lateral direction, resulting in material removal and other associated deformation mechanisms. Figure 3.8a illustrates the in-situ scratching of diamond-like carbon (DLC) thin films in SEM (Rabe et al. 2004). The scratch-induced deformation mechanisms vary depending on the applied loads or the penetration depths. For the case study shown in the figure, relatively lower penetration load (5 mN) resulted in the formation of small particles due to the indenter tip scratching the film surface. When the load was increased to 10 mN, flaking/chipping along the scratch path was observed. Even higher indentation loads (150 mN) led to coating delamination and indentation of substrate. Scratch testing can also be performed inside TEM for high-resolution and high-magnification imaging of tribological phenomena. Real-time TEM snapshots for nanoscratch testing are shown in Fig. 3.8b (Hintsala et al. 2017). This case study demonstrates scratch deformation of perpendicular magnetic recording (PMR) hard disc drive (HDD) film. TEM images show a pileup of material in front of the scratch probe. Prior to loading, the film comprises of aligned grains (one reference grain is encircled in Fig. 3.8b). However, as the probe penetrates and travels laterally to make a scratch, reorientation of the grain is observed. The change in the orientation can be followed by looking at the arrow-marked grain in the real-time snapshots 1 through 4. These mechanistic insights into tribological mechanisms are vital for numerous applications. In this specific case study of scratch deformation of HDD, plastic reorientation of grains can divert the magnetic moment resulting in data loss or failure to read. This example also attests that the microstructure-level mechanistic information can provide clues to engineer superior functional materials. Scratch testers typically involve 2D transducers, such that they have load-sensing capability in the lateral direction in addition to the normal direction. Lateral load sensing allows quantification of the resistance to scratch-induced deformation or wear. It is also useful to capture local phenomena or events, captured by jumps or drops in the force readings. An example of lateral force and displacement plots as a function of time is shown in Fig. 3.8c (Hintsala et al. 2017). The readers are referred to Video 3.3 for an example of simultaneous quantitative (F–h–t curves) and qualitative (real-time deformation mechanisms) evaluation of scratch-induced deformation by in-situ SEM testing.

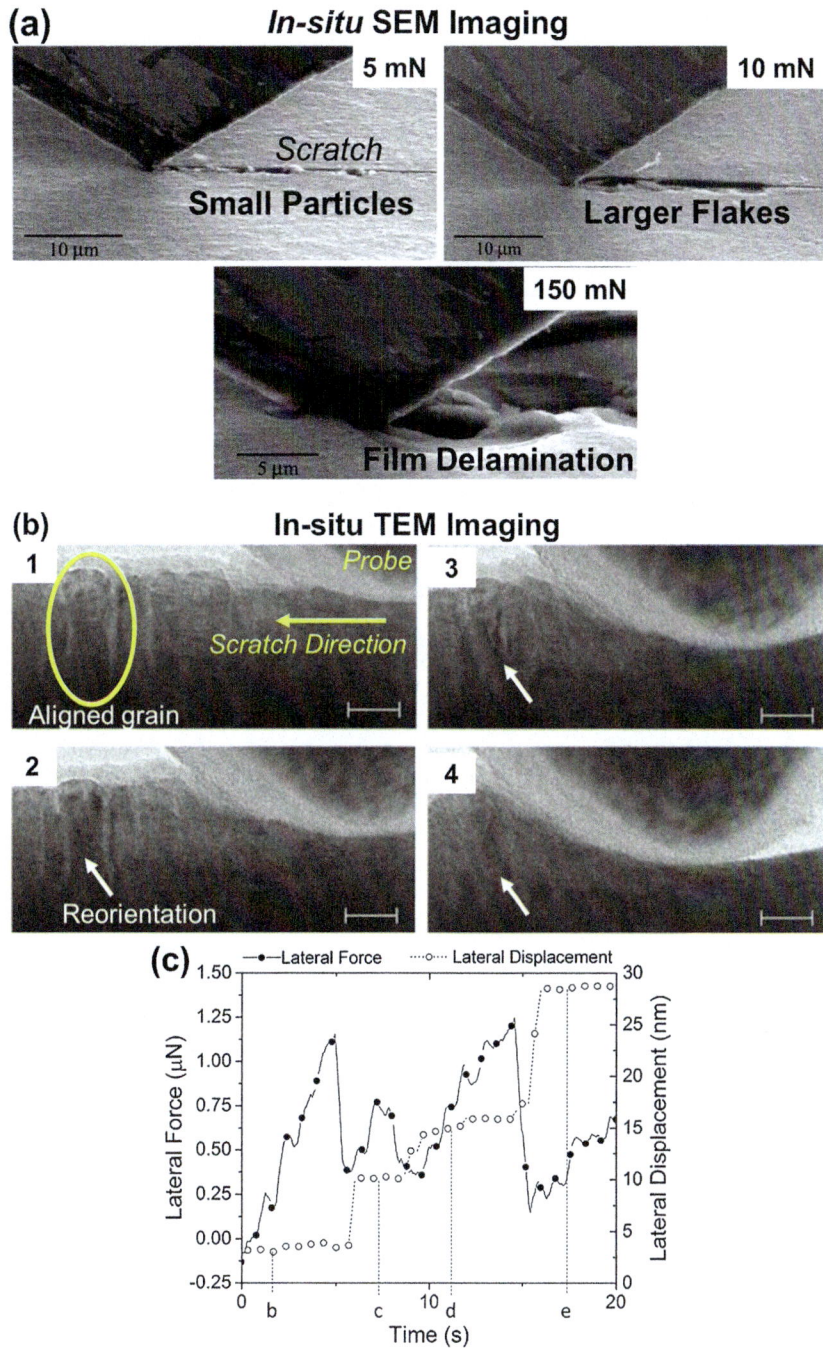

Fig. 3.8 Case studies showing real-time imaging during scratch testing in: (**a**) SEM (scratch testing of a diamond-like carbon film), and (**b**) TEM (perpendicular magnetic recording hard disc drive). (**c**) Lateral load and displacement readings during the scratch test enable the quantitative characterization of tribological phenomena. (Reproduced/adapted with permissions from Rabe et al. (2004), Hintsala et al. (2017))

A combination of high-resolution imaging, lateral force sensing, and targeted indentation can be exploited for probing local bonding strengths and mechanisms. Figure 3.9 illustrates the application of a scratch tester for evaluating the adhesion of micron-sized Al particles deposited on Al substrate by cold spray technique. The lateral motion of the indenter tip can cause particle de-bonding, which manifests as a load drop in the force–displacement profile (Fig. 3.9a). This critical force is then used for computing the adhesion strength using the formula:

$$\text{Adhesion Strength} = \frac{F_{cr}}{\pi r^2} \quad (3.3)$$

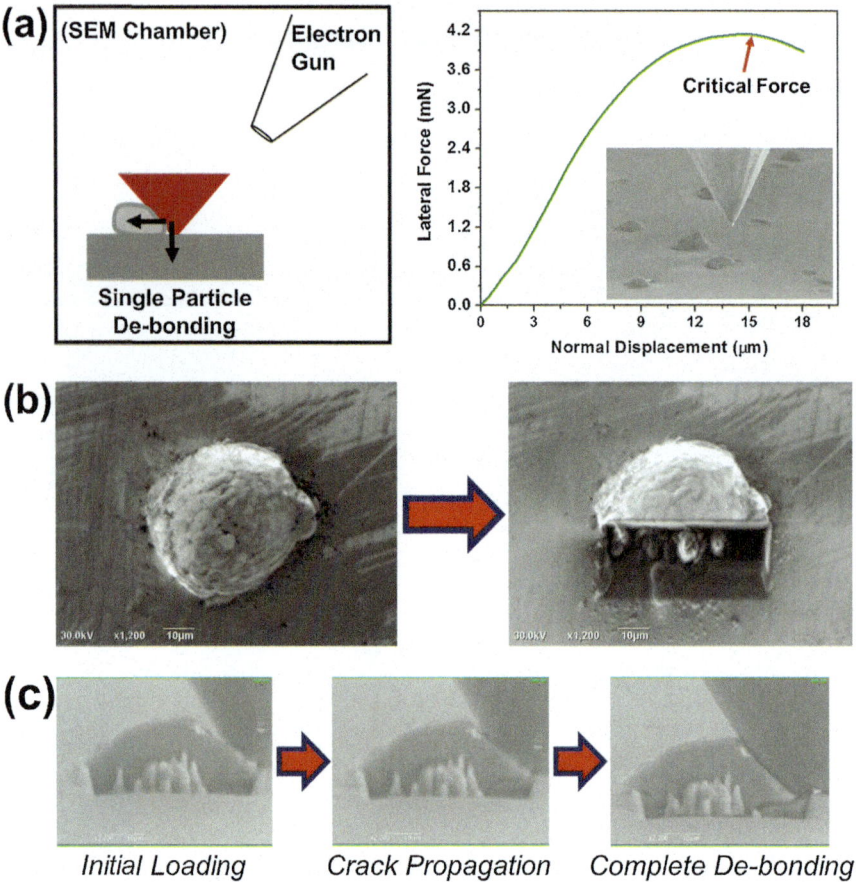

Fig. 3.9 (**a**) Application of in-situ scratch testing to evaluate adhesion of microparticles deposited on a substrate, (**b**) milling of Al particle to reveal the particle–substrate interface, and (**c**) real-time SEM images showing failure initiation, crack propagation, and complete de-bonding of the particle due to indenter-induced forces. (Authors' Unpublished Work)

3.2 In-Situ Tribology to Study Surface Interactions

where F_{cr} is the critical force required for particle de-bonding and r is the microparticle radius. In order to directly observe the de-bonding mechanisms, the particle can be sectioned by focused ion beam milling to expose the interface with the substrate (Fig. 3.9b). Performing targeted indentation-scratch testing of the milled particle allows the real-time imaging of interface de-bonding. Figure 3.9c shows the SEM snapshots as the indenter probe penetrates the substrate near the particle interface, causing crack initiation, propagation, and complete de-bonding of the deposited particle. In-situ interface imaging can be insightful for examining how different materials (metals/ceramics/polymers) bond during deposition. Therefore, this technique can have applications in understanding interface mechanics aspects in emerging additive manufacturing and surface engineering processes.

In-situ manipulation can be useful to study interfacial shear strength and friction associated with nanomaterials. Typically, nanomaterials are deposited on a substrate and the interfacial adhesion between the two is important from application and performance standpoint. Figure 3.10 demonstrates a case study where interfacial sliding characteristics are probed for the "*Ag nanowire/gold*" and "*ZnO nanowire/gold*" interfaces (Zhu et al. 2010). This is accomplished by using a nanomanipulator that picks the nanowire and brings it in contact with an AFM cantilever. Once in contact, the nanomanipulator is moved to apply loading on the nanowire, which results in buckling and eventually sliding at the nanowire/cantilever interface (Fig. 3.10a). Buckling theory can be used to measure the friction forces (Fig. 3.10b). The governing equation describing the shape post-buckling is expressed as:

$$y = \frac{M}{P}\left[\frac{x}{L} + 1.02\sin\left(4.49\frac{x}{L}\right)\right] \quad (3.4)$$

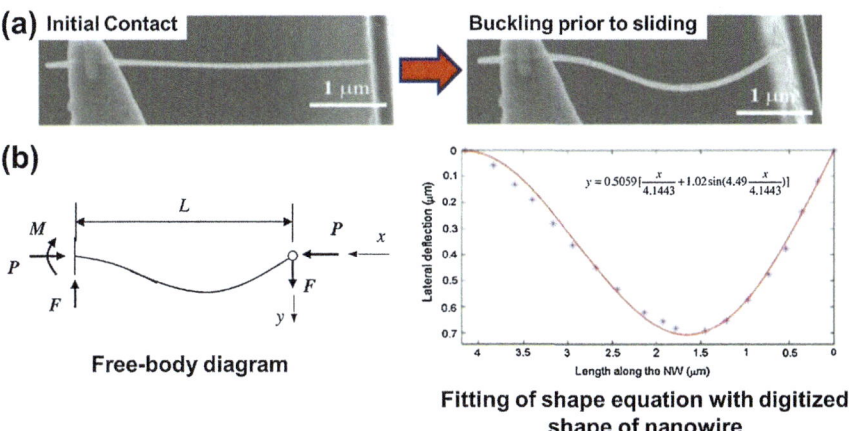

Fig. 3.10 (**a**) In-situ manipulation of nanowire in contact with AFM cantilever to induce buckling and sliding of nanowire, and (**b**) illustration of free-body diagram and fitting of a nanowire post-buckling to extract moment, *M*. (Reproduced/adapted from Zhu et al. (2010) (Open Access))

The terms in the question are marked in the free-body diagram in Fig. 3.10b. The shape of the nanowire observed by in-situ imaging can be fitted to the above equation to determine the moment, M (the fitting is illustrated in Fig. 3.10b). The friction force (F) can then be computed by the simple relation:

$$F = \frac{M}{L} \tag{3.5}$$

It was observed that the friction forces are higher for the contact between Ag nanowire and Au substrate, as compared to the ZnO nanowire/Au substrate. The interfacial shear strength (τ) can also be determined from these sliding experiments using the friction law:

$$F = \tau A \tag{3.6}$$

where A is the true contact area. Therefore, in-situ manipulation, imaging, and measurements provide remarkable ability to study and quantify the adhesion phenomena at nanometer length scales. These contact phenomena are critical to engineer and employ nanostructured materials and devices.

In addition to the nanomaterial–substrate interface, in-situ investigations can also provide insights into interfacial sliding between multiple layers of a 2D material upon external mechanical loading. Characterization of the interlayer de-bonding and sliding phenomena is possible by real-time high-resolution TEM imaging. Figure 3.11a demonstrates interlayer sliding for MoS_2 flakes (Oviedo et al. 2015). The application of shearing force on the top flake by a tip leads to smooth sliding as shown in the figure. However, after sliding half of its diameter, the top flake undergoes bending as half of the flake experiences attractive forces due to the adjacent layer, whereas the other half which is not in contact with the MoS_2 layer beneath does not experience any attraction. Eventually, the monolayer detaches or exfoliates. The shear stresses associated with sliding can be computed by measuring the forces and contact area at the point of exfoliation. An in-situ TEM setup design is shown in Fig. 3.11b, where MoS_2 is sandwiched between a Pt layer (deposited by FIB) and a SiO_2 film on Si substrate. The Si substrate is attached to a Cu lift-out grid, which can be moved using a micromanipulator for the sliding experiment. The MoS_2 is oriented such that it is aligned with the nanoindenter tip for precise measurement of the forces (Fig. 3.11c). The force is applied on the side of the Pt layer, which triggers shearing, captured in force curve (Fig. 3.11c) as well as real-time TEM image (Fig. 3.11d). This in-situ method enabled the determination of shear strength between MoS_2 layers, which is found to be around 25 MPa.

Nanoparticles have attracted attention for lubrication and anti-wear applications. They can act as nano-ball bearings. In-situ tests are employed to investigate rolling wear processes at nano-micrometer length scales. Figure 3.12 demonstrates the in-situ rolling of fullerenes nanoparticles imaged in a TEM (Lahouij et al. 2012). The rolling can be induced by applying a normal force on the particle using an indenter

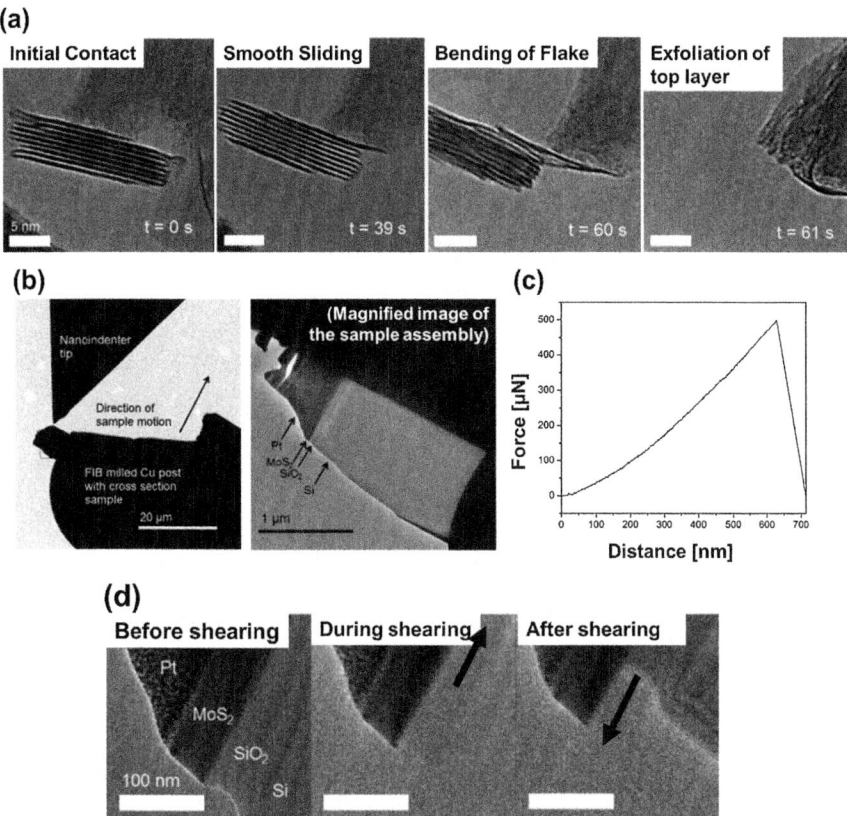

Fig. 3.11 (**a**) In-situ TEM images showing sliding, de-bonding, and exfoliation of MoS_2 flake, (**b**) in-situ TEM experimental setup for determining shear stress during sliding and exfoliation, (**c**) load–displacement curve recorded during the shearing test, and (**d**) TEM imaging to confirm shearing when force is applied on the specimen. (Reproduced/adapted with permission from Oviedo et al. (2015))

tip along with simultaneous tangential movement (in this case by moving the Si wedge). The rolling process typically involves pure sliding at low contact pressures (<100 MPa in the case study shown in Fig. 3.12). Higher contact pressures (~1 GPa) can also lead to exfoliation of the particle. Real-time imaging is useful to study the effect of particle sizes, morphologies, applied forces, and agglomerations on the rolling mechanisms of the particles (Lahouij et al. 2011, 2012). The sliding mechanisms are also influenced by the surfaces on which the particle slides/rolls and in-situ testing can be useful to identify the differences in the lubrication behavior of nanoparticles for different materials.

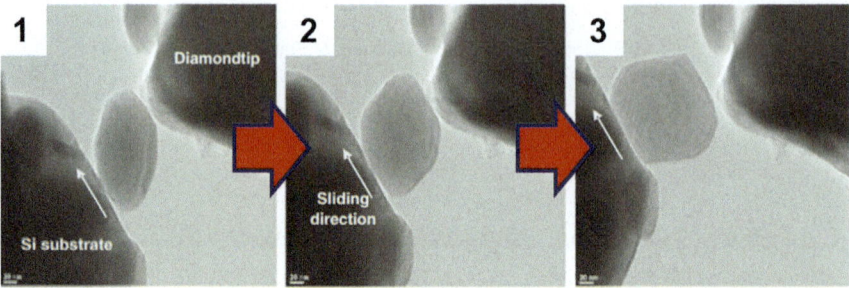

Fig. 3.12 In-situ TEM imaging of sliding of fullerenes nanoparticle: the nanoparticle is seen to slide on the substrate when a shear force is introduced. (Reproduced/adapted with permission from Lahouij et al. (2012))

3.3 In-Situ Nano- and Micro-Pillar Compression

Indentation loading creates complex stress-state under the tip, with very large strain gradients. Hence, a straightforward determination of mechanical properties, such as yield stress or failure strength is not possible. These limitations are overcome by the nano- and micro-pillar compression approach. Pillar compression is a popular and effective methodology for probing localized mechanical properties, which requires nano/micromachining at the regions of interest in the sample to obtain cylindrical pillars. These pillars are then subjected to compression loading by an indenter punch, allowing extraction of load–displacement readings. The fabrication of the pillars was discussed in the previous chapter in the section on focused ion beam machining. One major advantage of pillar compression is that the loading is uniaxial in nature. Therefore, the determination of contact geometries and areas is not required, which is a necessity in the indentation characterization method. A flat-ended probe is typically used for pillar compression. Figure 3.13a–c shows different deformation regimes for a Ni micro-pillar under compression, along with a corresponding load–displacement plot (Fig. 3.13d). The initial deformation is elastic in nature (Fig. 3.13a), characterized by a linear increase in compression force (up to ~30 mN) with indenter punch displacement (up to ~200 nm). The force response is not as steep in the plastic regime. The real-time SEM micrograph reveals the plastic deformation of the pillar is nonuniform in nature: plastic flow is enhanced at the top end of the pillar (close to the probe), manifesting as an outward bulge (Fig. 3.13c). The quantitative comparison of strain in different regions of the pillar (from bottom to top) can be seen in Fig. 3.13e. Real-time imaging during in-situ SEM compression enables this accurate characterization of strain distribution in the specimen, otherwise not possible in ex-situ testing. The stresses induced in the specimen during compression can also be correlated with the true strains determined from in-situ images. Figure 3.13f highlights the stress–strain response in different regions of the micro-pillar under compression (lines 1 through 4). The ability to resolve and decipher this information provides an enhanced understanding of deformation mechanisms. For instance, the true stress–true strain plot in Fig. 3.13f indicates initial

3.3 In-Situ Nano- and Micro-Pillar Compression

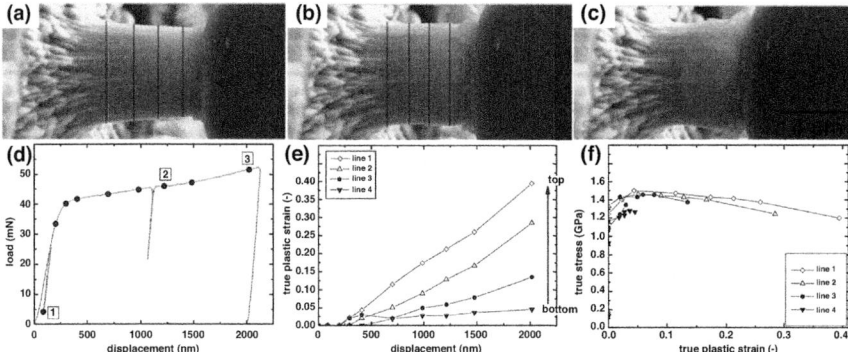

Fig. 3.13 Real-time SEM snapshots during micro-pillar compression of Ni, showing: (**a**) elastic, and (**b, c**) plastic deformation regimes. (**d**) Corresponding load–displacement response for pillar compression. (**e**) Comparison of plastic strain in different regions of the pillar (bottom to top). (**f**) Stress–strain response revealing work hardening and softening regimes during plastic deformation of the pillar specimen. (Reproduced/adapted with permission from Schwaiger et al. (2012))

work hardening (for lower strain values), where an increase in stress with plastic strain is seen; thereafter, softening is observed as the stresses drop when the strain imposed on the specimen increases from ~0.1 to 0.4. Therefore, the in-situ pillar compression methodology can be highly informative about deformation processes and mechanisms active in the material.

FIB fabrication process often introduces a slight taper in the pillar geometry. For instance, the pillars shown in Fig. 3.13 have a taper of ~4°. The variation in the taper angle is known to influence the load–displacement response captured during compression. This is illustrated in Fig. 3.14a. The graph shows the change in load response as the taper angle is increased from 0° (perfect cylinder) to 4°. The initial elastic region seems to be indistinguishable for different taper angles, indicating taper does not have a significant effect on elastic modulus determination. However, the plastic regime is more sensitive to taper (Fig. 3.14a). A higher taper angle can result in overestimation of yield stress since the load readings increase as the taper angle increases. Additionally, the slope in the plastic regime also varies with the taper angle. Therefore, the determination of strain hardening during pillar compression is expected to be susceptible to the taper angle. Another factor that influences mechanical response during pillar compression is the misalignment between the pillar and the indenter tip/punch. Unlike the taper angle, misalignment also affects the elastic portion of the load–displacement curve. A drop in the elastic slope of the curve can be observed in Fig. 3.14b for 3° misalignment. The slope in the plastic regime is not greatly influenced by tip-pillar misalignment. The effect of misalignment is also intertwined with the friction at the tip/pillar interface. The absence of friction, coupled with misalignment causes a significant deviation of the load response. For 3° misalignment and no friction, Fig. 3.14b demonstrates an abrupt transition from elastic to plastic deformation, as opposed to other curves where the

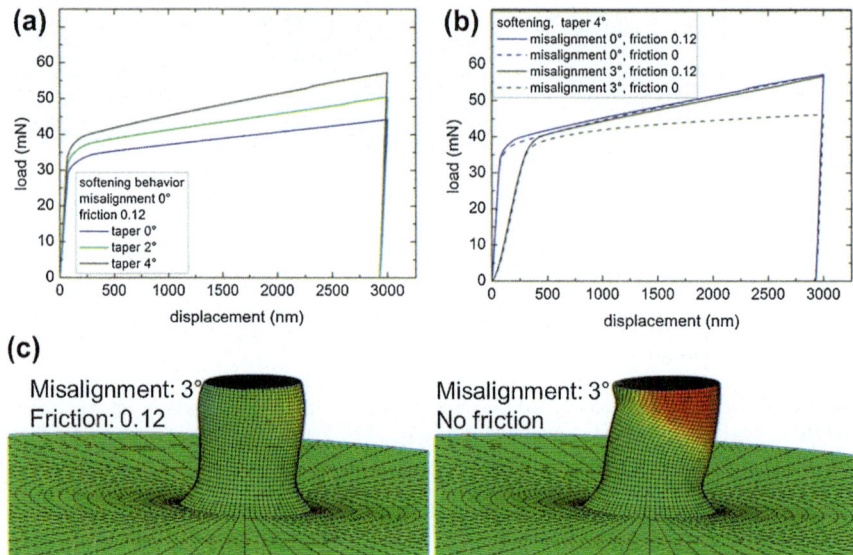

Fig. 3.14 Effect of pillar taper angle (**a**) and tip-pillar misalignment (**b**) on the load–displacement response of the specimen. (**c**) Comparison of plastic instability localization at pillar top with and without friction between the tip and the pillar. (Reproduced/adapted with permission from Schwaiger et al. (2012))

change in slope is not so sudden. This deviation is associated with localization of shear and plastic instability at the top of the pillar when friction is absent (Fig. 3.14c). These observations highlight the importance of experimental setup on the mechanical measurements obtained from pillar compression. In-situ imaging allows the observation and rectification of alignment issues prior to the test. It also facilitates a precise measurement of taper, which can then be factored in for mechanical property assessment.

Pillar compression technique is also used for probing sub-micrometer-sized specimens. In-situ testing of nanopillars inside TEM can be highly informative about plasticity mechanisms in the materials. A case study on in-situ TEM testing of single crystal Ni nanopillars is shown in Fig. 3.15. TEM imaging of the pillar reveals a high density of defects, such as small loop-like defects and long line defects, with a dislocation density of $\sim 10^{15}$ m^{-2}. Some of these defects are introduced due to sample irradiation during FIB. In-situ testing allows direct observation and measurement of dislocation density change during pillar compression by a flat punch. Figure 3.15a demonstrates the mechanical annealing phenomenon, where dislocation density dramatically drops due to compression. This provides the opportunity to measure the mechanical response of a dislocation-free pillar, demonstrated in Fig. 3.15b. The video and load–displacement curves captured during the in-situ test show clear demarcation of elastic deformation (up to 20 nm), discrete plastic events (20–140 nm) and pillar buckling resulting in decreasing loads (140–185 nm). A comparative stress–displacement plot is shown in Fig. 3.15c: Test

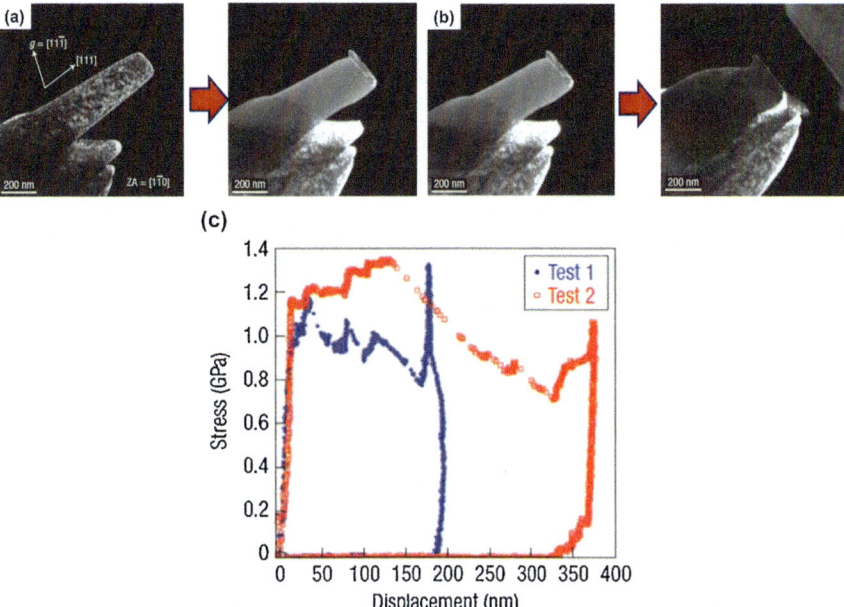

Fig. 3.15 In-situ TEM compression of a single crystal Ni pillar: (**a**) test 1 after FIB machining shows mechanical annealing to produce defect-free pillar, (**b**) test 2 on the defect-free pillar, and (**c**) comparison of stress–displacement response for the two tests. (Reproduced/adapted with permission from Shan et al. (2008))

1 and 2 correspond to before and after mechanical annealing, respectively. There is insignificant difference in yield stress value, which suggests the presence of FIB-induced defects does not adversely influence mechanical strength. Moreover, ion penetration during FIB is confined to 10–20 nm depth, limiting the influence of defects on the mechanical response. These examples illustrate the importance of in-situ pillar testing for fundamental mechanistic understanding of elastic and plastic phenomena in materials.

3.4 In-Situ Beam Deflection

As stated in the previous section, micromechanical testing enables the evaluation of small-volume specimens to understand the deformation and properties of localized features. However, proper alignment, fabrication errors/defects, and surface state are challenging factors to control, which can influence the test results (shown in Fig. 3.14 for micro-pillar testing) (Mara et al. 2013; Allison et al. 2014). Moreover, micro-pillar testing is appropriate for examining plasticity mechanisms and not effective to study fracture/failure behavior (Armstrong et al. 2009). Micro-tensile tests, on the other hand, are affected by stress concentration and slippage along the grips, and premature failure prior to yielding due to surface damage during microfabrication. Microbeam or microcantilever testing overcomes these challenges (test set-up is shown in Fig. 3.16a). The method can be effectively employed for studying elastic, plastic, and fracture behavior of a wide variety of materials (Armstrong et al. 2009, 2011; Kupka and Lilleodden 2012). Beams with different geometries and cross sections can be fabricated for the tests (Fig. 3.16). Triangular beam is useful for studying the deformation of thin films. Figure 3.16b illustrates a silicon beam, on which Cu thin films were deposited and the beam was then subjected to cantilever deflection. This triangular beam ensures that the thin film layer on top is in a uniform state of plain strain (Florando and Nix 2005). A pentagonal cross section shown in Fig. 3.16c is sometimes used for testing notched specimens (Mara et al. 2013), as it allows easy fabrication and selection of beam orientation (Deng et al. 2017). Figure 3.16d demonstrates an example of a notched beam (Kupka and Lilleodden 2012). A U-shaped notch is fabricated by FIB milling. The rectangular cross section used in this specific image is to ensure a constant crack width (Kupka and Lilleodden 2012). These examples show ample scope to modify the beam geometry to extract desired mechanical information from in-situ tests.

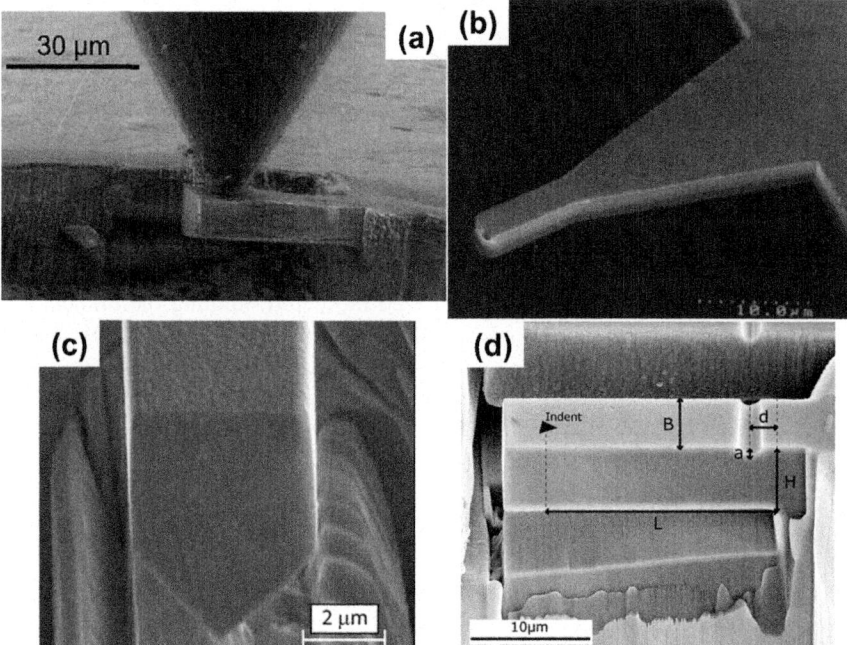

Fig. 3.16 (**a**) Illustration of an in-situ cantilever beam bending test (indenter tip-beam assembly in SEM). SEM images showing different beam geometries used for micro-bending tests: (**b**) triangular beam for probing thin films, (**c**) pentagonal cross section beams for testing notched specimens, and (**d**) a rectangular beam with a U-shaped notch to maintain constant crack width. (Reproduced/adapted with permissions from Allison et al. (2014), Florando and Nix (2005), Mara et al. (2013), Kupka and Lilleodden (2012))

3.4 In-Situ Beam Deflection

A case study showing in-situ SEM bending of a nanocomposite beam is shown in Fig. 3.17a (Allison et al. 2014). The deflection response is captured as the beam is subjected to mechanical loading. It can be seen that mechanisms such as crack initiation and ultimate fracture recorded via SEM can be correlated with the load response. This allows precise estimation of critical forces/stresses as well as deflections/strains for failure initiation in the material. Cantilever microbeams can be subjected to cyclic mechanical loading for studying fatigue characteristics of the material. Parameters such as maximum stress, elastic stiffness, yield strength, and energy dissipated can be determined as a function of loading cycles. This is demonstrated in Fig. 3.17b, where the response of Cu and steel specimens is recorded for up to 100 cycles (Howard et al. 2017). In-situ imaging allows real-time observation of fatigue mechanisms in the specimen. The SEM micrograph in Fig. 3.17b exhibits microcrack-induced surface damage in an ultrafine-grained Cu beam post-cyclic loading. The in-situ bending technique has also been applied to examine the mechanics of single crystal specimens. The case study in Fig. 3.17c shows in-situ SEM bending of Cu single crystal microbeams (Kiener et al. 2010). The figure shows the use of a dove-tailed grip for applying the bending forces, instead of an indenter probe that pushes down the beam. The gripper avoids undesired preloads while establishing tip-sample contact. It also allows easy adjustment of the sample length to be deflected. The figure shows the load–displacement curve with superimposed SEM snapshots at different instants. The real-time images show deformation state of the beam as the displacement is applied and reversed (cyclic loading/unloading/reloading scenario). Elastic modulus estimated from the load curve is found to be comparable with the bulk modulus of Cu. One main advantage of in-situ imaging is the ability to rectify compliances. For instance, the low unloading slope seen in the load–displacement curve in Fig. 3.17c is because of compliance during cantilever bending. It can be corrected by using the base displacement from real-time SEM images.

The examples discussed so far pertain to cantilever beam deflection. However, the in-situ beam bending technique can also be applied in three-point mode. Figure 3.18a shows the in-situ three-point bending of a pre-notched beam in SEM (Hintsala et al. 2015). The measurements consist of a combination of SEM imaging, EBSD mapping, and simultaneously captured load–displacement response from austenitic stainless steel. The example shown in Fig. 3.18a also demonstrates the use of in-situ analysis to characterize misorientation localization in the specimen. Grain rotation during plastic deformation is observed during the test. Load-displacement curves *i* through *v* stand for 5 loading cycles, which shows an increase in plastic deformation is correlated with grain rotation in this specimen. The technique can be extended for in-situ analysis in a TEM as well. Figure 3.18b shows a three-point bending test performed on a much thinner beam with ~100 nm thickness (as opposed to 500–2500 nm thick specimens for in-situ SEM testing, shown in Fig. 3.18a). The use of FIB for fabricating a fine-width notch in a slender beam specimen is challenging. In such cases, a fully converged TEM beam can be employed for localized sputtering of the material. The case study shown in Fig. 3.18b

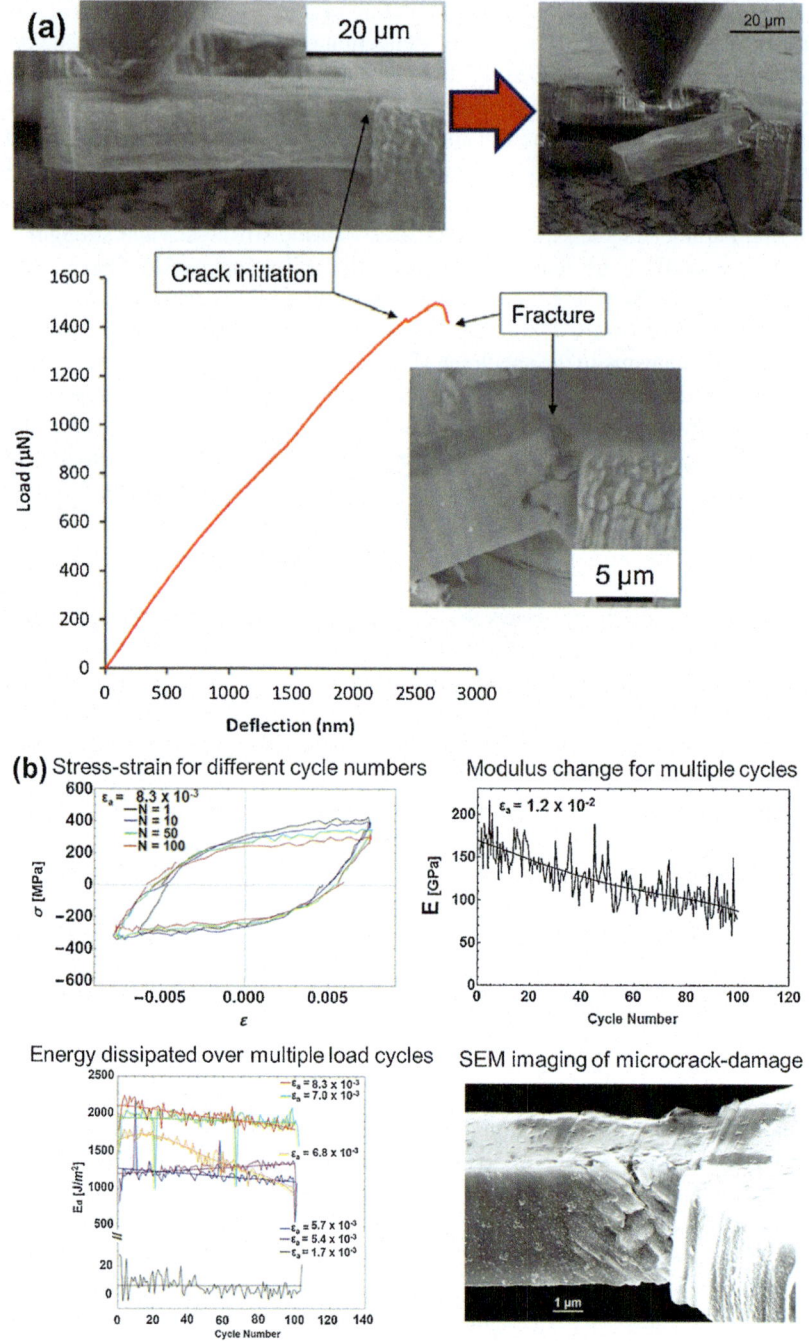

Fig. 3.17 In-situ SEM cantilever beam deflection showing: (**a**) crack initiation and fracture of a nanocomposite sample, (**b**) fatigue testing with stress–strain, modulus, energy dissipation as a

3.4 In-Situ Beam Deflection

A case study showing in-situ SEM bending of a nanocomposite beam is shown in Fig. 3.17a (Allison et al. 2014). The deflection response is captured as the beam is subjected to mechanical loading. It can be seen that mechanisms such as crack initiation and ultimate fracture recorded via SEM can be correlated with the load response. This allows precise estimation of critical forces/stresses as well as deflections/strains for failure initiation in the material. Cantilever microbeams can be subjected to cyclic mechanical loading for studying fatigue characteristics of the material. Parameters such as maximum stress, elastic stiffness, yield strength, and energy dissipated can be determined as a function of loading cycles. This is demonstrated in Fig. 3.17b, where the response of Cu and steel specimens is recorded for up to 100 cycles (Howard et al. 2017). In-situ imaging allows real-time observation of fatigue mechanisms in the specimen. The SEM micrograph in Fig. 3.17b exhibits microcrack-induced surface damage in an ultrafine-grained Cu beam post-cyclic loading. The in-situ bending technique has also been applied to examine the mechanics of single crystal specimens. The case study in Fig. 3.17c shows in-situ SEM bending of Cu single crystal microbeams (Kiener et al. 2010). The figure shows the use of a dove-tailed grip for applying the bending forces, instead of an indenter probe that pushes down the beam. The gripper avoids undesired preloads while establishing tip-sample contact. It also allows easy adjustment of the sample length to be deflected. The figure shows the load–displacement curve with superimposed SEM snapshots at different instants. The real-time images show deformation state of the beam as the displacement is applied and reversed (cyclic loading/unloading/reloading scenario). Elastic modulus estimated from the load curve is found to be comparable with the bulk modulus of Cu. One main advantage of in-situ imaging is the ability to rectify compliances. For instance, the low unloading slope seen in the load–displacement curve in Fig. 3.17c is because of compliance during cantilever bending. It can be corrected by using the base displacement from real-time SEM images.

The examples discussed so far pertain to cantilever beam deflection. However, the in-situ beam bending technique can also be applied in three-point mode. Figure 3.18a shows the in-situ three-point bending of a pre-notched beam in SEM (Hintsala et al. 2015). The measurements consist of a combination of SEM imaging, EBSD mapping, and simultaneously captured load–displacement response from austenitic stainless steel. The example shown in Fig. 3.18a also demonstrates the use of in-situ analysis to characterize misorientation localization in the specimen. Grain rotation during plastic deformation is observed during the test. Load-displacement curves i through v stand for 5 loading cycles, which shows an increase in plastic deformation is correlated with grain rotation in this specimen. The technique can be extended for in-situ analysis in a TEM as well. Figure 3.18b shows a three-point bending test performed on a much thinner beam with ~100 nm thickness (as opposed to 500–2500 nm thick specimens for in-situ SEM testing, shown in Fig. 3.18a). The use of FIB for fabricating a fine-width notch in a slender beam specimen is challenging. In such cases, a fully converged TEM beam can be employed for localized sputtering of the material. The case study shown in Fig. 3.18b

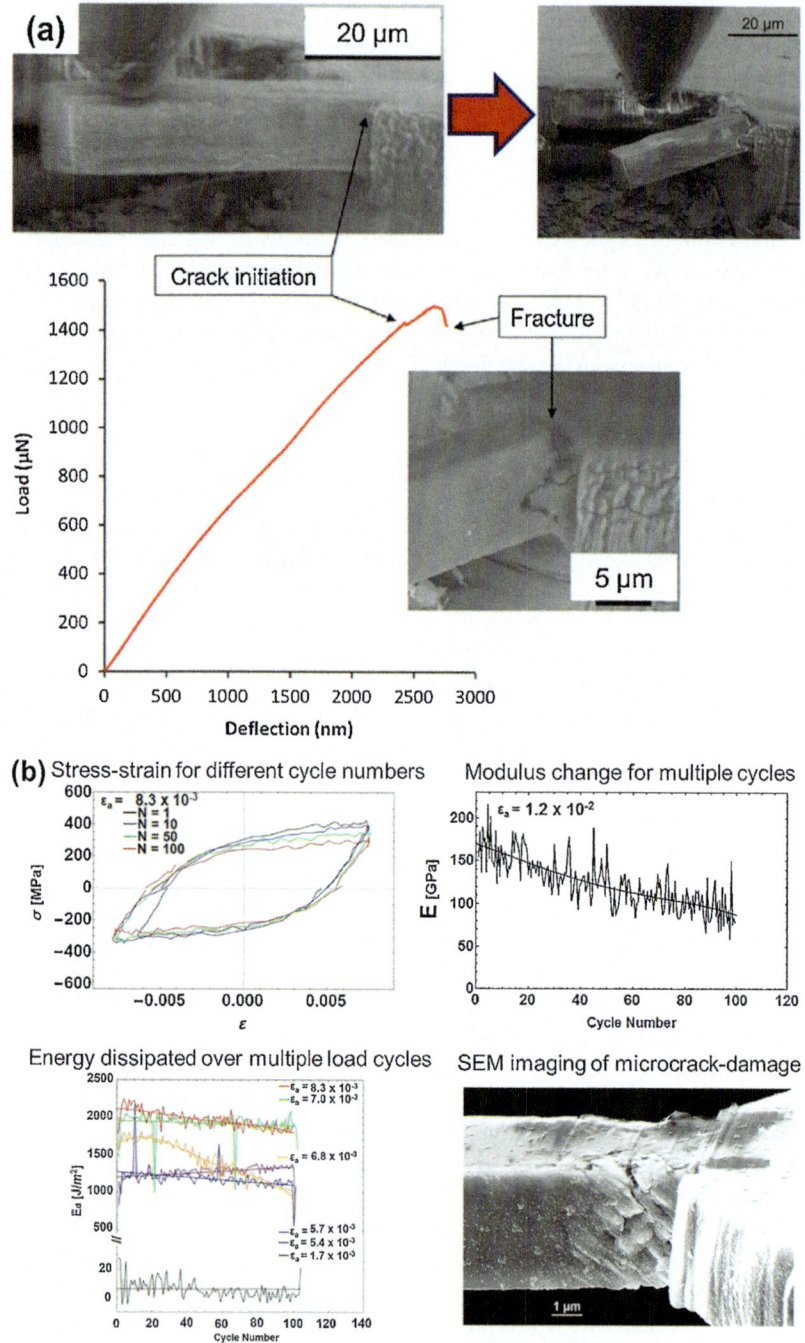

Fig. 3.17 In-situ SEM cantilever beam deflection showing: (**a**) crack initiation and fracture of a nanocomposite sample, (**b**) fatigue testing with stress–strain, modulus, energy dissipation as a

3.4 In-Situ Beam Deflection

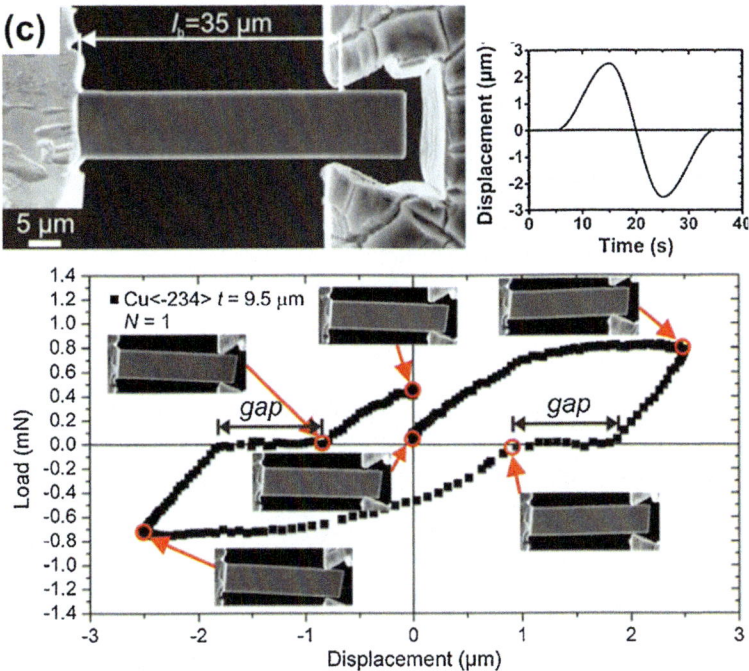

function of loading cycles and SEM image showing microcrack-induced sample damage, and (**c**) reversible bending of Cu microbeams with real-time SEM snapshots corresponding to different instants in the load–displacement curve. (Reproduced/adapted with permissions from Allison et al. (2014), Howard et al. (2017), Kiener et al. (2010))

represents an ultra-fine 5 nm radius notch. Due to the small beam size, the specimens prepared for in-situ TEM testing were single crystalline. The real-time snapshots and load–displacement curves show high ductility. TEM imaging also provides an estimation of dislocation activity during deformation: the dark region seen underneath the indenter tip in Fig. 3.18b is an indicator of higher dislocation activity. It is noteworthy that the methodology of fabricating and assuming a pre-notch to be a crack tip for fracture studies is not accurate because of the radius of curvature at the root of the notch (Armstrong et al. 2012). The stress intensity factor (K_{Ic}) can be corrected by the relation (Pugno et al. 2005):

$$K'_{Ic} = \sqrt{1 + \frac{\rho}{2d_o}} K_{Ic} \quad (3.7)$$

where d_o is expressed by the relation, $d_o = \frac{2}{\pi} \frac{K_{Ic}^2}{\sigma_u^2}$ (σ_u is the ultimate tensile strength of the material). This correction can be applied to overcome errors in K_{Ic} estimation for blunt pre-notches in three-point bending test specimens. The case study in

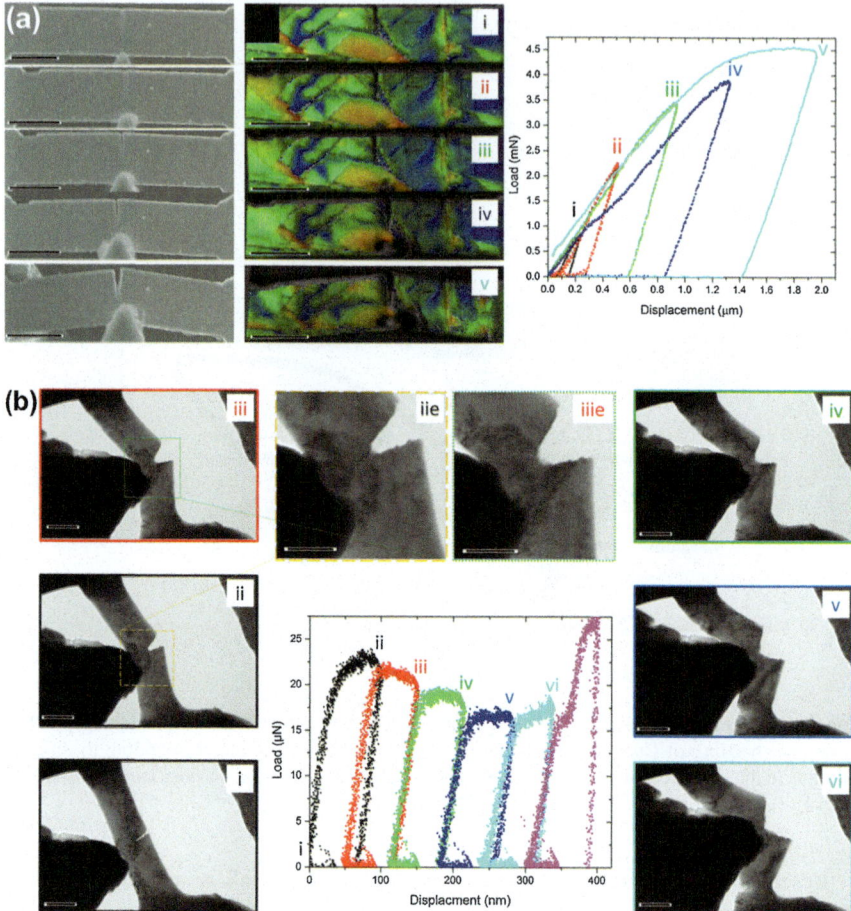

Fig. 3.18 In-situ imaging during three-point bending tests in: (**a**) SEM coupled with EBSD shows grain rotation due to plastic deformation, and (**b**) TEM for ultra-thin beams ($t \sim 100$ nm) shows dislocation activity underneath the indenter tip. (Reproduced/adapted with permissions from Hintsala et al. (2015))

Fig. 3.18 illustrates the importance of multi-scale, multi-resolution in-situ imaging to decipher size effects in fracture behavior, intrinsic mechanisms, and stress intensities.

3.5 In-Situ Tensile Characterization

Uniaxial tensile loading is one of the most common mechanical tests for extracting detailed deformation characteristics of materials, such as elastic modulus, elastic limit, yield stress, failure strength, and failure strain. Tensile testing involves gripping a sample at its two endpoints, followed by uniaxial stretching. The sample is stretched or loaded until the failure point. As the sample is stretched, the load required continues to increase and there is an abrupt drop to zero at failure. One key advantage of tensile testing over other mechanical tests is the absence of "substrate effect," which can influence the results during test methods such as pillar compression or nanoindentation of thin films/deposits. In a typical tensile test, the initial linear stress–strain regime represents elastic deformation, which is reversible in nature. After the elastic limit, the stress–strain curve is nonlinear in nature and corresponds to plastic or permanent deformation (Fig. 3.19a). It is noteworthy that not all materi-

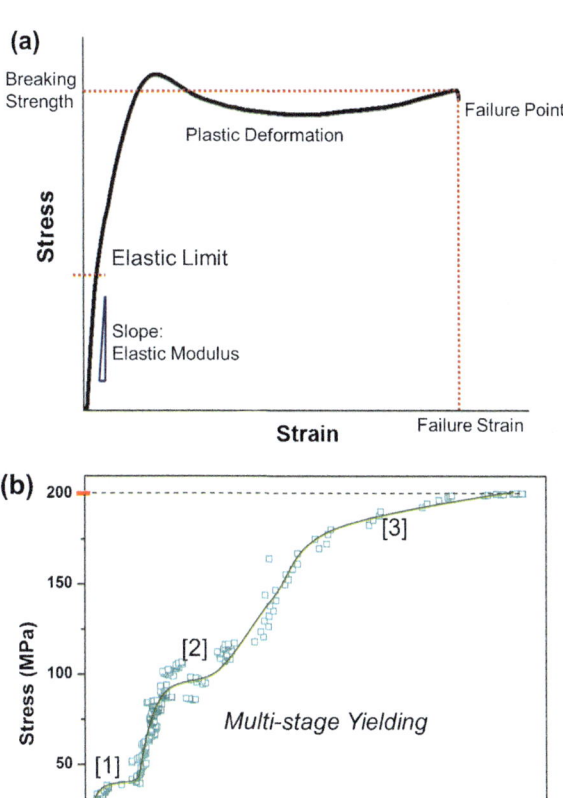

Fig. 3.19 (a) A typical tensile stress–strain curve consisting of elastic and plastic components, and (b) stress–strain response for a composite deviating from the conventional curve and consisting of multiple yield points prior to failure. (Reproduced/adapted with permissions from Nautiyal et al. (2016), Agarwal group (FIU))

als have a distinguishably large plastic regime. For instance, brittle ceramics tend to undergo limited plastic deformation prior to failure. Conventionally, material scientists rely on the stress–strain curves and postfracture surface imaging to decipher tensile deformation characteristics. However, this conventional analysis and interpretation approach leaves many unanswered questions:

1. The deformation mechanism/behavior in the elastic regime is difficult to understand by post-failure imaging since it is nonpermanent in nature. Considering the fact that most of the load-bearing structures typically experience loads in the elastic regime, it would be informative to understand the stress-transfer mechanisms prior to the yield point. The necessity of understanding stress-transfer is particularly pressing for engineering high-performance materials for critical applications, such as aircraft, space-vehicle bodies, armors, etc.
2. For brittle materials like ceramics, the failure is catastrophic in nature. Engineering ceramics with enhanced ductility requires close observation of mechanisms at play when the processing or compositions are tweaked. The stress–strain analysis provides merely quantitative information and lacks a mechanistic understanding of deformation.
3. Multicomponent materials, like composites, can exhibit complex stress–strain curves, such as multiple yield points (Fig. 3.19b). Understanding these additional phenomena during deformation is difficult by the conventional approach.

These challenges can be overcome by in-situ tensile testing. Real-time imaging provides the ability to resolve and distinguish these different mechanisms activated in the material upon mechanical loading.

Figure 3.20 shows an example of tensile testing of a nanofiber specimen (Li et al. 2017). The figure demonstrates the use of a MEMS device inside SEM for testing the fiber (Fig. 3.20a). In-situ testing is particularly useful to decipher deformation in a complex composite system. In this case study, the 2D graphene ribbons are embedded inside 1D carbon nanofibers. SEM imaging during mechanical loading reveals different mechanisms as the weight fraction of graphene filler in carbon fiber increases. For instance, carbon nanofiber with no graphene is seen to have a smooth fracture surface, which is characteristic of brittle failure. A single necking event is seen (Fig. 3.20b). After adding graphene, the SEM imaging reveals a relatively rougher fracture. At higher graphene loading, the sword-in-sheath mechanism is seen where the outer layer first deforms plastically followed by sliding and pullout of the inner layer resulting in fiber fracture (Fig. 3.20b). Real-time imaging enables this microscopic understanding of the strengthening mechanisms when engineering multi-material composites.

The example illustrated above showed brittle failure. In-situ imaging is also useful to study plasticity mechanisms. Figure 3.21a illustrates in-situ SEM tensile testing of a bulk metallic composite, with clearly distinguishable elastic, plastic, and necking regimes (Sun et al. 2018). As the displacement increases, the progressive reduction of specimen cross section is observed. SEM imaging shows the localization of plastic deformation. High-resolution, high-magnification imaging can be particularly useful to observe deformation characteristics in different regions of the

3.5 In-Situ Tensile Characterization

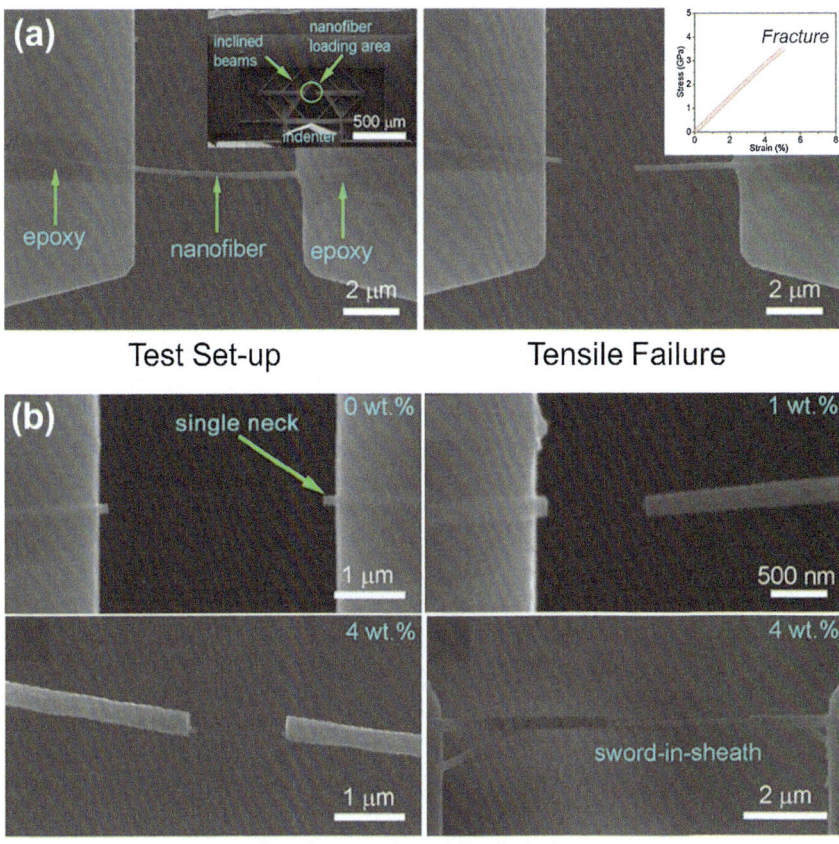

Fig. 3.20 In-situ SEM tensile testing of carbon nanofiber with embedded graphene ribbon: (**a**) images showing test setup and failure recorded by SEM, and (**b**) observed deformation mechanisms for different graphene weight fractions in carbon fibers. (Reproduced/adapted with permission from Li et al. (2017))

sample. Figure 3.21b illustrates shear bands in two different areas of the sample under deformation: necking region and the area away from the necking region. The density of shear bands is seen to increase as the sample is strained. In-situ observations clearly indicate inhomogeneous deformation in the bulk metallic glass composite as the number density of shear bands is quite different near and away from the necking region. The localized mechanistic information is useful to decode the microstructure characteristics of the specimen. For instance, the variable shear band density observed for the bulk metallic glass composite is indicative of the distribution and size range of crystalline phases in the specimen. Region-specific microscopic investigations during mechanical loading are particularly insightful for composite material systems, where the overall stress-transfer is dependent on the interplay of individual component characteristics.

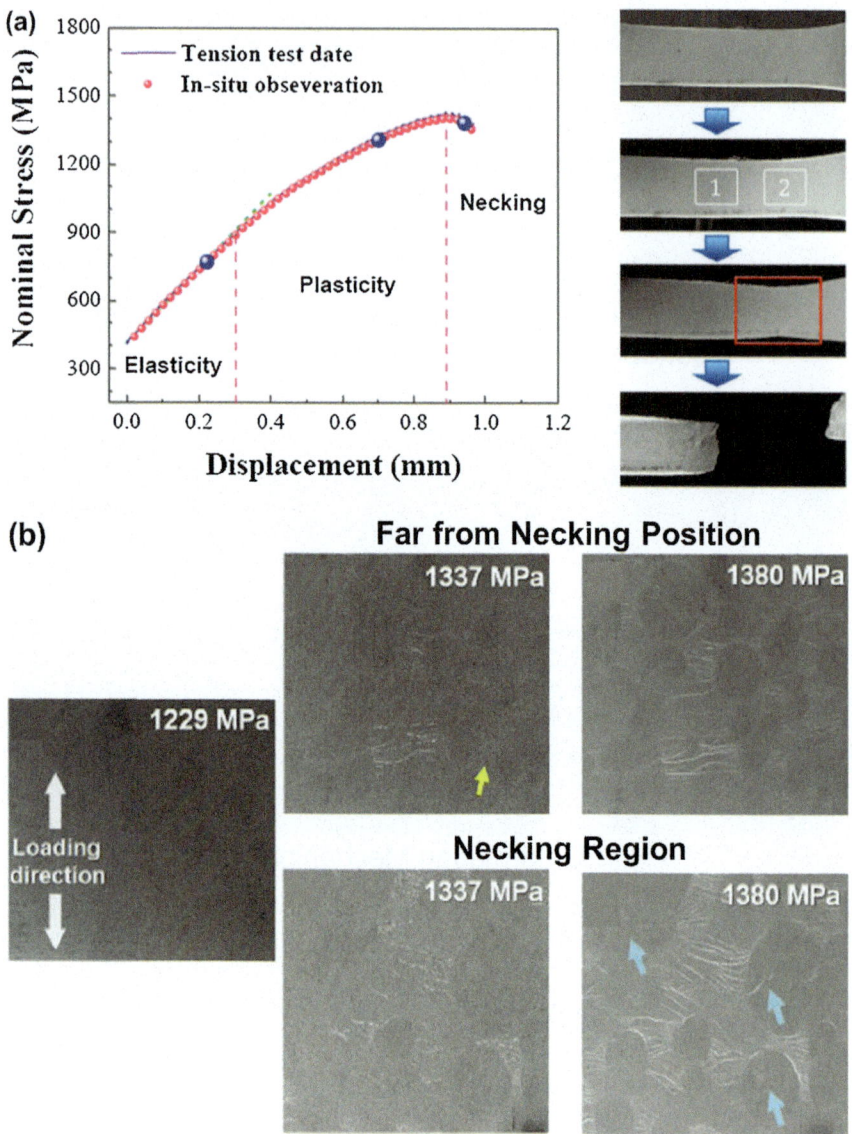

Fig. 3.21 In-situ SEM tensile testing of a bulk metallic glass composite, showing: (**a**) tensile stress–strain characteristics and real-time imaging of elastic, plastic, and necking regimes, and (**b**) high-magnification, local SEM imaging of sample under deformation to highlight differences in deformation mechanisms close to and away from the necking region. (Reproduced/adapted from Sun et al. (2018) CC BY 4.0)

3.5 In-Situ Tensile Characterization

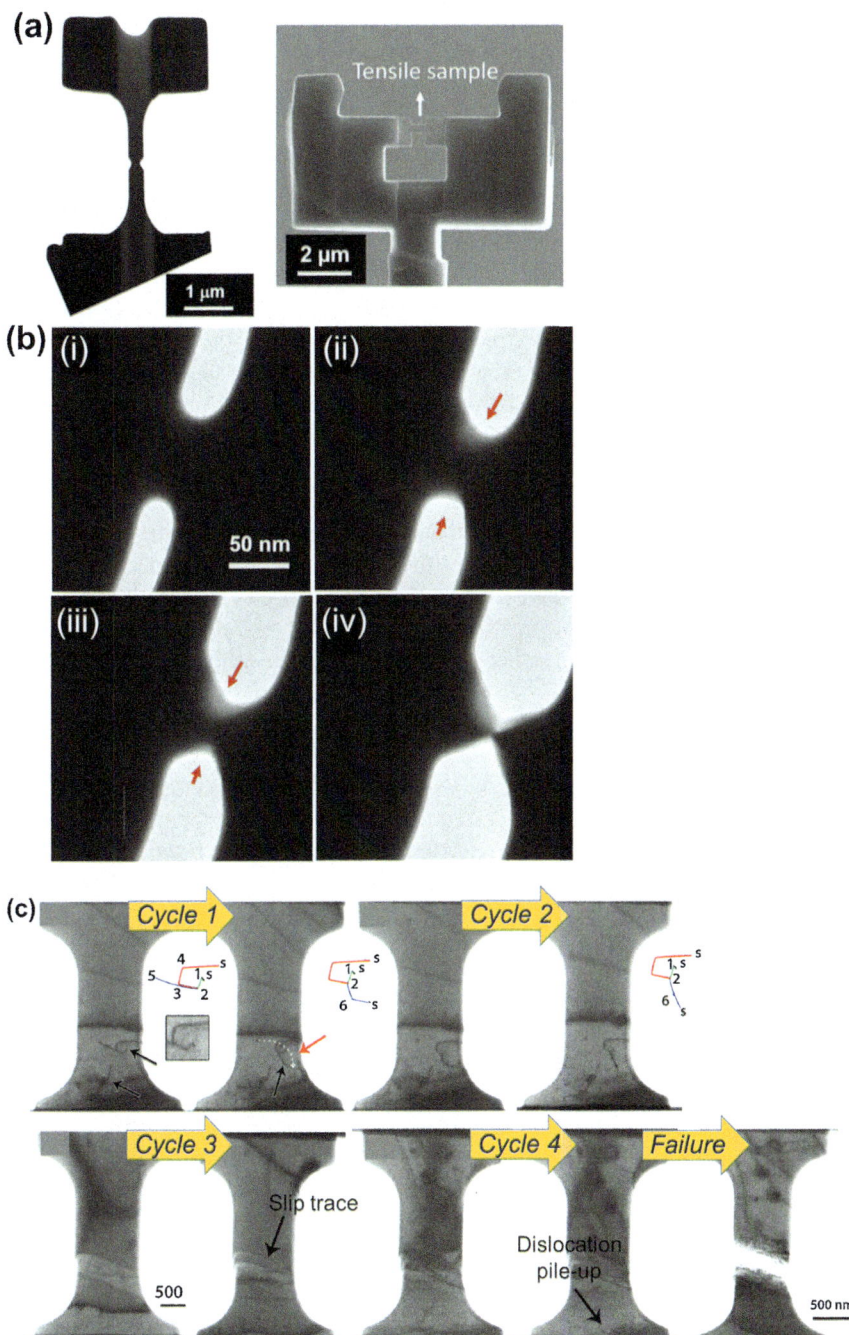

Fig. 3.22 In-situ TEM tensile tests: (**a**) Example of a submicron specimen and a tensile setup, (**b**) real-time snapshots showing necking in a specimen with notches, and (**c**) dislocation activity in a submicron Ni crystal specimen during tensile loading cycles, eventually leading to material failure. (Reproduced/adapted with permissions from Narayan et al. (2018), Samaee et al. (2018) CC BY 4.0)

Tensile characteristics of submicron specimens are typically probed in a transmission electron microscope to visualize the deformation behavior. TEM imaging can reveal the influence of flaws or notches on the mechanics of small-sized samples. Figure 3.22a illustrates in-situ TEM examination of a metallic glass sample (Narayan et al. 2018). The figure shows a notched free-standing dogbone sample and the tensile setup consisting of a push-to-pull device. One key concern during TEM testing of small-sized samples can be electron beam induced heating. For this reason, the beam current density should be carefully chosen. The case study illustrated in the figure confined the densities within 2×10^{-2} A cm^{-2}, as the sample deformation was not influenced when the current densities were lower than this value. The choice of grips is also important; heat-conducting W grips are helpful to negate heating effects. The real-time TEM snapshots during the tensile loading of the sample are shown in Fig. 3.22b. As the sample is strained, thinning of edges that connect the notches is observed. TEM imaging also reveals the formation of the elliptical band at notch edges. Eventually, necking is observed due to the merging of the elliptical bands from the two side notches. The influence of geometric parameters, such as specimen thickness, width and notch aspect ratio can be examined by this method. This example demonstrates the importance of in-situ mechanical testing to determine the role of nanoscale features and flaws on the mechanics of submicron specimens. Such investigations are highly relevant considering the recent trend toward miniaturization, where part designing can introduce sharp corners, notches or incisions.

In-situ TEM tensile testing provides information about dislocation activity during deformation. Dislocation structure evolution during successive tensile loading cycles is shown for a single crystal Ni specimen in Fig. 3.22c (Samaee et al. 2018). The in-situ snapshots are color-coded to follow the multiple activities in the specimen. Some key events resolved by real-time TEM imaging are listed below:

- Cycle 1: Unblocking of blue dislocation from the debris at point 5, which then glide around point 2 and pinned again at point 6.
- Cycle 2: Blue dislocation is seen to move in its slip plane, resulting in a reduction in the dislocation length.
- Cycle 3: A new slip trace is formed.
- Cycle 4: Dislocation pileup gliding in the $(11\bar{1})$ plane, near the bottom grip.

Eventually, the sample is seen to fail parallel to the $(11\bar{1})$ slip plane. Therefore, the application of in-situ TEM imaging during tensile tests provides insights into dislocation-driven plasticity mechanisms, otherwise difficult to speculate via post-mortem imaging.

3.6 In-Situ Double Cantilever Testing for Studying Fracture

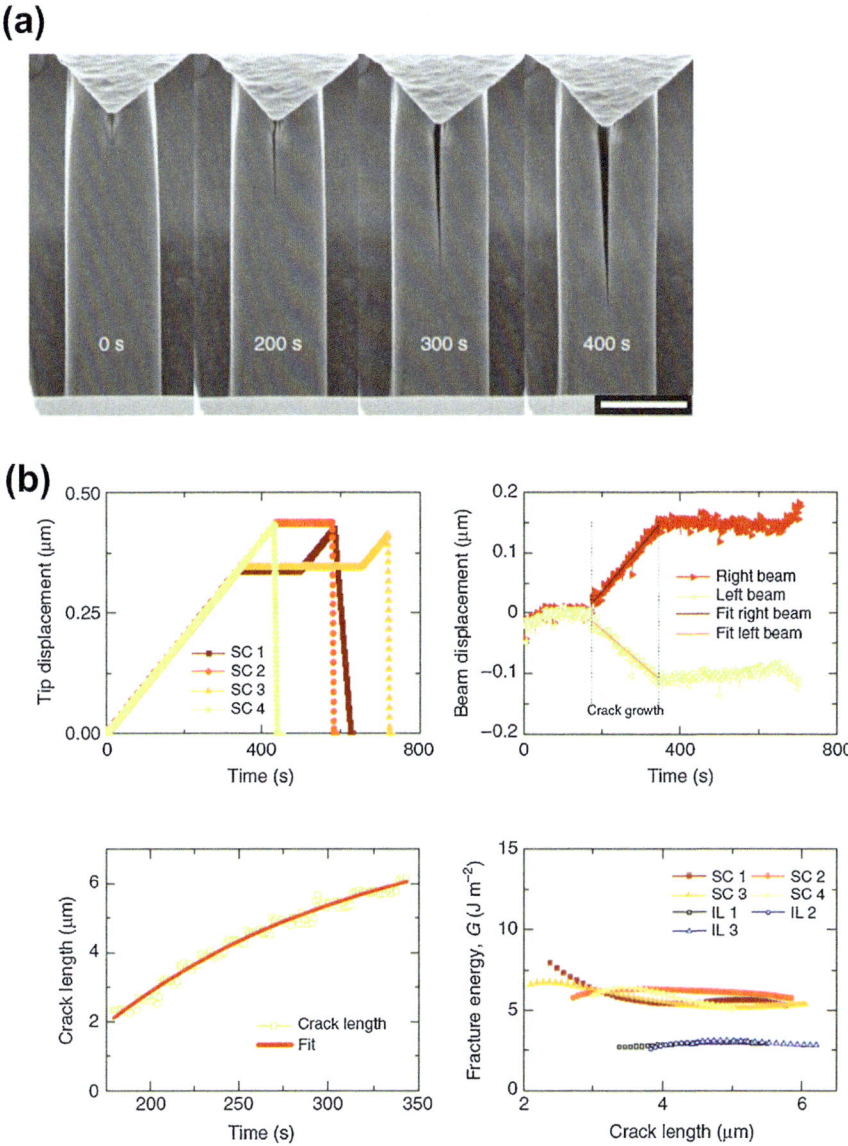

Fig. 3.23 In-situ double cantilever test for studying fracture of SiC: (**a**) Real-time SEM snapshots as the crack advances due to sliding of a wedge, and (**b**) determination of displacement, beam deflection, crack length, and fracture energy based on indenter test and in-situ SEM images. (Reproduced/adapted from Sernicola et al. (2017) CC BY 4.0)

3.6 In-Situ Double Cantilever Testing for Studying Fracture

The application of beam bending and tensile testing methods for studying fracture of materials was discussed in the previous sections. One major challenge with fracture studies is the uncontrolled and rapid crack propagation (in brittle materials), making failure visualization difficult. Double cantilever testing is an effective method for stable crack growth for in-situ imaging. The test setup consists of a double cantilever beam, which is loaded to cause bending of the two beams. Figure 3.23a shows a wedge sliding through the trough of the specimen of interest (Sernicola et al. 2017). In-situ SEM imaging allows observation and measurement of the crack as it grows. This setup enables the measurement of fracture energy since the energy stored within the two beams is primarily responsible for crack advancement. The energy stored within each beam is given by the Euler–Bernoulli theory:

$$U_M = \frac{Ed^3\delta^2}{8a^3} \tag{3.8}$$

where E is the elastic modulus, d is the beam width, δ is the maximum displacement, and a is the crack length. The strain rate release rate can be determined by the relation:

$$G = -\frac{dU_M}{da} = \frac{3Ed^3\delta^2}{8a^4} \tag{3.9}$$

This model is further modified for short cracks ($d >> a$) by taking into consideration the linear elasticity solution:

$$G = \frac{3Ed^3\delta^2}{8a^4} + \frac{3E(1+v)d^5\delta^2}{8a^6} = \frac{3Ed^3\delta^2}{8a^4}\left[1+(1+v)(d/a)^2\right] \tag{3.10}$$

This equation provides energy per unit area and hence can be employed for performing energy calculations during crack growth. Figure 3.23b shows the plot of quantitative information extracted from the double cantilever beam tests on a SiC specimen:

- The tip displacement is determined from the nanoindenter used to drive the wedge.
- The deflection (δ) is measured by the SEM image as the crack advances through the sample.
- Crack length (a) is determined from the SEM images (shown in Fig. 3.23a).
- Fracture energy can then be computed using the relation for G above.

A salient feature of this technique is that the crack growth is stable in nature: crack growth stops if the indenter driving the wedge is stopped. Therefore, this methodology is effective to characterize brittle specimens, where failure tends to be catastrophic in nature.

3.7 Summary

Mechanical test methods for in-situ characterization of materials were introduced and discussed. These methods have been employed for decades to measure various mechanical properties. However, this chapter provided the perspective of real-time imaging to develop mechanistic understanding as the materials deform and fail. The chapter covers test methods that span through the length scales: from localized region-specific investigations to bulk-scale deformation. The indentation method was discussed from the standpoint of multi-scale mechanics: low- to high-load indentation, and in-situ imaging in microscopes with different resolution capabilities, depending on the nature of desired information. Extension of the indentation approach for tribological characterization by the scratch method is useful for studying wear in materials. Other in-situ tribological test methods, such as rolling and sliding of surfaces in contact were discussed. Nano- and micro-pillar compressions are uniaxial test methods for local investigation, which provide stress–strain characteristics and allow quantitative as well as qualitative examination of elastic and plastic deformation. Beam bending, which is another micromechanical test method, is useful to study fracture behavior of materials. The tensile test method was discussed for uniaxial characterization of materials along with real-time imaging for deciphering mechanisms in different regions of the sample, under different conditions of stresses, and in the presence of additional flaws or features. The double cantilever beam method is useful for studying fracture behavior of brittle materials in real-time. These multiple mechanical test methods, coupled with in-situ imaging provide remarkably detailed and insightful information about the mechanics of materials. The comprehensive mechanical assessment enabled by these in-situ test methods contribute toward new material development with superior control over their properties and performance.

Questions and Assignments

1. Compare and contrast nanoindentation and micro-pillar compression testing. List key advantages, limitations, and challenges associated with each method. Take into consideration the following points:
   (a) Sample preparation.
   (b) Nature of loading.
   (c) Contact area.
   (d) Stress–strain relationship.
   (e) Mechanical results, outcome, analysis, and interpretation from each test.
   (f) Ease of in-situ imaging.

2. The study of fracture phenomena is important for the application of engineering materials. In-situ mechanical testing methods provide useful insights for failure analysis.

   (a) Discuss how in-situ microbeam bending is a more suitable technique to study the fracture of materials in comparison to the micro-pillar compression method.
   (b) What is the key advantage of double cantilever beam testing over microbeam bending during in-situ imaging of fracture behavior?

3. The chapter discusses several case studies showing real-time imaging of the deformation of pre-notched specimens. Write a 1000–1500 words mini critique on in-situ testing of notched samples by different methods (tensile, deflection, and double cantilever), elucidating insights and information that can be derived by real-time imaging. You may use references and examples not illustrated in this chapter. A maximum of four figures can be used.

4. As the sample size is reduced, especially in sub-micrometer length scales, mechanical properties are no longer size independent. Prominent size effects are often observed. These effects can be critical in small-scale technologies, like NEMS/MEMS. Write a brief note on the size effects observed for the following cases. Write no more than a paragraph for each point. You may use examples from literature or in the chapter to comment on the size effects associated with these test methods.

   (a) Tensile testing of nanofibers/nanotubes/nanowires: Consider the effect of their diameters.
   (b) Nano- and micro-pillar compression: Comment on *smaller is stronger* effect.
   (c) Indentation size effect (ISE): Consider the effect of penetration depth during nanoindentation testing.
   (d) The notch sensitivity of sub-micrometer-sized specimens during tensile testing: Refer to the following reference—
   Narayan et al. *Effect of notches on deformation of submicron sized metallic glasses: Insights from* in situ *expts*. Acta Mater. 2018, *154*, 172.

5. Identify the mechanical test method(s) suitable for achieving the objectives listed below:

   (a) Measuring the elastic modulus of very fine individual features (<600 nm in diameter) in the microstructure of a multiphase sample.
   (b) Measuring the aggregate or overall elastic modulus of the same sample described in (a) above.
   (c) Measuring yield strength of an individual grain in an alloy sample.
   (d) Measuring overall yield strength of the material/sample described in (c) above.
   (e) Measuring wear volume and coefficient of friction of a thin film ($t \sim 500$ nm) deposited on a substrate.
   (f) Measuring fracture strength of a ceramic matrix composite.

6. Choose a suitable mode of in-situ imaging from the list below to study the evolution of dislocation structure during tensile testing of a metallic sample:

    (a) Optical camera with high frame rate imaging capability.
    (b) Scanning electron microscope coupled with electron back scatter diffraction.
    (c) Transmission electron microscope with bright field and dark field imaging modes.

7. There are multiple approaches for micromechanical fracture study of interfaces. A list of methods is shown in the figure below (Sernicola et al. Nature Commun 8, 108). Compare and contrast these four different techniques for studying fracture by in-situ investigations.

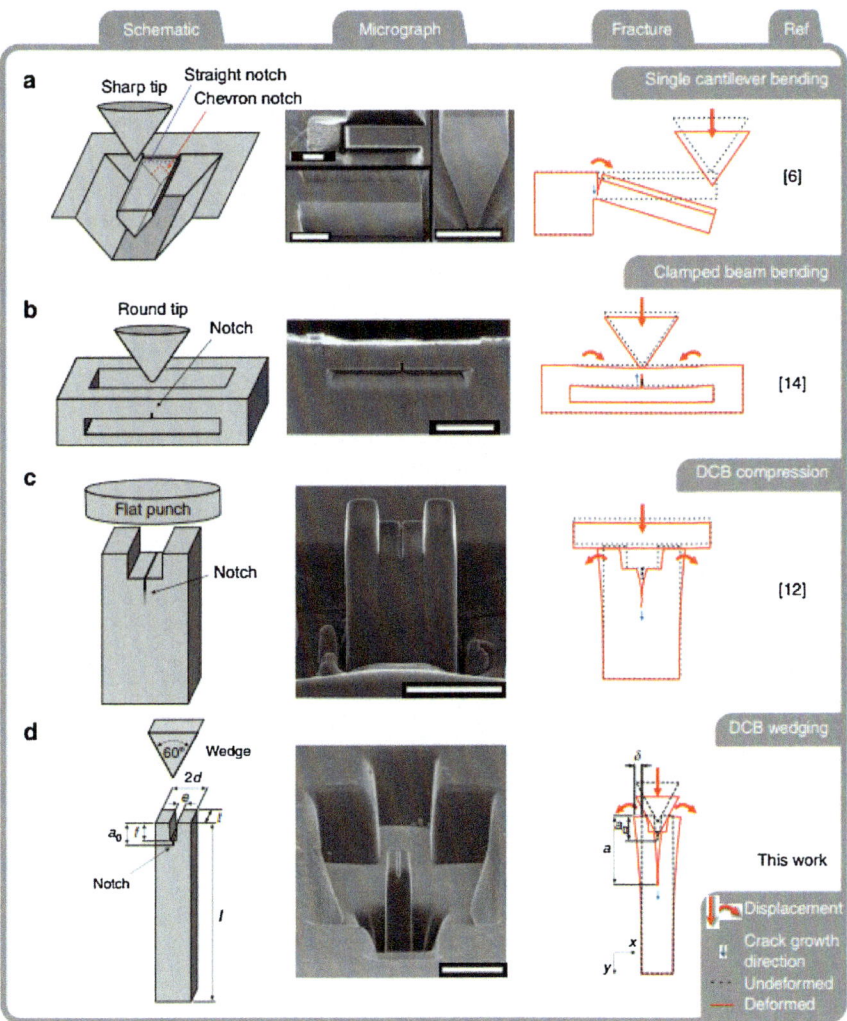

Reproduced from (Sernicola et al. 2017) CC BY 4.0

8. What kind of indenter tip(s) would be suited for the following applications (answer separately for each part)?
   (a) Performing micro-pillar compression.
   (b) Probing fine feature (~500 nm in diameter) in the microstructure.
   (c) Inducing cracks in a thin film of ceramic (deposited on a substrate).
   (d) Testing an ultrasoft hydrogel sample.
   (e) Performing a scratch test.
9. Which of the following two techniques are applicable for testing the mechanical response of a single graphene monolayer?
   (a) Low-load nanoindentation.
   (b) High-load microindentation.
10. Fatigue performance of a material is important for bearing mechanical loads over multiple loading cycles. Several examples of cyclic in-situ loading were presented in this chapter. Prepare a 1500–2000 words write-up on "In-situ Fatigue Characterization of Materials," focusing on different in-situ test methods which can be employed for fatigue testing (such as cyclic tension/compression/deflection/indentation). The write-up should emphasize on the relevance of in-situ imaging for understanding fatigue mechanisms in different classes of materials (metals/ceramics/polymers). Use case studies from literature.

References

Allison PG, Moser RD, Schirer JP et al (2014) In-situ nanomechanical studies of deformation and damage mechanisms in nanocomposites monitored using scanning electron microscopy. Mater Lett 131:313–316. https://doi.org/10.1016/j.matlet.2014.05.196

Armstrong DEJ, Rogers ME, Roberts SG (2009) Micromechanical testing of stress corrosion cracking of individual grain boundaries. Scr Mater 61:741–743. https://doi.org/10.1016/j.scriptamat.2009.06.017

Armstrong DEJ, Wilkinson AJ, Roberts SG (2011) Micro-mechanical measurements of fracture toughness of bismuth embrittled copper grain boundaries. Philos Mag Lett 91:394–400. https://doi.org/10.1080/09500839.2011.573813

Armstrong DEJ, Haseeb ASMA, Roberts SG et al (2012) Nanoindentation and micro-mechanical fracture toughness of electrodeposited nanocrystalline Ni-W alloy films. Thin Solid Films 520:4369–4372. https://doi.org/10.1016/j.tsf.2012.02.059

Bufford D, Liu Y, Wang J et al (2014) In situ nanoindentation study on plasticity and work hardening in aluminium with incoherent twin boundaries. Nat Commun 5:4864. https://doi.org/10.1038/ncomms5864

Chudoba T, Schwarzet N, Linss V, Richter F (2004) Determination of mechanical properties of graded coatings using nanoindentation. Thin Solid Films 469–470:239–247. https://doi.org/10.1016/j.tsf.2004.08.157

Deng Y, Hajilou T, Barnoush A (2017) Hydrogen-enhanced cracking revealed by in situ microcantilever bending test inside environmental scanning electron microscope. Philos Trans R Soc A Math Phys Eng Sci 375:20170106. https://doi.org/10.1098/rsta.2017.0106

Fisher-Cripps AC (2005) Nanoindentation, 3rd edn. Springer, New York

References

Florando JN, Nix WD (2005) A microbeam bending method for studying stress-strain relations for metal thin films on silicon substrates. J Mech Phys Solids 53:619–638. https://doi.org/10.1016/j.jmps.2004.08.007

Heiroth S, Ghisleni R, Lippert T et al (2011) Optical and mechanical properties of amorphous and crystalline yttria-stabilized zirconia thin films prepared by pulsed laser deposition. Acta Mater 59:2330–2340. https://doi.org/10.1016/j.actamat.2010.12.029

Hintsala E, Kiener D, Jackson J, Gerberich WW (2015) In-situ measurements of free-standing, ultra-thin film cracking in bending. Exp Mech 55:1681–1690. https://doi.org/10.1007/s11340-015-0069-2

Hintsala ED, Stauffer DD, Oh Y, Asif SAS (2017) In situ TEM scratch testing of perpendicular magnetic recording multilayers with a novel MEMS tribometer. JOM 69:51–56. https://doi.org/10.1007/s11837-016-2154-0

Howard C, Fritz R, Alfreider M et al (2017) The influence of microstructure on the cyclic deformation and damage of copper and an oxide dispersion strengthened steel studied via in-situ microbeam bending. Mater Sci Eng A 687:313–322. https://doi.org/10.1016/j.msea.2017.01.073

Kiener D, Motz C, Grosinger W et al (2010) Cyclic response of copper single crystal micro-beams. Scr Mater 63:500–503. https://doi.org/10.1016/j.scriptamat.2010.05.014

Kupka D, Lilleodden ET (2012) Mechanical testing of solid-solid interfaces at the microscale. Exp Mech 52:649–658. https://doi.org/10.1007/s11340-011-9530-z

Lahouij I, Dassenoy F, De Knoop L et al (2011) In situ TEM observation of the behavior of an individual fullerene-like MoS 2 nanoparticle in a dynamic contact. Tribol Lett 42:133–140. https://doi.org/10.1007/s11249-011-9755-0

Lahouij I, Dassenoy F, Vacher B, Martin J-M (2012) Real time TEM imaging of compression and shear of single fullerene-like MoS2 nanoparticle. Tribol Lett 45:131–141. https://doi.org/10.1007/s11249-011-9873-8

Lee JH, Holland TB, Mukherjee AK et al (2013) Direct observation of Lomer-Cottrell locks during strain hardening in nanocrystalline nickel by in situ TEM. Sci Rep 3:1061. https://doi.org/10.1038/srep01061

Li X, Yang Y, Zhao Y et al (2017) Electrospinning fabrication and in situ mechanical investigation of individual graphene nanoribbon reinforced carbon nanofiber. Carbon N Y 114:717–723. https://doi.org/10.1016/j.carbon.2016.12.082

Mara NA, Crapps J, Wynn TA et al (2013) Microcantilever bend testing and finite element simulations of HIP-ed interface-free bulk Al and Al-Al HIP bonded interfaces. Philos Mag 93:2749–2758. https://doi.org/10.1080/14786435.2013.786192

Narayan RL, Tian L, Zhang D et al (2018) Effects of notches on the deformation behavior of submicron sized metallic glasses: insights from in situ experiments. Acta Mater 154:172–181. https://doi.org/10.1016/j.actamat.2018.05.041

Nautiyal P, Jain J, Agarwal A (2015) A comparative study of indentation induced creep in pure magnesium and AZ61 alloy. Mater Sci Eng A 630:131–138. https://doi.org/10.1016/j.msea.2015.01.075

Nautiyal P, Rudolf C, Loganathan A et al (2016) Directionally aligned ultra-long boron nitride nanotube induced strengthening of aluminum-based Sandwich composite. Adv Eng Mater 18:1747–1754. https://doi.org/10.1002/adem.201600212

Oliver WC, Pharr GM (1992) An improved technique for determining hardness and elastic modulus using load and displacement sensing indentation experiments. J Mater Res 7:1564–1583. https://doi.org/10.1557/JMR.1992.1564

Oliver WC, Pharr GM (2010) Nanoindentation in materials research: past, present, and future. MRS Bull 35:897–907. https://doi.org/10.1557/mrs2010.717

Oviedo JP, Santosh KC, Lu N et al (2015) In situ TEM characterization of shear-stress-induced interlayer sliding in the cross section view of molybdenum disulfide. ACS Nano 9:1543–1551. https://doi.org/10.1021/nn506052d

Pugno N, Peng B, Espinosa HD (2005) Predictions of strength in MEMS components with defects - a novel experimental-theoretical approach. Int J Solids Struct 42:647–661. https://doi.org/10.1016/j.ijsolstr.2004.06.026

Rabe R, Breguet J-M, Schwaller P et al (2004) Observation of fracture and plastic deformation during indentation and scratching inside the scanning electron microscope. Thin Solid Films 469–470:206–213. https://doi.org/10.1016/j.tsf.2004.08.096

Rudolf C, Boesl B, Agarwal A (2015) In situ indentation behavior of bulk multi-layer graphene flakes with respect to orientation. Carbon N Y 94:872–878. https://doi.org/10.1016/j.carbon.2015.07.070

Rzepiejewska-Malyska K, Parlinska-Wojtan M, Wasmer K et al (2009) In-situ SEM indentation studies of the deformation mechanisms in TiN, CrN and TiN/CrN. Micron 40:22–27. https://doi.org/10.1016/j.micron.2008.02.013

Samaee V, Gatti R, Devincre B et al (2018) Dislocation driven nanosample plasticity: new insights from quantitative in-situ TEM tensile testing. Sci Rep 8:12012. https://doi.org/10.1038/s41598-018-30639-8

Schwaiger R, Weber M, Moser B et al (2012) Mechanical assessment of ultrafine-grained nickel by microcompression experiment and finite element simulation. J Mater Res 27:266–277. https://doi.org/10.1557/jmr.2011.248

Sernicola G, Giovannini T, Patel P et al (2017) In situ stable crack growth at the micron scale. Nat Commun 8:108. https://doi.org/10.1038/s41467-017-00139-w

Shan ZW, Mishra RK, Syed Asif SA et al (2008) Mechanical annealing and source-limited deformation in submicrometre-diameter Ni crystals. Nat Mater 7:115–119. https://doi.org/10.1038/nmat2085

Stach EA, Freeman T, Minor AMAM et al (2001) Development of a nanoindenter for in situ transmission electron microscopy. Microsc Microanal 7:507–517

Sun HC, Ning ZL, Wang G et al (2018) In-situ tensile testing of ZrCu-based metallic glass composites. Sci Rep 8:1–12. https://doi.org/10.1038/s41598-018-22925-2

Zhu Y, Qin Q, Gu Y, Wang Z (2010) Friction and shear strength at the nanowire-substrate interfaces. Nanoscale Res Lett 5:291–295. https://doi.org/10.1007/s11671-009-9478-4

Chapter 4
In-Situ Mechanical Characterization as a Function of Temperature

Mechanical properties of materials are highly sensitive to temperature changes. In-situ characterization approach can be utilized to probe temperature dependence of the mechanical properties at multiple length scales. This chapter introduces the principles and techniques for high- and low-temperature mechanical testing with simultaneous real-time imaging. Temperature-dependent characterization requires specialized instrumentation which allows for effective sample heating or cooling. Challenges associated with thermal drift, system compliances, thermal expansion, radiation, and image quality are addressed from the standpoint of high-temperature in-situ instrumentation design. Strategies to overcome or minimize these issues are discussed. The chapter presents in-situ temperature-dependent characterization by indentation, compression, flexural, and tensile testing techniques. Mechanical characterization at different temperatures is helpful to decipher plasticity parameters, such as Peierls stress or activation energy. The applications of high-temperature testing for studying creep mechanisms, fatigue deformation, and temperature-dependent fracture behavior of thin films are presented in the chapter. The high- and low-temperature in-situ testing is also shown to be useful for selectively activating and studying different slip systems in materials. Therefore, in-situ testing as a function of temperature is a promising approach for engineering superior materials with extreme environment applications.

Electronic supplementary material The online version of this chapter (https://doi.org/10.1007/978-3-030-43320-8_4) contains supplementary material, which is available to authorized users.

4.1 High-Temperature In-Situ Mechanical Testing

High-temperature in-situ investigations can be performed at multiple length scales to evaluate mechanical properties and observe deformation mechanisms activated in materials. A variety of mechanical testing methods can be employed, such as indentation, pillar compression, beam bending, or tensile loading to study the temperature-dependent response of materials under different stress states. The subsequent subsections discuss high-temperature in-situ characterization under different loading scenarios.

4.1.1 In-Situ High-Temperature Nanoindentation

High-temperature indentation typically involves heating of both the sample and the tip for stable testing. Lack of temperature match can result in thermal drift and influence the measurements. One of the concerns for in-situ imaging in SEM while heating is the deflection of the electron beam by alternating current supplied to the heaters (Wheeler and Michler 2013; Wheeler et al. 2015). This can result in a wobbly image. Because of this reason, direct current is preferred in in-situ SEM indenters. Additionally, the grounding of electrical components is important to avoid transmission of outside electronic noise into the microscope. In-situ nanoindenter stages are also equipped with water-cooling systems to keep the elevated temperatures localized to the tip and the sample. The use of materials with a low coefficient of thermal expansion is preferred to minimize compliances. In order to limit thermal drift in long-duration tests, an active surface referencing approach was recently reported which involves two vertical axes: one for the actual indentation and the other for surface referencing. These two axes are designed to be identical, and the differential displacement between the reference and the indenter is measured (Conte et al. 2019). Mounting of samples for high-temperature nanoindentation requires careful consideration as the glues used for room-temperature indentation might not be suitable beyond a certain temperature. Special types of cement are often used for mounting samples for high-temperature testing, but they are prone to cracking due to elevated temperatures. Moreover, the use of a nonconducting adhesive can be detrimental to in-situ SEM imaging. Therefore, mechanical clamping is preferred to mount the samples during high-temperature testing (Trenkle et al. 2010). Indentation systems with both resistive and infrared heaters have been demonstrated for high-temperature testing. A schematic representation of some of the key components of a high-temperature indenter stage is shown in Fig. 4.1. It is noteworthy that this is not a universal system, and there are several variations of the instrumentation design (Wheeler et al. 2015).

Coupling high-temperature indentation with high-resolution real-time imaging is useful to investigate specific deformation mechanisms as a function of temperature.

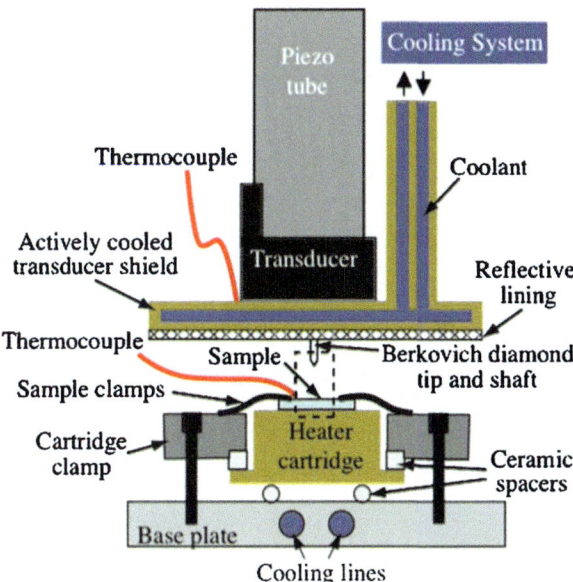

Fig. 4.1 Schematic representation of components and working principles of high-temperature indenter. (Reproduced with permission from Trenkle et al. (2010))

Figure 4.2a shows the deformation mechanisms captured by SEM imaging as the indenter tip penetrates a Zr-based bulk metallic glass (Wheeler et al. 2011). It is seen that relatively higher penetration depths are required to activate the shear offsets at 200 °C in comparison to room-temperature indentation. The nature of shear offsets is also seen to be temperature sensitive: they have pointed or sharp features during the room-temperature indentation, but are smooth, semicircular in shape at 200 °C. High-temperature indentation also allows a comparison of load–displacement response vis-à-vis deformation mechanisms. For the same Zr-based bulk metallic glass, serrations in the load–displacement curves were found to be a function of temperature and a trend of increasing serrations with the increasing indentation temperature was observed (Fig. 4.2b). This finding is the opposite of SEM observation, which shows a decrease in surface shear at higher temperatures. These in-situ investigations reveal that surface shear and load drops are unrelated mechanical phenomena.

The high-temperature in-situ indentation can be useful to study creep in materials. Figure 4.3 demonstrates a case study on high-temperature creep in a cold-sprayed aluminum alloy (Nautiyal et al. 2019a). The creep displacement–time plots in Fig. 4.3a indicate an order of magnitude higher creep at 300 and 400 °C as compared to room-temperature creep deformation. The SEM images comparing temperature-sensitive creep mechanisms are shown in Fig. 4.3b. It can be seen that pileup is the prominent mechanism at relatively lower temperatures. At 300 and 400 °C,

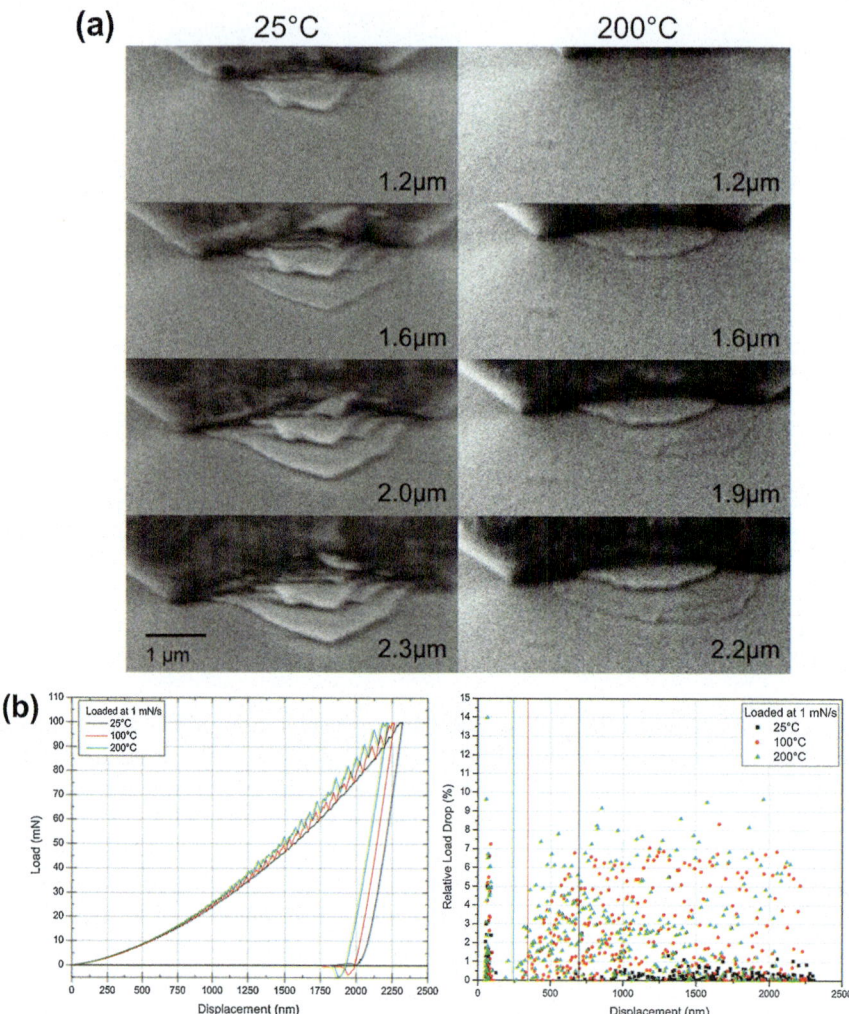

Fig. 4.2 (**a**) The comparison of real-time indentation deformation mechanisms in a Zr-based bulk metallic glass at room temperature and 200 °C captured by SEM imaging, and (**b**) load–displacement plots showing enhanced serrations/load drops at elevated temperatures which is also confirmed by plotting relative load drops as a function of indentation depth at room temperature, 100 °C and 200 °C. (Reproduced/adapted with permission from Wheeler et al. (2011))

there is no pileup and the elasto-plastic zone is much larger in size. In-situ SEM imaging at 400 °C reveals crack formation, propagation, and delamination mechanisms in the vicinity of the indenter tip (shown in Fig. 4.3c). This is a rather surprising observation for an aluminum alloy, which is not expected to demonstrate the crack propagation mechanism. This observation is attributed to the cold spray manufacturing process, which involves the supersonic impact and deposition of powder particles to form the bulk alloy structure. High-temperature exposure creates stress

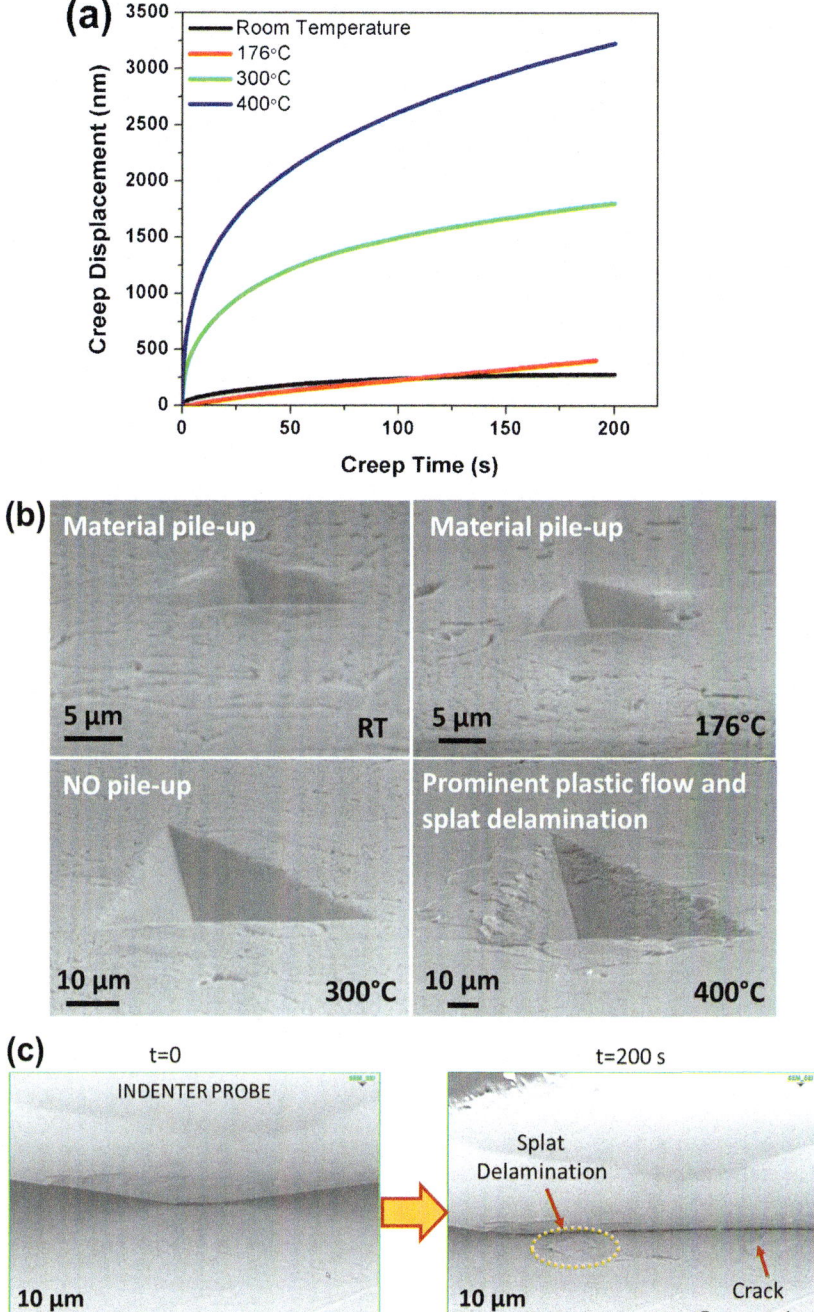

Fig. 4.3 In-situ high-temperature creep investigations by indentation technique: (**a**) comparative creep curves at different temperatures, (**b**) comparison of temperature-sensitive deformation mechanisms by SEM imaging, and (**c**) real-time indentation reveals crack propagation and delamination mechanisms at particle interfaces in a cold-sprayed aluminum alloy. (Reproduced/adapted with permission from Nautiyal et al. (2019a))

concentration along the particle interfaces. These stressed boundaries then act as sites for crack initiation. Readers are referred to Supplementary Video, Video 4.1 where crack propagation and splat delamination events are captured by real-time SEM imaging. Therefore, the in-situ high-temperature indentation can be a useful approach to study the mechanics of materials processed by additive manufacturing techniques. The processing-mechanics correlation is vital for wide-scale acceptability of newly emerging techniques and in-situ characterization can provide useful insights for establishing such correlations.

High-temperature indentation is a useful approach to probe deformation in nanomaterial assemblies. Figure 4.4a shows the in-situ SEM indentation of a boron nitride nanotube "buckypaper," consisting of a dense and entangled network of 1D nanotubes (Nautiyal et al. 2019b). Load–displacement plots showed similar deformation for room temperature and 250 °C indentation, but higher displacement for indentation at 500 °C (Fig. 4.4b). The elastic modulus showed a modest drop from ~1.2 GPa to ~0.9 GPa as the test temperature was increased from room temperature

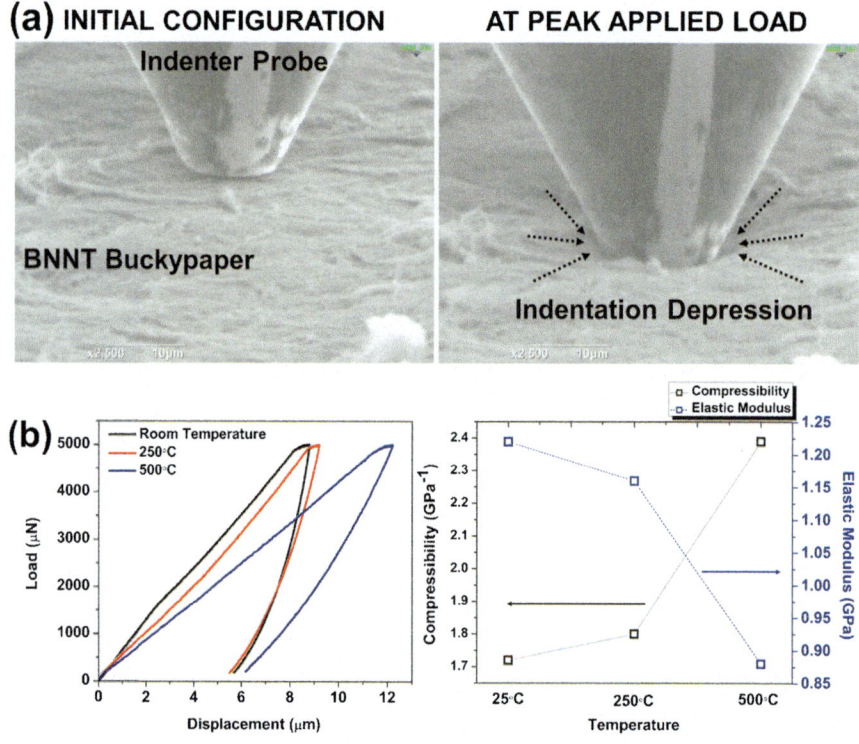

Fig. 4.4 The in-situ SEM indentation of a boron nitride nanotube buckypaper: (**a**) SEM micrographs showing indenter-induced stretching/deformation of buckypaper, and (**b**) load-displacement and elastic modulus response as a function of test temperature. (Reproduced with permission from Nautiyal et al. (2019b))

to 500 °C. Buckypaper response to compressive loading at different temperatures are shown in Supplementary Videos, Videos 4.2, 4.3, and 4.4. SEM imaging demonstrated that the buckypaper is highly flexible and damage tolerant. The videos show remarkable strain-redistribution ability due to interconnected microstructure, which prevents stress–concentration and local failure. The mechanical response was captured for 50 loading-unloading cycles and there were no signs of failure, indicative of fatigue-resilience at elevated temperatures. Supplementary Videos, Videos 4.5, 4.6, and 4.7 show cyclic deformation of BNNT buckypaper at room temperature, 250 °C and 500 °C, respectively. These mechanistic insights into high-temperature deformation of nanomaterials obtained are highly informative for engineering advanced structural composites.

4.1.2 In-Situ High-Temperature Micro-pillar Compression

For the localized investigation of high-temperature properties, the micro-pillar compression technique is useful as it eliminates strain gradients which can be prominent during nanoindentation. It also enables direct measurement of temperature-dependent yield stress and failure strength from stress–strain plots, which is not possible by the indentation technique. Figure 4.5a illustrates deformation Zr-metallic glass pillars at temperatures ranging from room temperature to 387 °C (Wheeler et al. 2012). The sample deformation is characterized by shear banding below the glass transition temperature, T_g (351.85 °C). However, the deformation is seen to be completely homogeneous above T_g with no shear bands, as shown in Fig. 4.5a for the pillar compressed at 387 °C. The comparison of the stress–strain response of the pillars revealed the serrated flow below T_g, but no variation in yield stress even as the compression temperature is raised from room temperature to 289 °C (Fig. 4.5b). However, the yield stress is significantly lower above T_g. This case study demonstrates the applicability of in-situ high-temperature testing to decipher the material response in different regimes and understand the mechanistic significance of critical temperatures (T_g in this case).

In-situ micro-pillar compression is useful to decipher brittle-to-plastic transformation exhibited by ceramic materials. For instance, a flash-sintered YSZ ceramic shows very different stress–strain characteristics at room temperature and elevated temperatures (above 400 °C) as shown in Fig. 4.6a (Cho et al. 2018). Compression at room temperature shows continuously increasing stress and ultimately abrupt failure at ~9% strain. On the other hand, the stress response is not monotonic for compression at 400 °C. In-situ SEM imaging during the test reveals a higher crack density at 400 °C, and these cracks tend to propagate downward through the pillar as the strain is applied (Fig. 4.6b). This mechanism is very different from room-temperature response, where failure is catastrophic in nature. Pillar compression technique can also be used to evaluate temperature-sensitive "fatigue response" of materials. Figure 4.6c illustrates multiple loading-unloading-reloading cycles

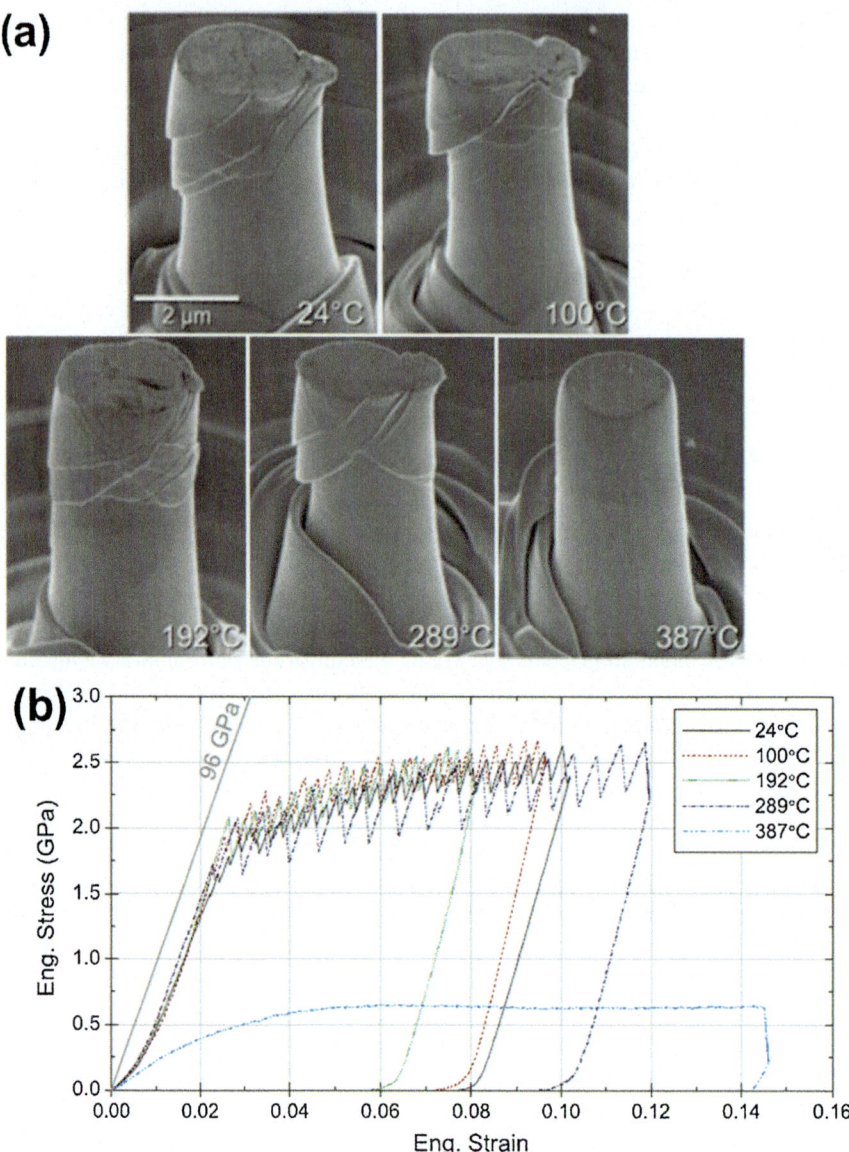

Fig. 4.5 (**a**) A comparison of micro-pillar compression mechanisms in a Zr-metallic glass for different test temperatures, and (**b**) stress–strain plots corresponding to the pillar compression. (Reproduced/adapted with permission from Wheeler et al. (2012))

performed on the same YSZ ceramic specimen at room temperature and 400 °C. While no cracks were observed at room temperature after performing 30 cycles, the fatigue loading at 400 °C resulted in the formation of multiple cracks in the pillar near the point of loading. The case study shown in the figure also

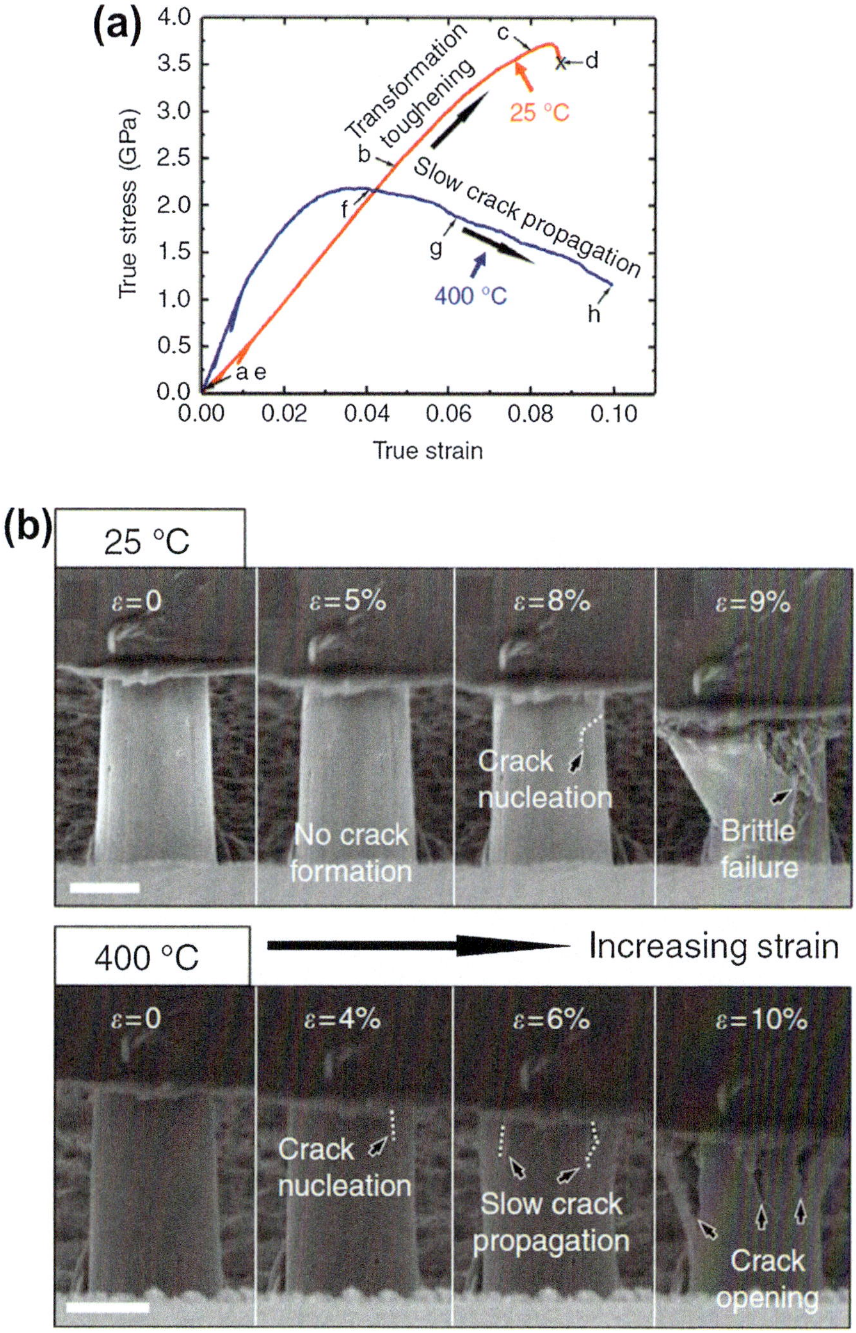

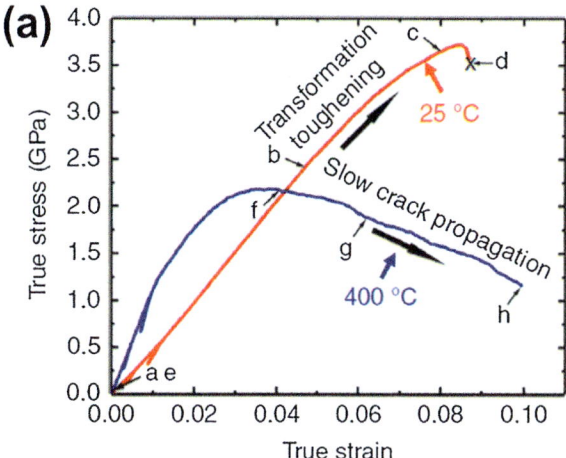

Fig. 4.6 (**a**) A comparison of stress–strain curves for micro-pillar compression of YSZ ceramic at room temperature and 400 °C, and (**b**) corresponding mechanisms observed in real time. (**c**) Cyclic pillar compression tests as a function of temperature—comparison of stress–strain response and mechanisms observed via SEM. (Reproduced/adapted from Cho et al. (2018) CC BY 4.0)

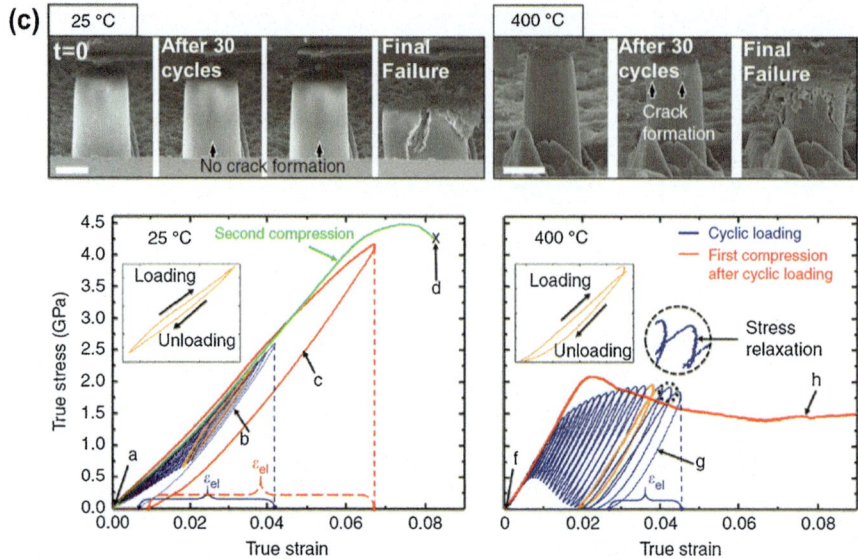

Fig. .6 (continued)

demonstrates a compression test post cyclic deformation to assess the effect of fatigue on the material's mechanical response. While room temperature test results in compression strength exceeding 4 GPa and catastrophic failure, high-temperature compression (at 400 °C) is characterized by much lower strength of ~2 GPa and extensive cracking.

Multilayered or multicomponent materials can have complex and difficult to decode deformation behavior. The issue becomes even more challenging with systems comprising of hard and soft phases, as there can be an interplay of multiple deformation mechanisms. For instance, multilayer metal/ceramic specimens are interesting from the standpoint of combining desirable mechanical properties, such as superior fracture toughness, high formability and work hardening ability. The activation of these mechanisms can be highly sensitive to temperature in such structures. Figure 4.7 demonstrates an example of the micro-pillar compression of a Cu/TiN multilayer system from room temperature up to 400 °C (Raghavan et al. 2015). The case study in the figure also assesses the role of layer thickness on the mechanics of the material system. The stress–strain curves and deformation mechanisms are shown for three different thickness ratios of Cu/TiN: 5/10, 50/100 and 700/1000 nm. A progressive reduction in yield stress was seen with the increase in test temperature (Fig. 4.7a). SEM imaging showed the plastic flow of Cu, as they form microcrystals on the cylinder surfaces. This stress-assisted diffusion also resulted in the coalescence of Cu crystals to form larger crystals. The plastic flow behavior is more pronounced during compression at higher temperatures (Fig. 4.7b). This example illustrates the importance of high-resolution real-time

4.1 High-Temperature In-Situ Mechanical Testing

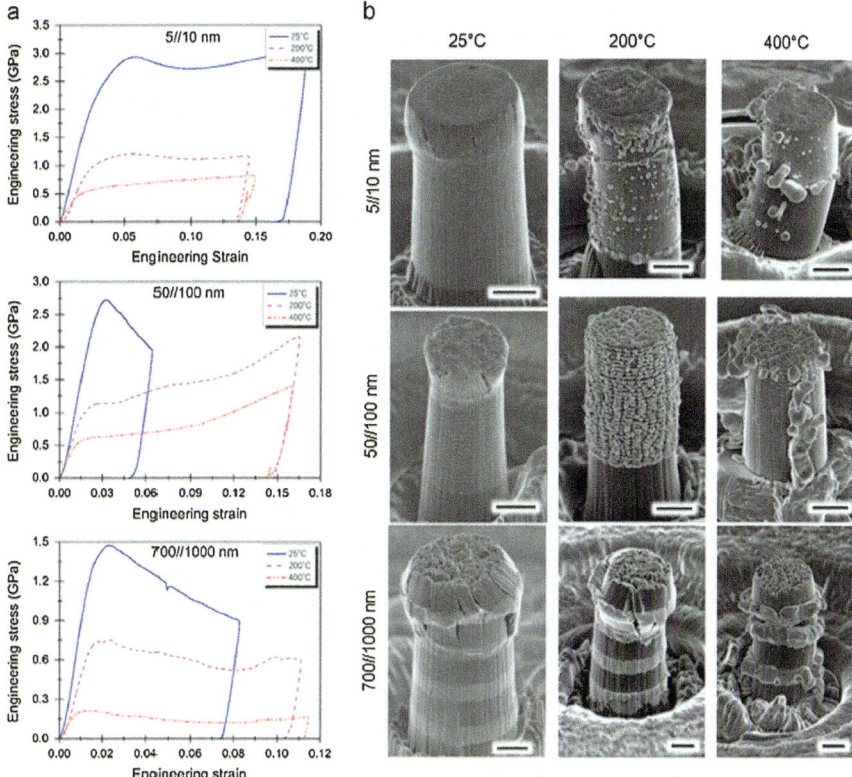

Fig. 4.7 In-situ characterization of a multilayer Cu/TiN material system: (**a**) comparative stress–strain plots for micro-pillar compression tests, and (**b**) SEM imaging of deformation mechanisms at different temperatures and for different thickness ratios of Cu/TiN. (Reproduced with permission from Raghavan et al. (2015))

imaging for a precise understanding of temperature-dependent mechanisms in composite material systems.

A variation of micro-pillar compression, named micro-pillar splitting, is a useful technique for evaluating fracture toughness at micrometer length scales. In this technique, a thin ceramic film is deposited over a substrate, which is then machined to form a pillar. The indentation of the pillar is performed to induce cracks in the coating for toughness calculation. By adding high-temperature instrumentation and real-time imaging, the pillar splitting technique can also be used to study fracture behavior as a function of temperature. The example in Fig. 4.8 demonstrates fracture investigation for different ceramic coatings (Best et al. 2019). The SEM images show the differences in fracture mechanisms from room temperature to 500 °C (Fig. 4.8a). The real-time imaging indicated splitting fracture (for AlTiN and AlCrSiN films) and cleavage fracture (CrN and AlCrTiN films) mechanisms. Quantitative analysis showed the toughness increased and the fracture strength

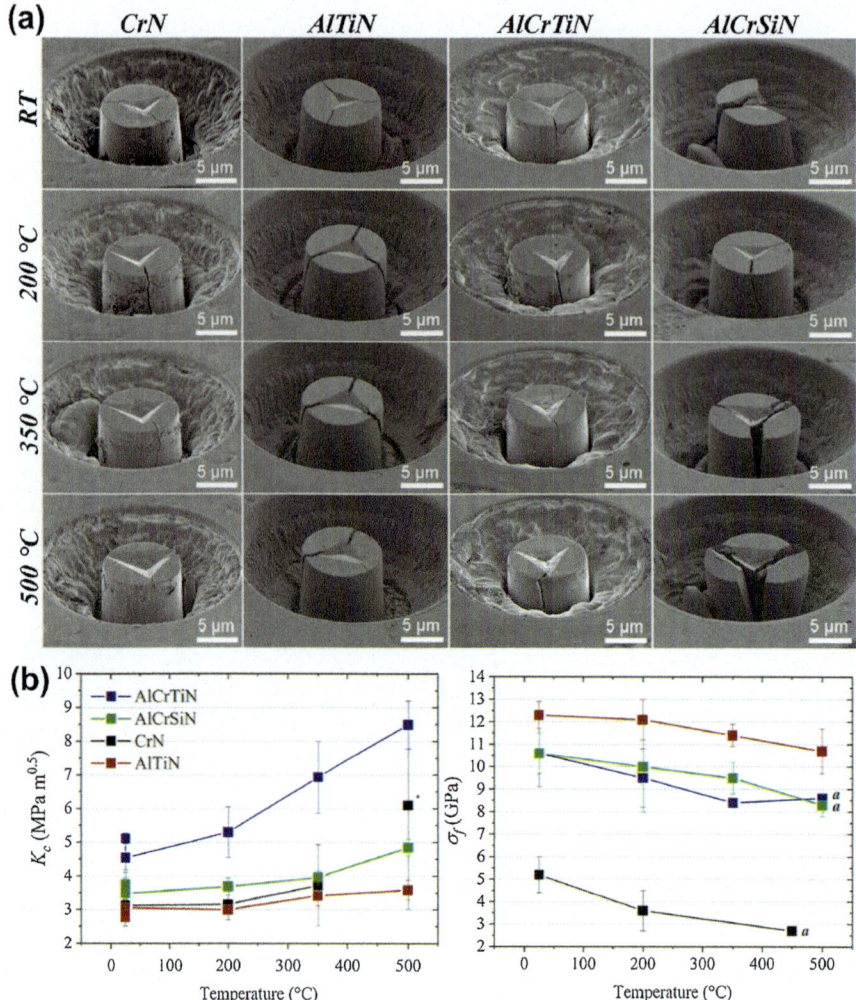

Fig. 4.8 Investigation of fracture in different ceramic coatings by micro-pillar splitting as a function of temperatures: (**a**) comparison of fracture mechanisms by SEM imaging, and (**b**) comparison of fracture toughness and fracture strength values. (Reproduced with permission from Best et al. (2019))

dropped at elevated temperatures (Fig. 4.8b). The main advantage of this test over notched specimen testing for toughness estimation is the reduction in ion impregnation. Fabricating notches with controlled dimensions by the milling process can be tedious and difficult to control. The ion beam induced defects can also influence the test outcome. The possibility of inducing milling artifacts is reduced in micro-pillar splitting test setup.

High-temperature micro-compression technique has been demonstrated for extracting plasticity parameters for brittle materials. In order to study plasticity in brittle materials, it is important to prevent early catastrophic failure due to mechanical loading. Bending and tension tend to induce brittle failure and it's not always possible to visualize deformation due to abrupt failure. Because of this reason, in-situ microscale compression is a viable technique for studying plasticity in brittle materials. When combined with high-temperature testing, micro-compression can enable the calculation of activation energy parameters. The compression of single crystals is useful for probing slip systems. The example in Fig. 4.9 demonstrates the application of high-temperature microscale compression on GaN {0001}-oriented euhedral prisms in SEM (Wheeler et al. 2013). The effect of temperature on deformation is shown by comparative SEM images before and after the test in the temperature range of ~24 to 480 °C (Fig. 4.9a). It was observed that the selection of the prism aspect ratio is critical to study crystal plasticity. For instance, the prisms with aspect ratios greater than 2.2 exhibited early failure along the prismatic planes due to stress concentration at the top of the prisms. This stress concentration was ascribed to geometric imperfections. On the other hand, the slip was not always activated in the prisms with an aspect ratio less than 1.5 because of constraints imposed on the top and the bottom ends of the specimen. This constrained state was seen to induce fracture or tearing. An optimum aspect ratio of 1.8 for the prism with uniform hexagonal geometry was identified to be desirable for activating pure slip. The effect of aspect ratios on deformation behavior is illustrated in Fig. 4.9b through

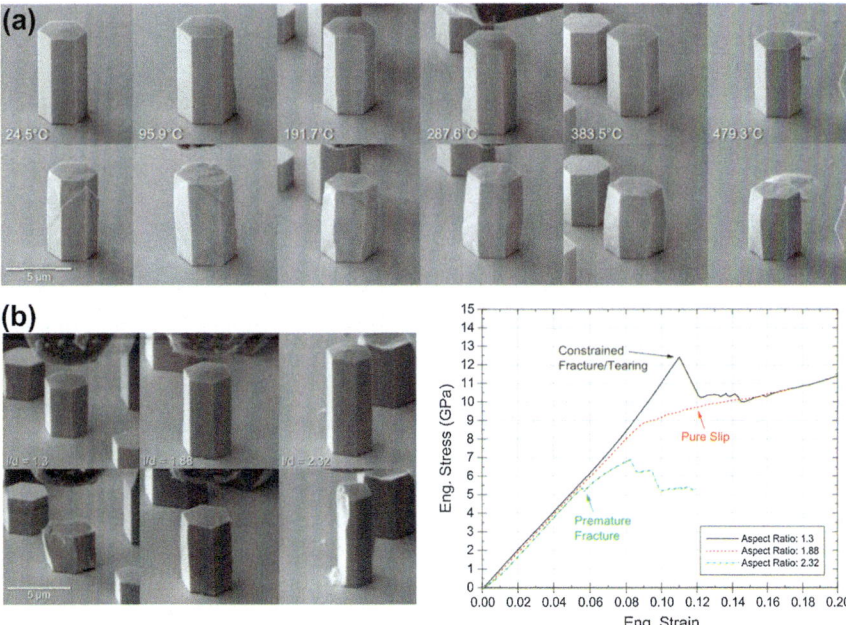

Fig. 4.9 In-situ micro-compression of GaN prisms: (**a**) deformation as a function of temperature imaged by SEM, and (**b**) the effect of prism aspect ratios on the nature of deformation. (Reproduced with permission from Wheeler et al. (2013))

the SEM images and stress–strain curves. In-situ imaging allowed the observation of the angle between the slip and the loading axis (as can be seen from Fig. 4.9a). This direct determination of slip angle from the facets enables the identification of the slip system by comparing with the theoretically computed angles from the mathematical relation:

$$\cos\theta = 3/4\left(\frac{a}{c}\right)^2 l \left\{ 3/4\left(\frac{a}{c}\right)^2 \left(h^2 + k^2 + hk + \left(\frac{a}{c}\right)^2 l^2 \right) \right\}^{-1/2} \quad (4.1)$$

where a and c are the lattice constants, and θ is the angle between the basal plane and an arbitrary HCP plane. For the case study shown in Fig. 4.9, slip on the second-order pyramidal {11-22} plane in the <11-23> direction was found to be active at all the temperature conditions. By performing the tests as a function of temperature (T), Peierls stress (τ_p) can be determined by linear extrapolation (based on the critical resolved shear stress equation):

$$\tau = \frac{k_B T}{V} \sinh^{-1}\left[\frac{\dot{\gamma}}{2.\rho_m.v.b^2} \exp\left(\frac{\tau_p V}{k_B T}\right) \right] \quad (4.2)$$

where τ is the resolved shear stress, v is the frequency, $\dot{\gamma}$ is the shear strain rate, V is the activation volume, b is the Burger's vector, and ρ_m is the mobile dislocation density. The value of Peierls stress was computed to be 3.76 GPa for the slip system in the prisms observed from in-situ imaging of the compression tests. This example highlights the importance of high-temperature in-situ compression for determining plasticity parameters, which is particularly challenging for brittle materials.

4.1.3 In-Situ High-Temperature Microbeam Bending

Thermal drift is a major challenge during high-temperature mechanical testing and can influence the mechanical properties and the deformation mechanisms observed during in-situ imaging. The temperature gradients between the sample and the probe (used for either micro-compression or indentation) are a major source of drift. A mode of testing where the effect of drift can be minimized is microbeam bending. In micro-bending, the region of the specimen experiencing major deformation is away from the area where the probe makes contact with the sample. Figure 4.10a illustrates the in-situ micro-bending in SEM. It was demonstrated in the previous section that real-time high-temperature testing is useful for studying slip in materials. Microbeam bending is another useful mode of testing to investigate crystal plasticity. Beams with desired crystal orientations can be machined for examining specific slip phenomena. In the case study shown in Fig. 4.10, Cu samples were fabricated to study single-slip (SS) and multiple-slip (MS) deformation. For the

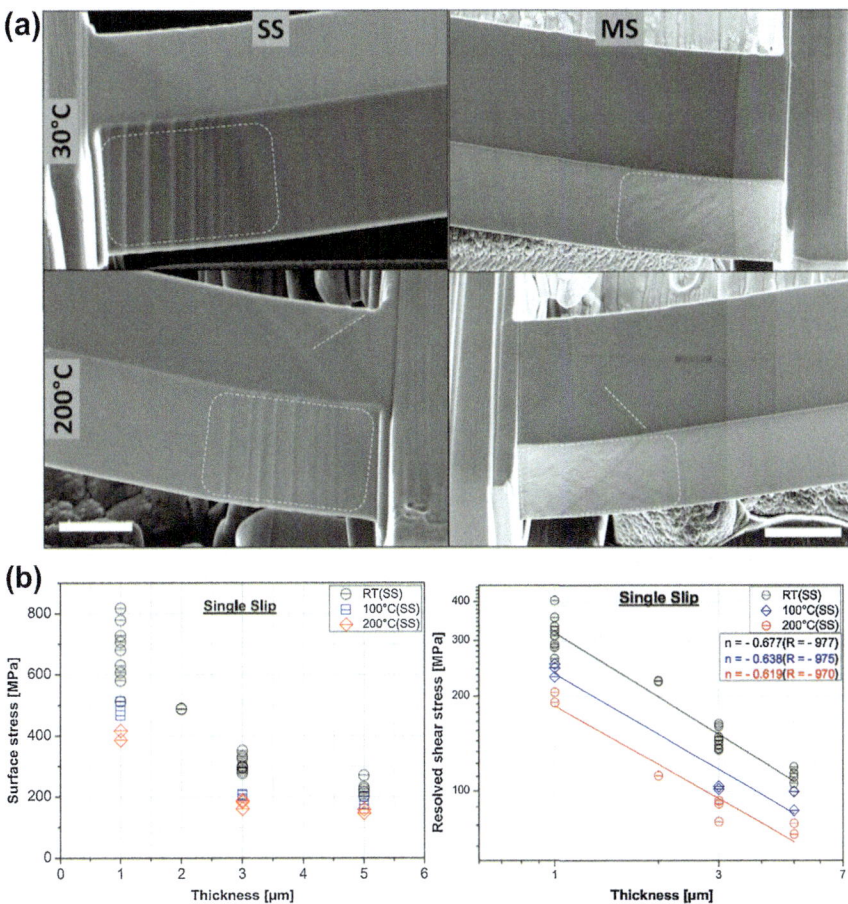

Fig. 4.10 (a) SEM imaging of in-situ micro-bending of single crystal Cu beams as a function of temperature under different slip conditions, and (b) surface stress and resolved shear stress–beam thickness plots to evaluate the correlation between temperature and size effect. (Reproduced/adapted with permission from Rafael Velayarce et al. (2018))

former, beams are machined to have the orientation such that the neutral plane is parallel to the [153] orientation. Whereas for the latter, the neutral plane is favored to be parallel to [100]. The application of in-situ bending for selectively activating SS and MS scenarios by fabricating the beams with the desired orientations is shown in Fig. 4.10a. On performing in-situ bending as a function of temperature, slip steps were seen to be more pronounced at elevated temperatures (SEM images of the bent beams at 30 vs. 200 °C shown). High-temperature in-situ micro-bending provides the ability to understand the correlation between temperature and size effects in mechanical properties. Figure 4.10b demonstrates the variation of surface stresses and resolved shear stress for beams with different thicknesses and at different temperatures (ranging from room temperature up to 200 °C). It was observed that the size effect is somewhat arrested at elevated temperatures, as evidenced by the calculated slope (n) of the resolved shear stress–thickness plot in Fig. 4.10b. Therefore, in-situ high-temperature micro-bending approach is powerful to obtain qualitative as well as quantitative insights into the deformation of materials.

4.1.4 In-Situ High-Temperature Tensile Testing

Tensile testing is a widely followed method to characterize the mechanical behavior of materials, in the form of stress–strain curve, elastic modulus, yield strength, failure strength, and failure strain. In-situ tensile testing at elevated temperatures enables the characterization of these properties as a function of temperature. Tensile testing is a versatile technique to probe the mechanics of materials at multiple length scales: from submicron "miniature" samples to bulk/macro-scale samples. The imaging mode, the testing device, and heating methodology are determined by the size of the samples being examined. Both miniature and bulk size sample testing are discussed in this section.

Miniature Sample Characterization

Miniature samples, such as nanomaterials, thin films, nano-micrometer sized dog-bone specimens require microheaters for sample heating. There are two approaches to heat the sample during the test. One is the resistive heating method, where current or voltage is applied to induce sample heating. But this method requires the sample to be electrically conductive, and would not be feasible for insulators. Another challenge with resistive heating is maintaining a constant temperature during deformation, because the resistance changes as the sample's cross-section area changes due to tensile loading. This issue can be overcome by using an external heater to maintain a uniform temperature. External heaters can also be used for high-temperature testing of electrical insulators. Figure 4.11a, b show an example of a microheater design for an in-situ SEM tensile tester (Sim et al. 2013). It consists of two micro-machined heaters that support the specimen, with a Tungsten heating element layer on top (Fig. 4.11c). The sample is heated by passing the current through the element. The sample temperature is controlled and monitored during the test by calibrating the element resistance as a function of temperature. The schematic shows a four-point probe which constantly measures the element resistance for monitoring real-time temperatures.

In-situ heating of the specimen during tension can be exploited to study the failure mechanisms at different temperatures. Figure 4.12a illustrates the fracture mechanisms of gold thin films captured during in-situ SEM tensile testing (Sim and Vlassak 2014). In this case study, a transition in failure behavior was observed when the films were tested above 200 °C. The specimens exhibited grain boundary voiding above 200 °C, which was absent at relatively lower test temperatures. In addition

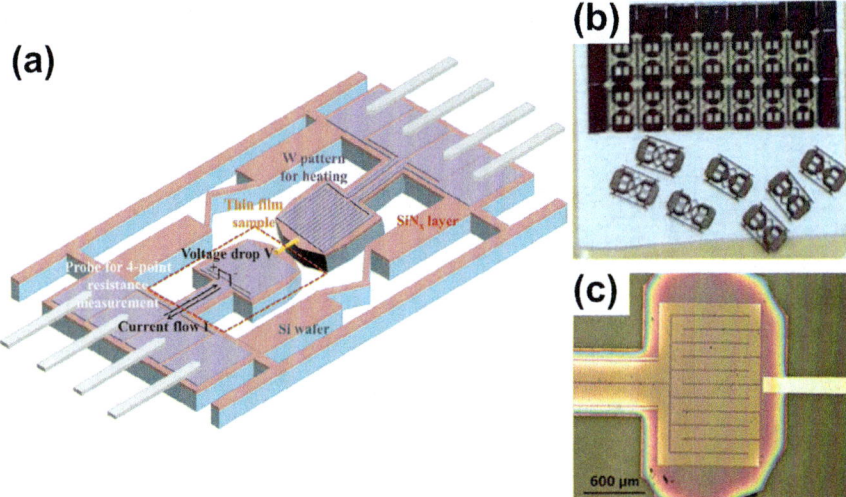

Fig. 4.11 (**a**) Schematic representation of the design and operation of a microheater for high-temperature tensile testing inside SEM, (**b**) images of microfabricated heaters, and (**c**) tungsten heating element on the microheater surface. (Reproduced with permissions from Sim and Vlassak (2014), Sim et al. (2013))

to real-time imaging, stress–strain response is also recorded to obtain insights into high-temperature mechanics of thin films (shown in Fig. 4.12b). The quantitative assessment provides information such as the change in loading/unloading slope as well as flow stress with test temperature. These insights are vital for high temperature application of thin films.

Macroscale Sample Characterization

High-temperature testing of larger specimens (mm- to cm-sized samples) is performed using the mechanical stage equipped with sample heater, as shown in Fig. 4.13a (Dessolier et al. 2018). Sample heating can be performed either by using a contact heating system or heating jaws. It should be noted that contact heating is tricky, as deformation during the test can result in loss of intimate contact between the heating element and the sample. Heating jaws maintain contact throughout the duration of the test, avoiding the issue of contact loss. However, there can be non-homogeneous sample heating, particularly for nonconducting samples. Therefore, the heating method should be chosen by taking into consideration the type of sample being tested. As can be seen from the figure, the cooling system comprising of a

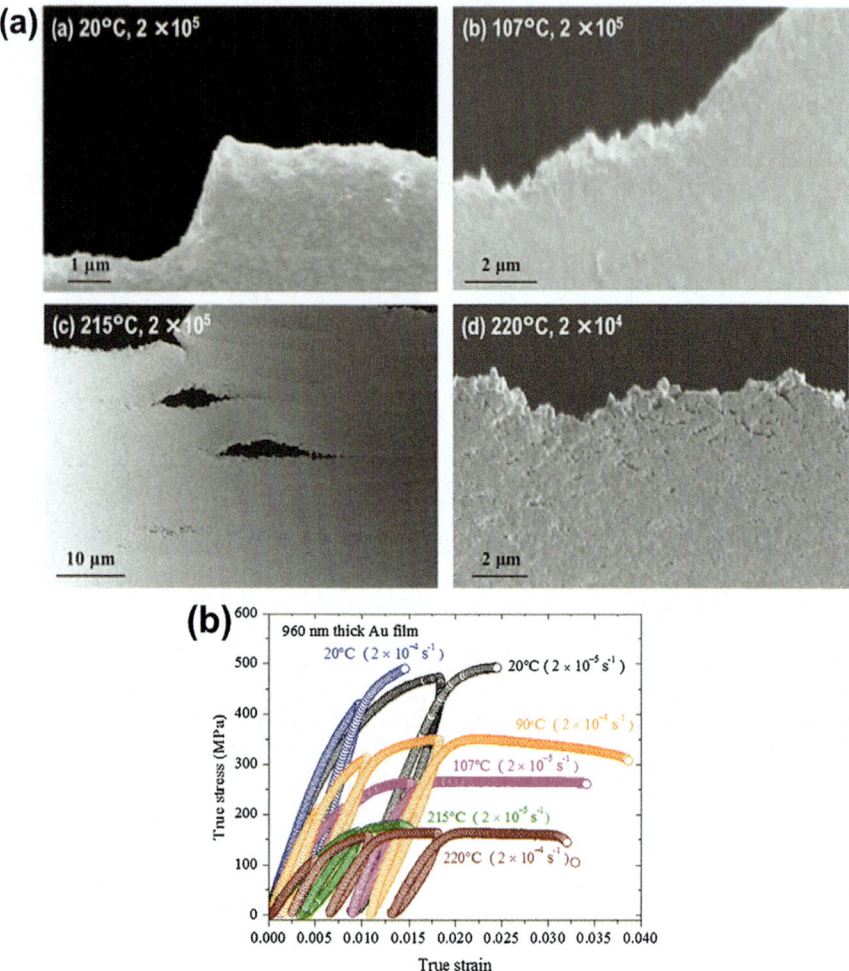

Fig. 4.12 In-situ high-temperature tensile testing of gold thin films: (**a**) SEM imaging to study fracture behavior, and (**b**) stress–strain response at different test temperature conditions. (Reproduced with permission from Sim and Vlassak (2014))

set of pipes with de-ionized water/coolant is typically employed in high-temperature mechanical stages.

A case study on in-situ SEM high-temperature testing of Ni-based superalloy is shown in Fig. 4.13b (Summers et al. 2016). The strength and ductility values do not necessarily follow a definitive trend as test temperature is varied. In-situ testing at 400 °C resulted in enhanced strain-to-failure (4.2%) as compared to room-temperature testing (2.7%). The yield stress remained unchanged (~900–910 MPa) on increasing the temperature from room temperature to 400 °C. However, tensile testing at 750 °C showed reduced yield strength (630 MPa) and failure strains (1.8%). Real-time SEM imaging revealed slip deformation bands under room-temperature

4.1 High-Temperature In-Situ Mechanical Testing

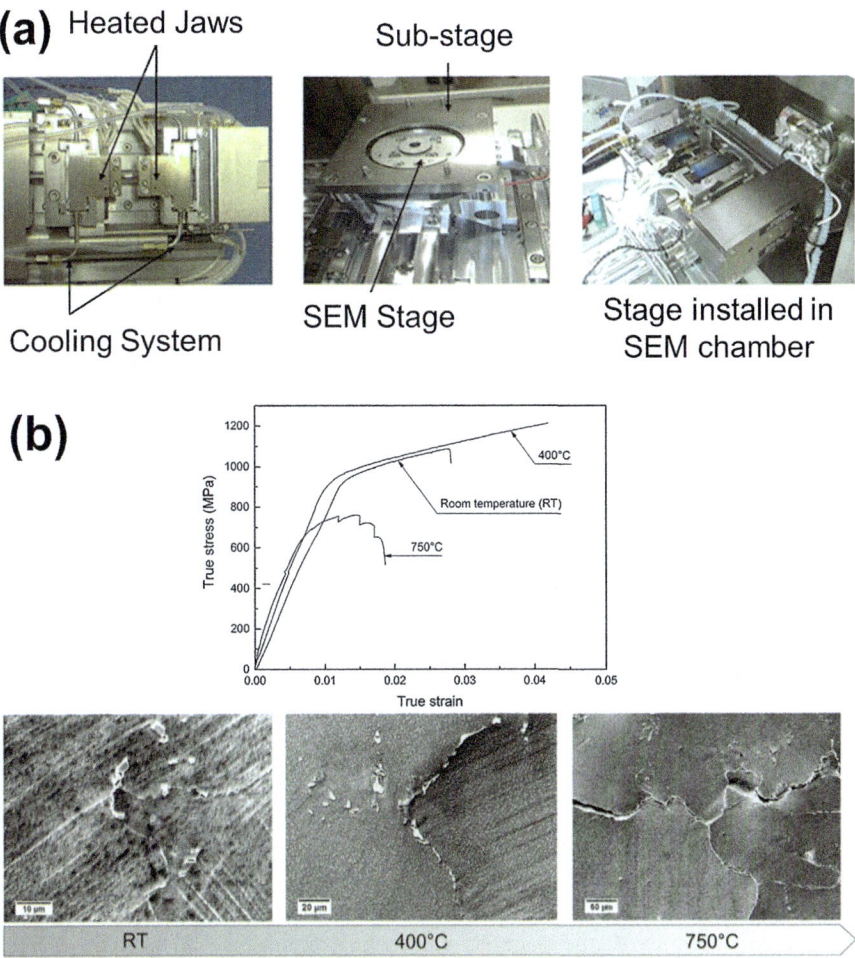

Fig. 4.13 In-situ tensile testing of macro-scale samples: (**a**) in-situ tensile stage for high-temperature testing (inside SEM), and (**b**) tensile testing of Ni-based superalloy (comparative stress–strain curves and real-time SEM micrographs). (Reproduced with permissions from Summers et al. (2016), Th. Dessolier, G. Martin, P. Lhuissier, C. Josserond, F. Roussel, F. Charlot, J-J. Blandin (2018) Microscopy and Microanalysis)

to mid-temperature testing conditions (Fig. 4.13b). Interestingly, failure at higher temperature was not characterized by slip band formation. The formation of voids at the grain boundaries was noticed at 750 °C, which ultimately grew due to void coalescence, leading to intergranular failure of the superalloy. On the other hand, tensile loading at room-temperature induced cracking of carbides in the microstructure, which acted as failure sites.

Case studies discussed so far clearly illustrate the strong influence of elevated temperature conditions on the mechanical response of materials. Diffusion processes, oxidation, and plastic deformation mechanisms are accentuated at higher temperatures. Material properties, such as yield stress, elastic modulus, hardness, or fracture toughness show significantly different values as temperatures are increased. These variations are not always monotonic and in-situ imaging is beneficial to understand the cause of such anomalies. This temperature dependence is also expected to come into play as temperatures are reduced below room temperature. However, altogether different processes are expected to play a significant role. For instance, thermal diffusion is not anticipated to be prominent and the ductile-to-brittle transition phenomenon is going to be more important in low-temperature regimes. The next section introduces in-situ cryogenic characterization for probing low-temperature mechanics of materials.

4.2 Cryogenic In-Situ Mechanical Investigations

This segment of the chapter discusses the mechanical characterization of materials below freezing temperatures (up to temperatures as low as ~ −113 °C). In-situ imaging allows observation of low-temperature deformation mechanisms. Low-temperature characterization is accomplished by using a cryogenic cooling system in conjunction with the mechanical testing instrumentation (Lupinacci et al. 2014; Hagen and Thaulow 2016). Some of the key components of a cryogenic system include liquid nitrogen dewar, cold finger, coolant transfer line, temperature sensors, and temperature controller (Lee et al. 2014). The schematic in Fig. 4.14a shows some of these components. Liquid nitrogen-assisted cooling has been successfully employed to attain temperatures as low as ~130 K (Lee et al. 2014). In order to attain low temperatures effectively, it is important to prevent the radiation of heat from the microscope chamber wall to the sample stage. The use of thermal insulators helps to minimize the radiation, and the use of titanium has been demonstrated to arrest heat flow because of its low thermal conductivity (21.9 W/m.K). Figure 4.14b shows the photographs of a low-temperature nanoindenter for in-situ characterization inside SEM. These images show the assembly outside and inside of the microscope chamber for performing cryogenic mechanical testing. During low-temperature nanoindentation, both the tip and the sample should be cooled down to the desired temperature to avoid thermal drift.

Figure 4.15a demonstrates the in-situ SEM indentation at low temperatures (up to 160 K) (Lee et al. 2013). Cryogenic indentation is employed in this example to examine the change in material hardness as a function of temperature. This is shown in Fig. 4.15b, where the values of a hardness parameter (H_o) are plotted as a function of indentation test temperatures. This case study shows the hardness variation captured for W, Nb, Au, and Al single crystals. It was seen that Au and Al have very weak or minimal temperature dependence, whereas W and Nb showed a stronger correlation with temperature (in 160 K to room-temperature regime).

4.2 Cryogenic In-Situ Mechanical Investigations

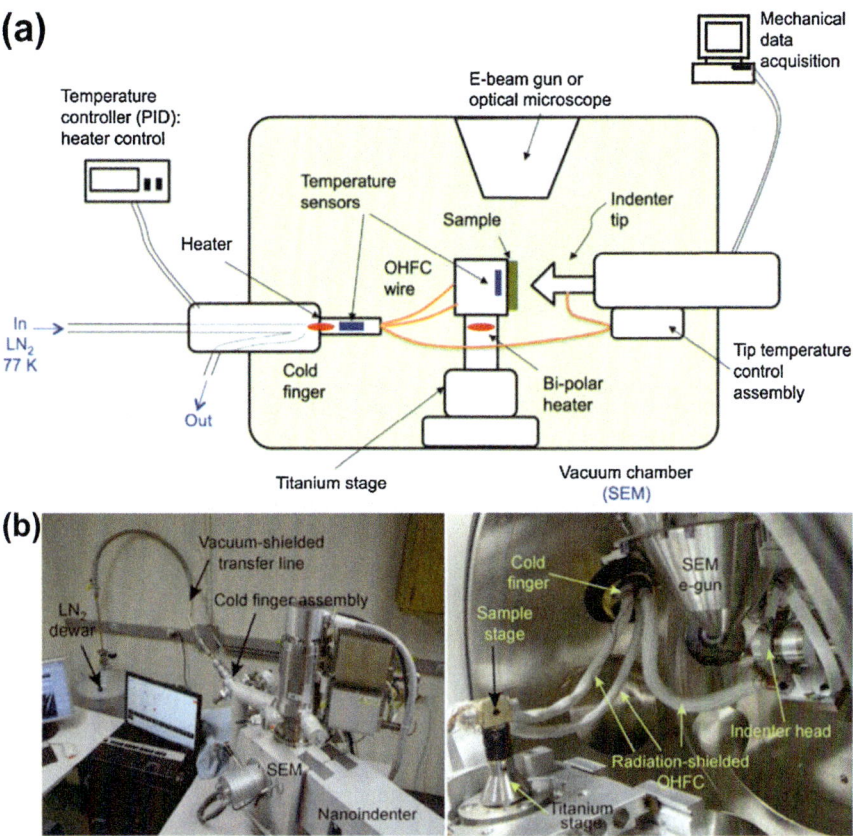

Fig. 4.14 (**a**) Schematic representation of mechanical testing instrumentation coupled with a cryogenic cooling system for low-temperature testing, and (**b**) images showing a scanning electron microscope equipped with a cooling system and in-situ nanoindenter. (Reproduced with permission from Lee et al. (2014))

In-situ low-temperature testing can be employed for uniaxial loading as well. Some of the examples of in-situ SEM imaging during cryogenic compression are shown in Fig. 4.16. These examples illustrate real-time deformation of Niobium, Tungsten and Tin. For instance, the micro-compression of a Nb pillar at 160 K produced sharp, concentrated slip traces as shown in Fig. 4.16a (Lee et al. 2014). These slip traces are associated with screw dislocations. Contrary to this, compression of W pillar at the same temperature showed uniform deformation and no local slip traces (Fig. 4.16b). This behavior is related to the higher critical temperature of W compared to Nb, resulting in slower screw dislocation mobility. On the other hand, compression of Sn pillars revealed a sudden and catastrophic failure at maximum strength at ~131 K (Fig. 4.16c). This failure behavior is quite different from room-temperature response, where no abrupt failure was observed (Lupinacci et al. 2014). These differences are dictated by the competition between different mechanisms:

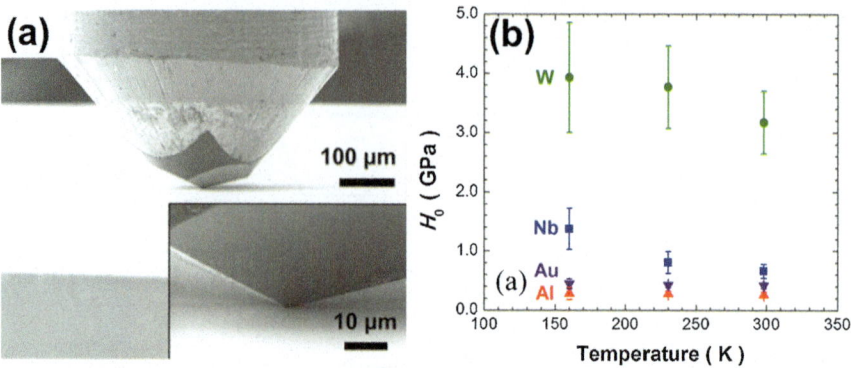

Fig. 4.15 (a) In-situ SEM imaging of indenter penetrating a single crystal sample, and (b) variation of a hardness parameter (H_o) at low temperatures (up to 160 K). (Reproduced with permission from Lee et al. (2013))

dislocation plasticity dominates at room temperature, whereas twinning is prominent during cryo-compression. An advantage of in-situ testing is to selectively study the deformation of single-slip orientations to establish temperature dependence of deformation. Figure 4.16d illustrates micro-compression of bcc α-Fe for [$\bar{2}35$] and [$\bar{1}49$] orientations. SEM imaging allowed the observation and analysis of slip traces. Different slip systems were activated at different temperatures. For instance, compression of [$\bar{2}35$] pillar was dominated by <$\bar{1}01$>[111] slip system at room temperature, but <$\bar{2}11$>[111] was the prominent slip system at 198 K. For [$\bar{1}49$] pillar, SEM imaging revealed <101>[$\bar{1}11$] slip system at room temperature and <$\bar{1}01$>[111] at 198 K. These examples highlight the significance of real-time imaging during cryogenic testing to understand the underlying deformation mechanisms in metals.

In-situ cryogenic tests can be employed to systematically probe the variation in strength and ductility of materials in the subfreezing temperature regime. The example in Fig. 4.17a shows enhanced strength and arrested ductility of Fe pillars when tested at 198 K, as compared to room-temperature mechanical response. Additional phenomena, such as strain bursts, load drops, and changes in curve slopes can be resolved and analyzed. In-situ SEM testing also enables the investigation of size effects in mechanical properties at low temperatures. A comparison of stress–strain curves for Nb pillars is shown in Fig. 4.17b at room temperature and 165 K. As the pillar diameter was reduced from ~1000 nm to ~400 nm, a prominent enhancement in flow stress was observed. But the comparison of the curves reveals the size effect is somewhat suppressed during cold compression (165 K).

In-situ cryogenic testing unravels deformation of materials in harsh environmental conditions, which has implications for low-temperature engineering applications such as in superconducting power generator, nuclear fusion reactor, superconducting magnetic levitated train and space shuttle thermal protection system (Ogata 2014).

4.2 Cryogenic In-Situ Mechanical Investigations

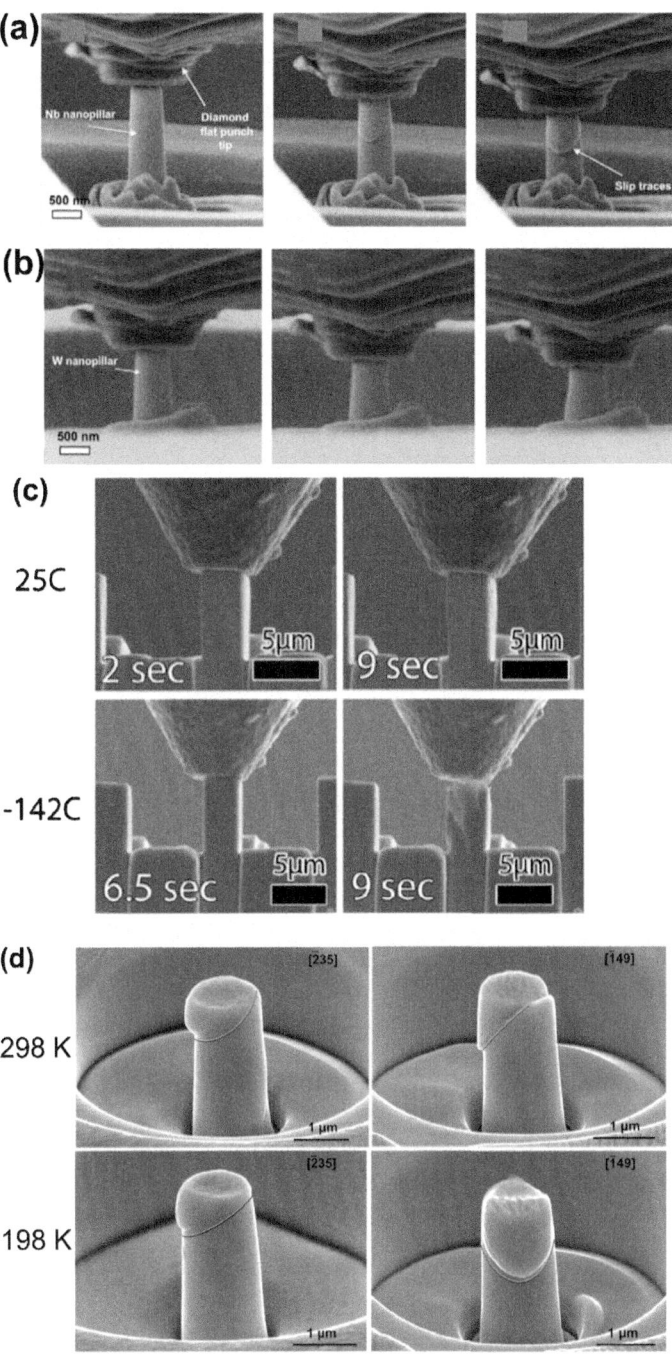

Fig. 4.16 SEM imaging during cryo-compression of micro-pillars: (**a**) Niobium (at 160 K), (**b**) Tungsten (at 160 K), (**c**) Tin (298.15 K vs. 131.15 K), and (**d**) observation of slip traces by SEM imaging of different pillar orientations. (Reproduced/adapted with permissions from Lee et al. (2014), Lupinacci et al. (2014), Hagen and Thaulow (2016))

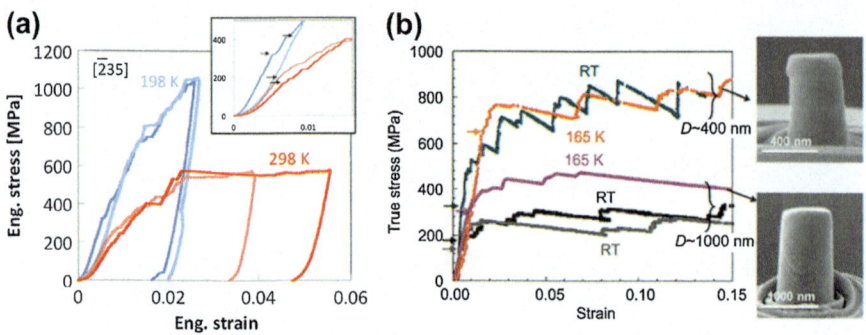

Fig. 4.17 Comparison of stress–strain curves for room-temperature and cold-temperature compression of: (**a**) Fe, and (**b**) Nb. (Reproduced with permissions from Hagen and Thaulow (2016), Lee et al. (2014))

4.3 Summary

In-situ mechanical characterization as a function of temperature provides mechanical properties as well as insights into deformation mechanisms across the temperature spectrum (from 130 K to over 1000 K). Heating elements and cryogenic systems are used for high- and low-temperature mechanical testing, respectively. Thermal drift is a major challenge during such tests and instrumentation design is vital to avoid drift effects in measurement. During nanoindentation, compression, or micro-bending where a probe or tip is involved, heating or cooling of both the sample and the tip is preferred to minimize the temperature mismatch and the resulting drift. Prevention of thermal expansion in the equipment components during high-temperature testing is important to prevent compliances. Additionally, water cooling is employed to localize high temperatures to the sample and the tip. For in-situ SEM imaging, the current supplied to heaters can deflect the electron beam, resulting in wobbly imaging. Because of this reason, the use of direct current based heaters is preferred over the ones designed for alternating current. During cold-temperature testing, heat radiation can hinder the ability to achieve ultralow temperatures. The selection of effective thermal insulator material (such as Ti) for the sample stage is helpful to prevent heat transfer. In-situ high- and low-temperature testing can be performed at multiple length scales and for a variety of loading scenarios, such as indentation, uniaxial compression, microbeam bending, or uniaxial tension. Case studies on in-situ testing of different materials were presented. A comparison of stress–strain characteristics and other properties at different temperatures is important for various engineering applications.

Questions and Assignments

1. Which of the following practices can help minimize thermal drift during high-temperature nanoindentation?

   (a) Set the sample temperature to the desired test temperature, while the tip is maintained at room temperature.
   (b) Set the tip temperature to the desired test temperature, while the sample is maintained at room temperature.
   (c) Gradually increase the temperature of the tip and the sample as the test is in progress.
   (d) Set the tip and the sample temperature to the desired test temperature before nanoindentation begins.

2. Discuss the problem associated with using glues or cement to mount samples for high-temperature nanoindentation. How do the current indentation systems overcome these issues?

3. Which of the following statement(s) are true regarding high-temperature mechanical testing?

   (a) Water-cooling systems are used to remove the debris or flakes created due to oxidation of the indenter tip.
   (b) Alternating current supply to the heaters is preferred as opposed to direct current supply for stable imaging.
   (c) Electrical components are grounded to avoid transmission of electronic noise from outside to inside the microscope chamber.
   (d) Materials with high coefficient of thermal expansion are preferred to minimize compliances.

4. Suggest some real world scenarios/applications where the structural components can be exposed to low- or high-temperature conditions, compromising their mechanical integrity.

   Think of the engineering material examples along the following lines:

   (a) Ductile-to-brittle transition can induce catastrophic failure.
   (b) High temperatures can cause creep.

5. Read the design of in-situ cryogenic nanoindenter system by Seok-Woo et al. (in Sci China Tech Sci, Vol. 57, No.4: 652–662). What is the significance of using titanium for designing the sample stage? Suggest some other alternate materials for designing the stage which will help accomplish and maintain low temperatures during cryogenic mechanical tests.

6. What are resistive heating and external heating approaches during high-temperature tensile test? List any two problems associated with resistive heating of samples during in-situ tensile deformation.

7. List five key components of a cryogenic system for conducting low-temperature mechanical tests using an in-situ nanoindenter.

8. Metals and alloys tend to demonstrate ductile-to-brittle transition at low enough temperatures. Discuss how this transition in deformation characteristics can influence the ability to image and capture cryogenic deformation mechanisms in real time. What imaging tools/techniques can be more effective to capture the mechanisms in such a scenario?

9. Conte and coworkers (Rev. Sci. Instrum. 90, 045105) reported active surface referencing approach for stable measurements over long test durations. Elaborate on the principles and applications of active referencing for high-temperature nanoindentation. List some test scenarios where this technique could be useful for long-duration tests at elevated temperatures. You may refer to the schematic representation of the nanoindenter system below.

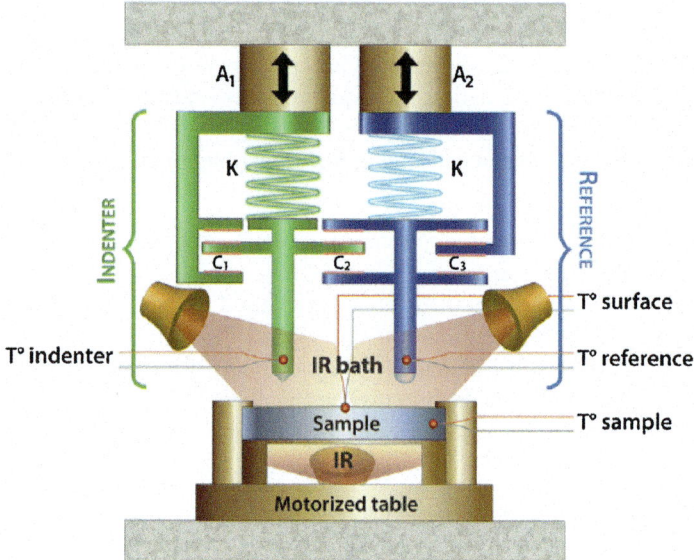

A high-temperature nanoindentation system [Reproduced with permission from Conte et al. (2019)]

10. Prepare a brief two-page report on in-situ high-temperature tensile testing with real-time TEM imaging. Use examples/published articles from literature for explaining the concepts and application. You may use any material system as an example. Your report should cover following aspects:

    (a) In-situ TEM-based high-temperature mechanical testing instrumentation.
    (b) Real-time imaging observations: what mechanisms can be seen?
    (c) Correlation between stress–strain curves and electron micrographs.
    (d) Comparison with room-temperature response (if available).

References

Best JP, Wehrs J, Polyakov M et al (2019) High temperature fracture toughness of ceramic coatings evaluated using micro-pillar splitting. Scr Mater 162:190–194. https://doi.org/10.1016/j.scriptamat.2018.11.013

Cho J, Li Q, Wang H et al (2018) High temperature deformability of ductile flash-sintered ceramics via in-situ compression. Nat Commun 9:2063. https://doi.org/10.1038/s41467-018-04333-2

Conte M, Mohanty G, Schwiedrzik JJ et al (2019) Novel high temperature vacuum nanoindentation system with active surface referencing and non-contact heating for measurements up to 800 °C. Rev Sci Instrum 90:045105. https://doi.org/10.1063/1.5029873

Dessolier T, Martin G, Lhuissier P, Josserond C, Roussel F, Charlot F, Blandin J-J, Maniguet L (2018) Conducting controlled in situ high temperature tensile tests within a SEM. In: Microscopy analysis. https://microscopy-analysis.com/article/july_18/High_Temperature_Tensile_Testing_in_situ

Hagen AB, Thaulow C (2016) Low temperature in-situ micro-compression testing of iron pillars. Mater Sci Eng A 678:355–364. https://doi.org/10.1016/j.msea.2016.09.110

Lee S-W, Meza L, Greer JR (2013) Cryogenic nanoindentation size effect in [0 0 1]-oriented face-centered cubic and body-centered cubic single crystals. Appl Phys Lett 103:10196. https://doi.org/10.1063/1.4820585

Lee S-W, Cheng Y, Ryu I, Greer JR (2014) Cold-temperature deformation of nano-sized tungsten and niobium as revealed by in-situ nano-mechanical experiments. Sci China Technol Sci 57:652–662. https://doi.org/10.1007/s11431-014-5502-8

Lupinacci A, Kacher J, Eilenberg A et al (2014) Cryogenic in situ microcompression testing of Sn. Acta Mater 78:56–64. https://doi.org/10.1016/j.actamat.2014.06.026

Nautiyal P, Zhang C, Champagne V et al (2019a) In-situ creep deformation of cold-sprayed aluminum splats at elevated temperatures. Surf Coatings Technol 372:353–360. https://doi.org/10.1016/j.surfcoat.2019.05.045

Nautiyal P, Zhang C, Loganathan A et al (2019b) High-temperature mechanics of boron nitride nanotube "Buckypaper" for engineering advanced structural materials. ACS Appl Nano Mater 2:4402–4416. https://doi.org/10.1021/acsanm.9b00817

Ogata T (2014) Evaluation of mechanical properties of structural materials at cryogenic temperatures and international standardization for those methods. AIP Conf Proc 1574:320–326. https://doi.org/10.1063/1.4860643

Rafael Velayarce J, Zamanzade M, Torrents Abad O, Motz C (2018) Influence of single and multiple slip conditions and temperature on the size effect in micro bending. Acta Mater 154:325–333. https://doi.org/10.1016/j.actamat.2018.05.054

Raghavan R, Wheeler JM, Esqué-de los Ojos D et al (2015) Mechanical behavior of Cu/TiN multilayers at ambient and elevated temperatures: stress-assisted diffusion of Cu. Mater Sci Eng A 620:375–382. https://doi.org/10.1016/j.msea.2014.10.023

Sim G-D, Vlassak JJ (2014) High-temperature tensile behavior of freestanding Au thin films. Scr Mater 75:34–37. https://doi.org/10.1016/j.scriptamat.2013.11.011

Sim G-D, Park J-H, Uchic MD et al (2013) An apparatus for performing microtensile tests at elevated temperatures inside a scanning electron microscope. Acta Mater 61:7500–7510. https://doi.org/10.1016/j.actamat.2013.08.064

Summers WD, Alabort E, Kontis P et al (2016) In-situ high-temperature tensile testing of a polycrystalline nickel-based superalloy. Mater High Temp 33:338–345. https://doi.org/10.1080/09603409.2016.1180857

Trenkle JC, Packard CE, Schuh CA (2010) Hot nanoindentation in inert environments. Rev Sci Instrum 81:073901. https://doi.org/10.1063/1.3436633

Wheeler JM, Michler J (2013) Elevated temperature, nano-mechanical testing in situ in the scanning electron microscope. Rev Sci Instrum 84:045103. https://doi.org/10.1063/1.4795829

Wheeler JM, Raghavan R, Michler J (2011) In situ SEM indentation of a Zr-based metallic glass at elevated temperatures. Mater Sci Eng A 528:8750–8756. https://doi.org/10.1016/j.msea.2011.08.057

Wheeler JM, Raghavan R, Michler J (2012) Temperature invariant flow stress during microcompression of a Zr-based bulk metallic glass. Scr Mater 67:125–128. https://doi.org/10.1016/j.scriptamat.2012.03.039

Wheeler JM, Niederberger C, Tessarek C et al (2013) Extraction of plasticity parameters of GaN with high temperature, in situ micro-compression. Int J Plast 40:140–151. https://doi.org/10.1016/j.ijplas.2012.08.001

Wheeler JM, Armstrong DEJ, Heinz W, Schwaiger R (2015) High temperature nanoindentation: the state of the art and future challenges. Curr Opin Solid State Mater Sci 19:354–366. https://doi.org/10.1016/j.cossms.2015.02.002

Chapter 5
Application of In-Situ Mechanics Approach in Materials Science Problems

The in-situ characterization approach can be utilized to develop insights into the mechanics of a wide diversity of materials. This chapter introduces the application of in-situ mechanical investigations for different materials science problems. The applicability of in-situ measurement techniques over a wide range of length scales (from atomic- to macroscale) and load scales (nano- to Mega-Newton) makes them highly suited for developing a comprehensive understanding of deformation phenomena activated in materials. Testing in high-resolution microscopes is beneficial to manipulate and examine the mechanical response of extremely fine nanomaterials. Real-time imaging of deformation is advantageous to establish morphology-mechanics correlation for different nanomaterials, such as 0D nanoparticles, 1D nanotubes or nanowires, and 2D nanosheets or nanoribbons. Superior mechanical properties of nanomaterials are harnessed by using them as reinforcement fillers in metal, ceramic, or polymer matrices. The in-situ mechanics approach is useful to decipher the strengthening mechanisms in these composites. The real-time imaging approach can provide insights into load-transfer and failure mechanisms in 3D architectures and metamaterials. The effect of tweaking geometrical parameters, such as unit cell shape, size, and angles of inclination on deformation characteristics can be determined with real-time imaging. Since the individual features in metamaterials can have nano- and microscale dimensions, high-resolution imaging provides insights into the local response which affects the bulk, overall response. Coupling mechanical measurements with the thermal stimulus or electrical output during in-situ testing is an advantageous strategy to probe the mechanics and transformations in smart materials, such as shape memory or piezoelectric materials. The in-situ approach is increasingly gaining attention for biological applications. Real-time imaging can provide insights into mechanisms activated in different regions and along different orientations of the tissues. The application of the in-situ technique for probing the deformation of irradiated materials is discussed toward the end of

Electronic supplementary material The online version of this chapter (https://doi.org/10.1007/978-3-030-43320-8_5) contains supplementary material, which is available to authorized users.

Table 5.1 Application of in-situ measurement approach for different materials science problems

Material classes	In-situ imaging	Quantitative assessment	Examples
Nanomaterials (*0D, 1D, and 2D*)	Fracture, dislocations	Size effect in mechanical properties. Morphology-mechanics correlation	Nanoparticles, nanocubes, nanotubes, nanowires, nanofibers, nanosheets, nanoribbons
Composites (*low dimension and macro-sized*)	Interface mechanics. Strengthening mechanisms	Interfacial shear strength. Prediction of composite properties	Nanofiller-reinforced metal, ceramic and polymer matrix composites, Multilayer composite systems
Architected materials (*metamaterials, porous structures*)	Load transfer behavior	Geometry-mechanics correlation	Sponge, foam, octahedron lattice, composite lattice, 3D nanotube arrays
Smart materials (*shape memory, piezoelectric*)	Phase transformations. Stimulus-deformation coupling	Critical stress for transformations	Shape memory ceramic, shape memory alloy, piezoelectric ceramic
Biological materials (*tissues, animal parts*)	Response of heterogeneous microstructures. Effect of interfaces	Directionality in properties. Location-specific properties	Bones, insect wings

Images in the table are reproduced with permissions from Shan et al. (2008), Yang et al. (2016), Yee et al. (2019), Schwiedrzik et al. (2014)

the chapter. Table 5.1 summarizes the application of in-situ mechanics approach for different classes of materials, mechanistic understanding obtained by multi-scale imaging, and the associated mechanical properties. This table covers some of the key areas where the in-situ approach has gained popularity, but the applications of in-situ characterization are not limited to these categories and there are multiple avenues where the technique is utilized or has promising future potential. The subsequent sections discuss different materials science applications in greater detail. The challenges, solutions, advantages, and limitations are highlighted through case studies. The readers are also referred to several supplementary videos throughout the chapter for better visualization and understanding of these applications.

5.1 In-Situ Characterization of Nanomaterials

In-situ testing is extremely useful for probing the mechanics of nanomaterials. Because of their ultrafine dimensions, nanomaterials can be tricky and difficult to observe, manipulate, and use with conventional mechanical testers which are built for much larger samples. Microscopes are required in order to resolve and observe the fine nanomaterials, hence coupling high-resolution imaging with mechanical testing is an obvious route to determine their mechanical properties. This segment discusses in-situ testing of nanomaterials with different morphologies, that is, 0D, 1D, and 2D nanomaterials.

5.1.1 0D Nanomaterials

Nanomaterials which have all the dimensions in the nanoscale are classified as zero-dimensional (0D) nanomaterials. For instance, spherical nanoparticles, nanocubes, irregular nanoclusters, and nanocrystals or quantum dots with dimensions in the nanometer length scale (typically less than 100 nm) can be termed as 0D nanomaterials. Due to ultrafine feature sizes involved, electron microscopes are employed for real-time imaging. Figure 5.1 shows the in-situ compression of Si nanoparticle in TEM. Typically, a diamond indenter can be used to compress the particle deposited on a substrate. TEM imaging allows observation of transformations in the particle due to compressive loading. In Fig. 5.1, the strain field can be seen in the first image where the particle makes contact with the tip and the substrate because of the formation of bend contours during the test (Deneen Nowak et al. 2007). The bands of contrast seen are ascribed to elastic deformation of the nanoparticle arising due to local displacement of lattice planes. TEM imaging can be useful to capture fracture events. The second TEM micrograph in Fig. 5.1 shows the state of the particle just before the particle fractured, where the release of the strain is seen. Therefore, real-time TEM imaging allows to distinguish the response of the nanoparticle to mechanical loading in elastic and plastic regimes.

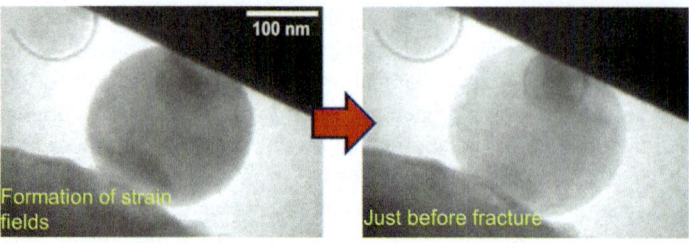

Fig. 5.1 In-situ TEM compression showing the state of the nanoparticle in elastic and plastic regimes. Release of strain is observed prior to fracture. (Adapted with permission from Deneen Nowak et al. (2007))

In-situ TEM imaging is useful to correlate stress–strain characteristics with intrinsic mechanisms. Stress–strain curves for compression of MgO nanocubes are shown in Fig. 5.2a (Issa et al. 2015). Contrast band tilted at ~45° can be seen to form corresponding to the strain burst recorded in the stress–strain curve. In-situ testing allows quantification of exact dimensions to study size effects in the mechanical response of nanomaterials. A comparison of stress-strain curves for nanocubes with different sizes is shown in Fig. 5.2a, indicating size effect in stress response. Due to their extremely fine size, the understanding of the size effects is not possible by ex-situ testing. TEM imaging is also helpful to study dislocation mechanisms during deformation. Figure 5.2b shows a series of TEM images as MgO nanocube is compressed, with progressively increasing strains from 9.1% to 11.2%. The nucleation-exhaustion of dislocations was observed as an active mechanism. In these TEM snapshots, the green lines show the contrast bands due to the formation of a dislocation network whereas the red lines show the contrast bands that escape the sample surface. The local peaks observed in stress–strain response is attributed to dislocation nucleation events in the nanomaterial observed by real-time imaging.

In-situ testing provides insights into the role of defects on the mechanics of nanoparticles. A case study on the deformation of twinned nanoparticles is shown in Fig. 5.3 (Casillas et al. 2012). The figure demonstrates TEM imaging of decahedral gold nanoparticle in dark field mode as it is compressed by the indenter probe. Tetrahedra 1 through 5 are marked in the micrographs in the figure. The sequence of real-time snapshots (A through D) showed the bending of the fringes upon mechanical loading. However, this behavior was not observed to be homogeneous: tetrahedra 1 and 2 exhibited significant bending of fringes. Tetrahedron 4 did not experience significant strains. The change in bending of contours in tetrahedron 5 was observed to be rather modest during compression. This example highlights the role of twin boundaries as "strain filers" in twinned nanoparticles. Controlling defects in the nanoparticles influence their mechanical response, and in-situ mechanical testing enables precise mapping of mechanical properties with the nature and extent of defects.

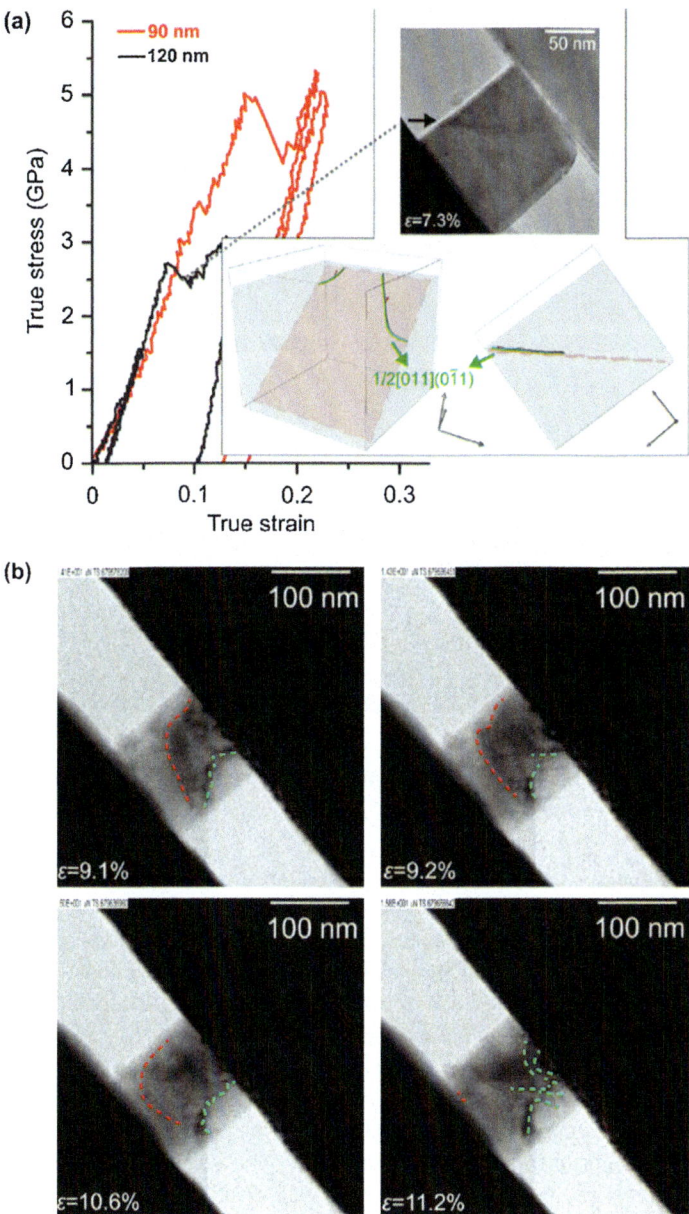

Fig. 5.2 (**a**) Stress–strain curves for compression of nanocubes with different dimensions (the inset shows contrast band formation corresponding to strain burst event), and (**b**) real-time TEM snapshots showing dislocation nucleation-exhaustion mechanism at different compressive strains. (Reproduced with permission from Issa et al. (2015))

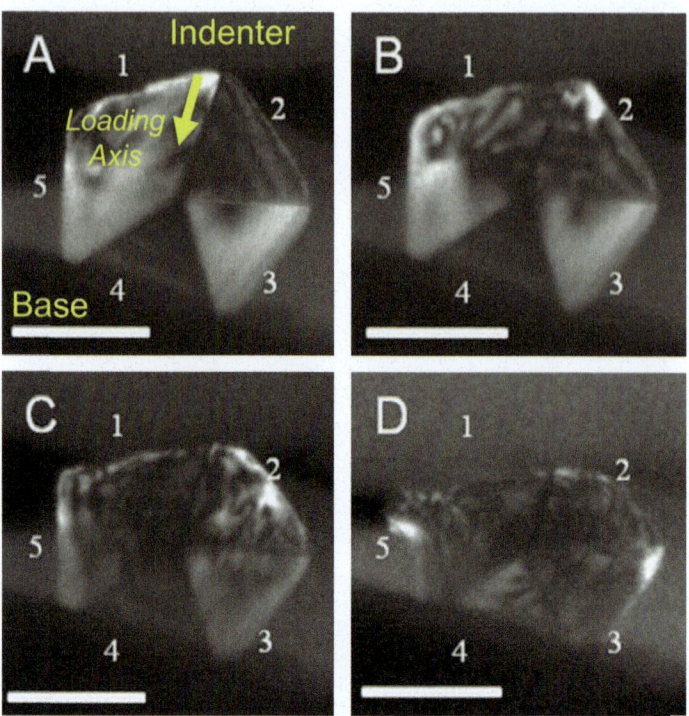

Fig. 5.3 Real-time TEM micrographs showing compression of twinned gold nanoparticle. In-situ imaging shows the bending of contours in different tetrahedral in the nanoparticle. (Reproduced with permission from Casillas et al. (2012))

Tweaking the structural hierarchy of the nanomaterials is a promising approach to tailor their mechanical properties. For instance, hollow nanoparticles exhibit higher strain to failure values (Shan et al. 2008). Figure 5.4a–c illustrates real-time compression of a nanocrystalline hollow CdS spheres in TEM. This case study demonstrates the compression-to-failure test on a 210 nm diameter particle, with a shell thickness of 30 nm. Real-time imaging allows the determination of "instantaneous" contact diameter (as shown in Fig. 5.4b), which is required for computing the contact pressure experienced by the nanoparticle. In the example shown in Fig. 5.4, the contact pressure prior to failure was calculated to be ~1 GPa. In-situ TEM imaging during compression and before fracture showed plastic deformation of the hollow particle. These plasticity events were also captured in the load–displacement curve between positions "a" and "b," as shown in Fig. 5.4d. The fracture event resulted in a big drop in the load response. After the major fracture event, the smaller fragments were compressed by the indenter and produced serrations in the load response (shown in Fig. 5.4d after point "c").

5.1 In-Situ Characterization of Nanomaterials

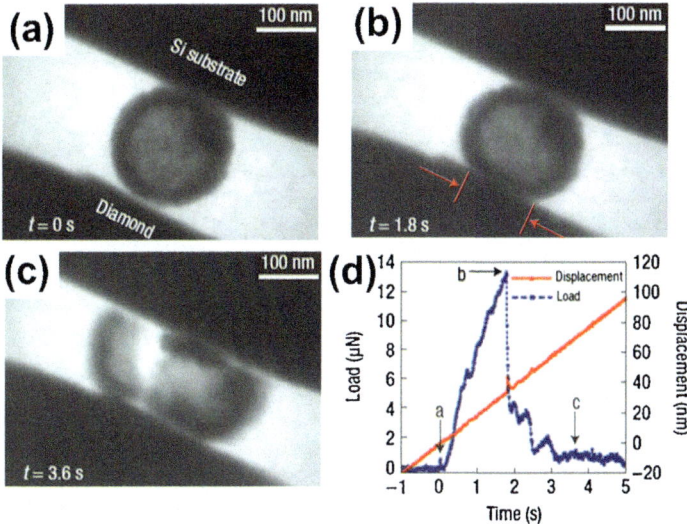

Fig. 5.4 Real-time TEM micrographs for compression of a hollow nanoparticle: (**a**) before compression, (**b**) plastic deformation during compression, and (**c**) fracture event. (**d**) Corresponding load–displacement response for the compression test. (Reproduced/adapted with permission from Shan et al. (2008))

In-situ TEM imaging is also useful to interrogate buckling and post-buckling mechanisms associated with hollow nanomaterials. Figure 5.5a shows the load–displacement response of a hollow amorphous carbon nanoparticle (Yang et al. 2018). The real-time TEM snapshots corresponding to different regions in the curve are shown. It was seen that the initial linear response (for $h < 230$ nm) arises due to the elastic flattening of the outer shell. However, at higher displacements, TEM imaging showed a variable contact area between the sphere and the indenter, as well as between the sphere and the substrate. This manifested as a nonlinear load response. The buckling mechanism observed during real-time imaging was responsible for delayed failure or enhanced failure strains. The in-situ mechanics approach is particularly advantageous to quantify the effect of critical dimensions, such as shell thickness in this case, on the failure behavior. Figure 5.5b shows in-situ TEM imaging of compression of nanospheres with different shell thickness to particle diameter ratio (t/D). While a solid nanoparticle failed at relatively lower ~18% strain, hollow spheres demonstrated delayed failure. The failure strains were observed to be a function of t/D ratio, where smaller t/D resulted in much higher failure strains. For instance, the nanosphere with $t/D \sim 0.064$ resisted failure up to 39% strain (Fig. 5.5b)! This case study illustrates the criticality of choosing feature dimensions in nanomaterials to achieve a desired mechanical response. In-situ characterization enables a systematic assessment of how the mechanical properties vary with these critical dimensions.

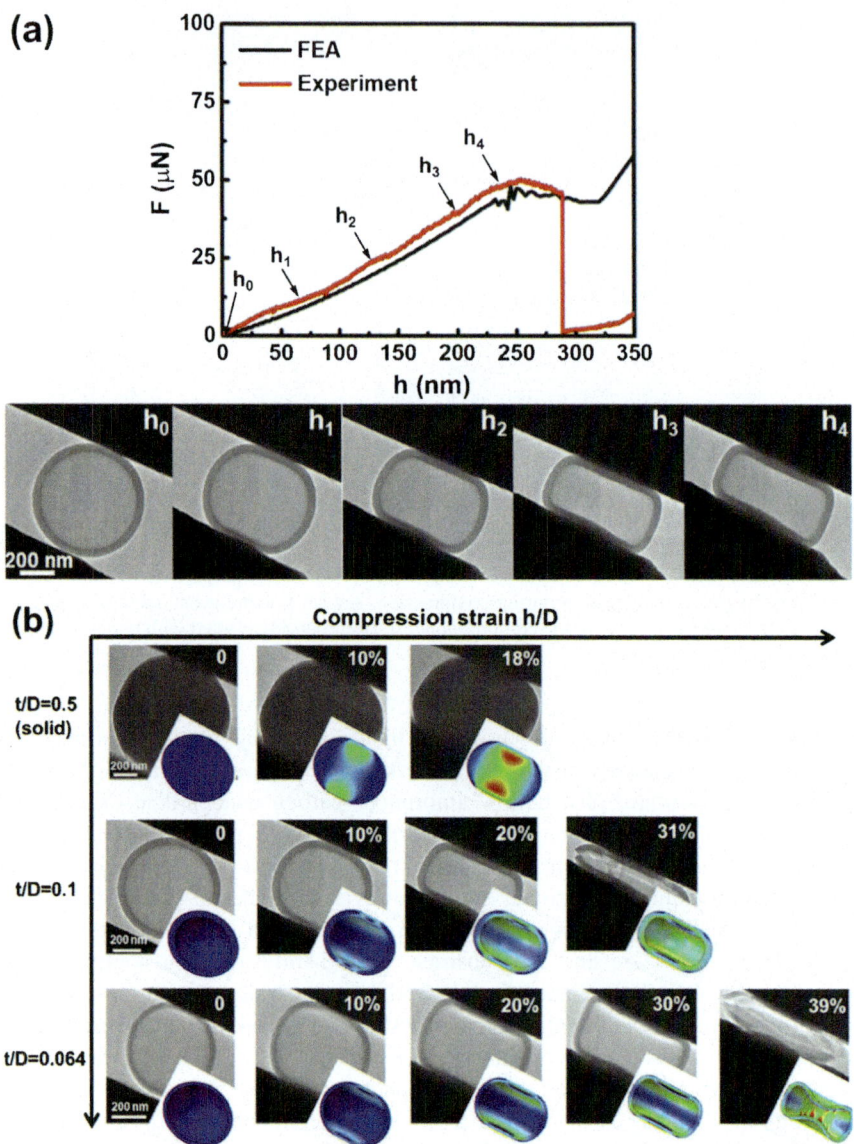

Fig. 5.5 (**a**) Force–displacement curve and corresponding real-time TEM images showing flattening, buckling and post-buckling mechanisms in a hollow carbon nanosphere subjected to compression, and (**b**) variation in failure strains as a function of shell thickness to sphere diameter ratio. (Reproduced with permission from Yang et al. (2018))

5.1.2 1D Nanomaterials

1D nanomaterials are high aspect ratio nanomaterials that have nanoscale thickness/diameter, but microscale length. For instance, nanowires, nanofibers, or nanotubes are all examples of 1D nanomaterials. Typically, 1D nanomaterials display superior load-bearing ability along the axial length. Therefore, axial mechanical loading is preferred for characterizing their mechanical properties. Figure 5.6a demonstrates the tensile loading of a ZnO nanowire with real-time SEM imaging (Xu et al. 2010). The tensile test is conducted by clamping the two ends of the nanowire on a nanomanipulator and an AFM cantilever. The tensile loads are determined based on cantilever deflection captured by SEM, whereas strain is computed from the

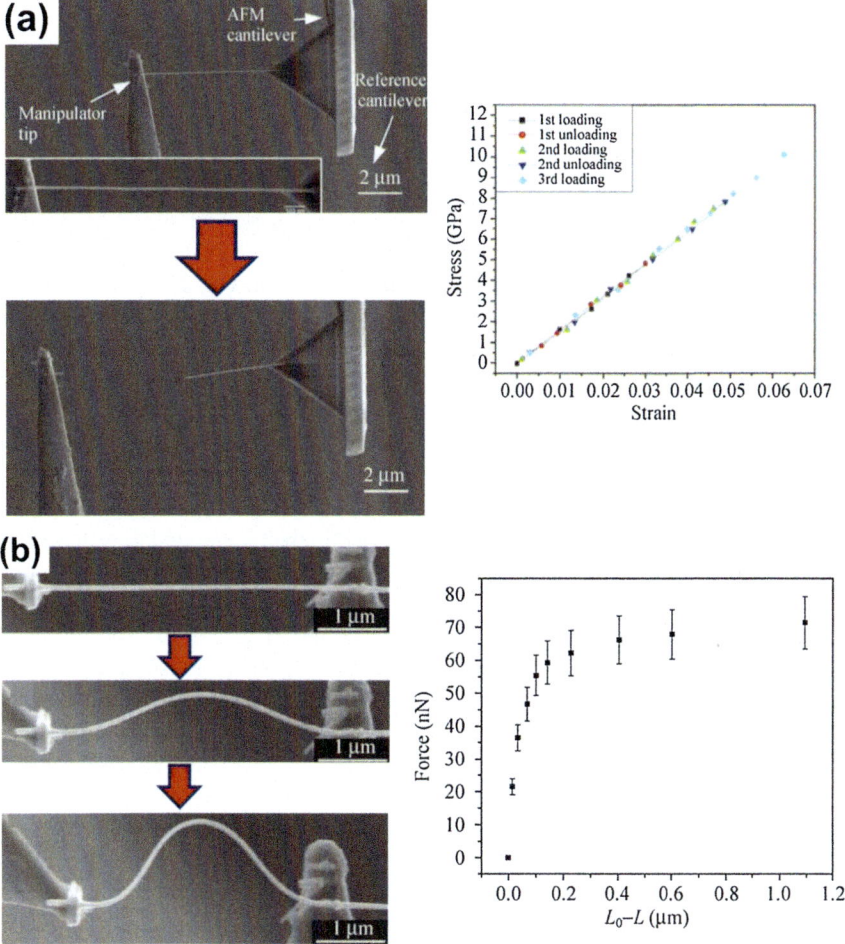

Fig. 5.6 In-situ SEM mechanical testing of ZnO nanowire in: (**a**) tension mode, and (**b**) buckling mode (Reproduced/adapted with permission from Xu et al. (2010))

high-magnification SEM images of the nanowire experiencing elongation. From the stress–strain response, Young's modulus of the nanomaterial can be determined (linear slope). Multiple loading-unloading-reloading cycles can also be investigated, as shown in the stress–strain response in Fig. 5.6a.

Opposite or compressive loading of the nanowire results in buckling (Fig. 5.6b). The corresponding load response is also shown. The load rapidly increases initially, until buckling happens at a critical load. Thereafter, in-situ imaging captures post-buckling deformation and the load remains steady after buckling of the nanowire. The buckling force of an ideal column is determined by the Euler's formula:

$$P_{cr} = \frac{\pi^2 EI}{L_e^2} \qquad (5.1)$$

where E is Young's modulus and I is the moment of inertia of the nanowire. From the in-situ experiment, P_{cr} is known and Young's modulus can be determined by using this expression. For the case study demonstrated in Fig. 5.6b, the bending modulus was computed to be 151 GPa.

High-resolution in-situ mechanics is desirable for interrogating properties of nanomaterials with specific geometries. A case study on the deformation of a "pentagonal" silver nanowire is shown in Fig. 5.7a (Vlassov et al. 2014). High-magnification SEM and TEM micrographs in Fig. 5.7a show the cross section of the nanorod with the special fivefold twinning structure (Chen et al. 2004). In-situ loading inside SEM was performed to observe deflection and deformation mechanisms. A sequence of real-time SEM images during cantilever deflection is shown in the figure. The pentagonal nanorod experienced some plastic deformation and eventually failed in a brittle fashion. The corresponding force response captured during the test shows the critical points for the onset of plasticity and failure initiation. Nonzero loads were observed after failure initiation and before complete failure, a sign of necking. Nevertheless, the plasticity was much lower than what is expected from a plastic metal like silver. This deviation is ascribed to the twin boundaries present in a pentagonal-structured nanorod in this case study. Another example is demonstrated in Fig. 5.7b, where real time TEM imaging of a "bamboo-like" boron nitride nanotube is performed during tensile loading (Tang et al. 2011). The quasi-1D nanobamboos consist of multiple short BN segments. The stress–strain curve captured elastic, plastic and failure regimes. The strength and Young's modulus of the bamboo-like nanotube were found to be 8 GPa and 225 GPa, respectively. The mechanical properties were inferior to the crystalline, non-bamboo-shaped nanotubes, which demonstrate strength and elastic modulus exceeding 30 GPa and 1 TPa, respectively (Wei et al. 2010). The difference in properties arises because of the discontinuity of lattice in bamboo-shaped nanotubes, which consist of multiple joints. Therefore, in-situ characterization provides important insights for correlating structural features/characteristics of nanomaterials with their mechanical response. Understanding these correlations is important for engineering nanomaterials with desired mechanical attributes.

5.1 In-Situ Characterization of Nanomaterials

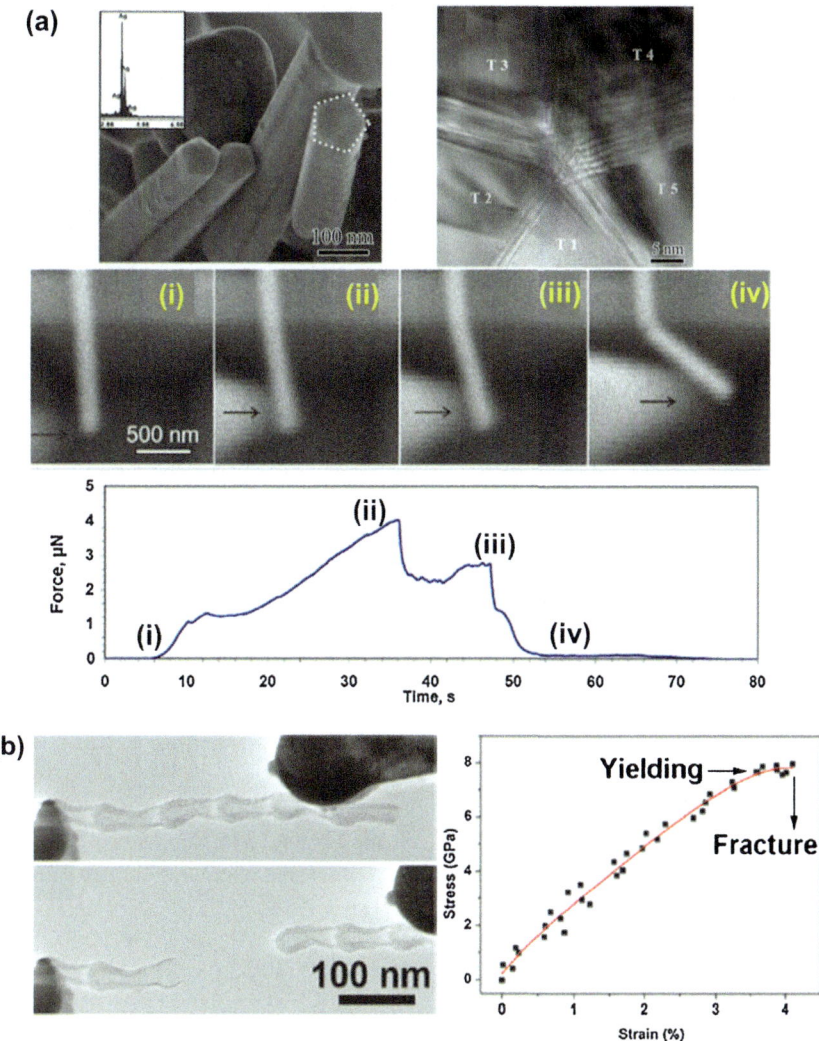

Fig. 5.7 (**a**) In-situ SEM cantilever bending of "pentagonal" Ag nanorod along with corresponding force response, and (**b**) In-situ TEM tensile failure of "bamboo-like" BN nanotube and the corresponding stress–strain curve showing yielding and fracture points. (Reproduced/adapted with permission from Chen et al. (2004), Vlassov et al. (2014), Tang et al. (2011))

The in-situ mechanics approach is useful to determine the fracture toughness of 1D nanomaterials. Figure 5.8 shows the three-point bending of Ag nanowire in TEM to study failure and fracture (Alducin et al. 2016). Three-point bending of a nanowire can be performed using an AFM-TEM holder, such that the nanomaterial is supported on its two ends while the AFM sensor contacts and induces deformation. The real-time snapshots shown in Fig. 5.8 captured elastic and plastic deforma-

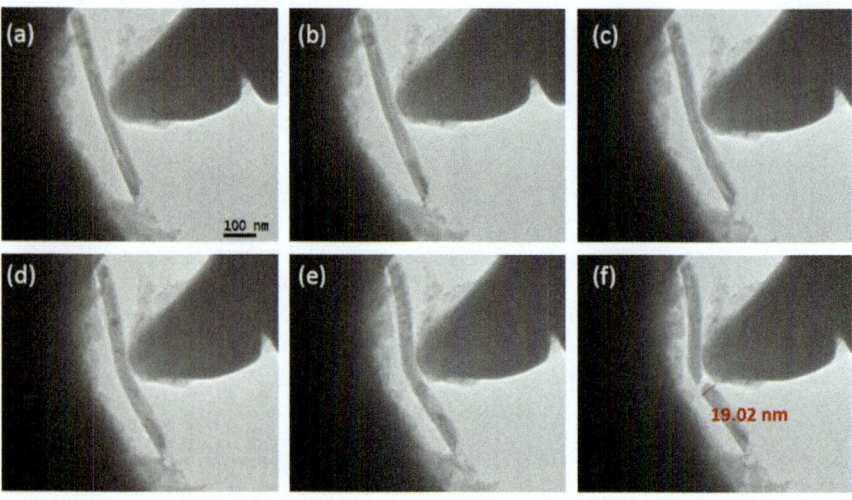

Fig. 5.8 Real-time TEM snapshots showing deformation and fracture of an Ag nanowire subjected to three-point bending. (Reproduced with permission from Alducin et al. (2016))

tion, and finally failure of the nanowire. From the initial snapshots (Fig. 5.8a, b), the elastic modulus was computed to be 108 GPa. The critical stress at failure can be computed based on the in-situ mechanical test:

$$\sigma_F = \frac{F}{\pi a} \tag{5.2}$$

where F is the force at the point of failure and a is nanowire diameter. The fracture toughness is a function of the ratio of crack length and nanowire diameter and can be determined by the relation:

$$K_{IC} = \sigma_F \sqrt{\pi c}\, f\!\left(\frac{c}{a}\right) \tag{5.3}$$

where c is the length of the fracture. For the example shown in Fig. 5.8, the fracture toughness was calculated to be 18.5 MN/nm$^{3/2}$. Therefore, real-time imaging during fracture provides the ability to quantify toughness of ultrafine nanomaterials which is otherwise difficult to calculate.

Some nanomaterials can be susceptible to oxide layer formation on their outer surface. In-situ testing approach is helpful to quantify mechanical properties, like elastic modulus, by isolating the contributions of the oxide layer from the original nanotube. For instance, an amorphous boron oxide layer is known to exist on boron nanowires, as shown in the TEM micrograph in Fig. 5.9a (Liu et al. 2013). The real-time TEM images for bending and breaking of the oxide-coated nanowire are shown in Figure 5.9b. As the load was applied on the nanowire, it transformed its shape from a straight rod to a curved hook. In-situ observations revealed a maximum

5.1 In-Situ Characterization of Nanomaterials

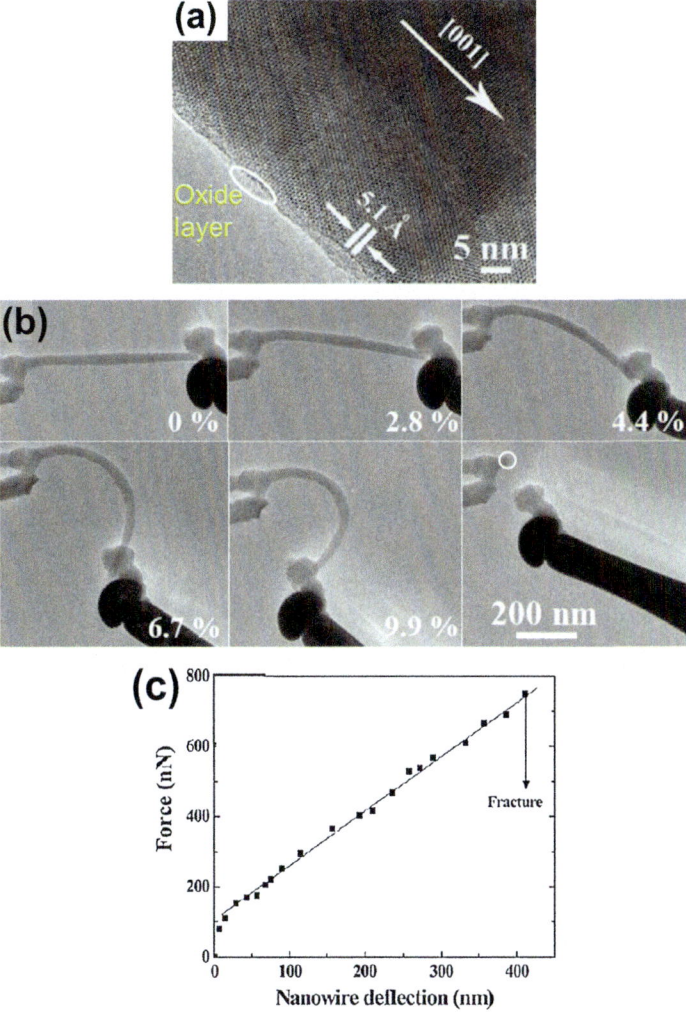

Fig. 5.9 (**a**) TEM micrograph showing a fine oxide layer deposited on B nanowire, (**b**) real-time TEM snapshots captured during bending deformation of B nanowire, and (**c**) force-deflection curve for the nanowire under deflection. (Reproduced/adapted with permission from Liu et al. (2013))

bending angle of ~129.6° until breakage, indicating excellent resistance to bending deformation. The force–deflection response was linear in nature (Fig. 5.9c). The vertical deflection of the nanowire is given by the elastic theory expression:

$$\Delta d = \frac{F l_0^3}{3EI} \tag{5.4}$$

where F is the force applied, E is the elastic modulus, l_o is the length of the nanowire (prior to deflection), and I is the moment of inertia, which can be expressed by the following relation for a rod-shaped nanowire:

$$I = \frac{1}{4}\pi r^4 \tag{5.5}$$

Elastic modulus can be computed by using Hooke's law ($F = k\Delta d$), and substituting Eqs. (5.4) and (5.5) to Hooke's law equation:

$$E = \frac{4l_o^3}{3E\pi r^4} \tag{5.6}$$

Bending moment of the nanowire is expressed by the equation:

$$M = \frac{EI}{\rho} \tag{5.7}$$

where ρ is the curvature radius of the nanowire subjected to bending deformation. In order to account for and isolate the effect of oxide, Eq. (5.7) can be modified as:

$$M = \frac{EI}{\rho} = M_B + M_{oxide} = \frac{E_B I_B}{\rho} + \frac{E_{oxide} I_{oxide}}{\rho} \tag{5.8}$$

Moment of inertia for the oxide (I_{oxide}) can be calculated by the expression:

$$I_{oxide} = I - I_B = \frac{1}{4}\pi r_{out}^4 - \frac{1}{4}\pi r_{in}^4 \tag{5.9}$$

Substituting this equation to Eq. (5.8), elastic modulus equation is obtained to be:

$$E = E_B \left(\frac{r_{in}}{r_{out}}\right)^4 + E_{oxide}\left[1 - \left(\frac{r_{in}}{r_{out}}\right)^4\right] \tag{5.10}$$

In this equation, E can be calculated from the stress–strain curve (obtained from Fig. 5.9c), r_{in} and r_{out} can be determined from the TEM micrographs and E_{oxide} value is known for B_2O_3. Therefore, E_B can be calculated, which is the modulus of the pristine nanowire (with no oxide contribution). This example demonstrates the suitability of the high-resolution in-situ mechanics approach for characterizing and isolating the properties of multicomponent or core-shell 1D nanomaterials.

High-resolution imaging during deformation is useful to capture dislocation activity in nanomaterials. Figure 5.10 shows a GaN nanowire subjected to compression by a flat-ended scanning tunneling microscopy probe (Huang et al. 2011). The real-time TEM imaging revealed a diameter change, resulting in a surface step

5.1 In-Situ Characterization of Nanomaterials

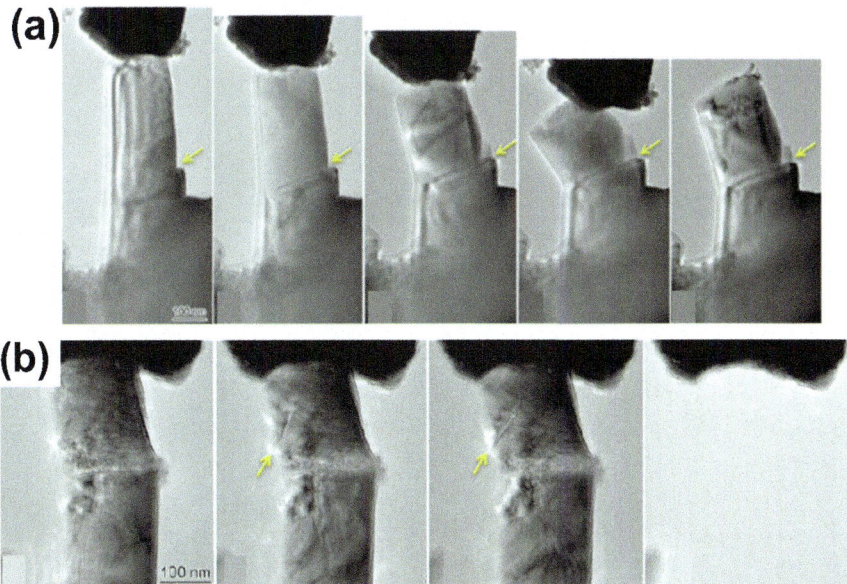

Fig. 5.10 In-situ TEM imaging of GaN nanowire showing: (**a**) initiation and propagation of slip due to compression, and (**b**) dislocation activity leading to nanowire fracture. (Reproduced with permission from Huang et al. (2011))

(marked in Fig. 5.10a by a yellow arrow). The slip initiated from the step and propagated along the prismatic plane. TEM imaging is helpful to decipher dislocation activity leading to fracture. In-situ imaging prior to failure is demonstrated in Fig. 5.10b. Nanowire compression resulted in dislocation nucleation, shown by the yellow arrow in the figure. Once nucleated, the dislocation was seen to cross the nanowire along a prismatic plane. Eventually, it escaped through the other free surface of the nanomaterial. The in-situ observations enabled the conclusion that dislocation activity was a precursor of fracture for GaN nanowire. Additionally, high-resolution imaging also revealed that most of the dislocations near the fracture surface were 1/3[11–20] type and the slip plane was (1–100). This example highlights the importance of in-situ testing to decipher deformation mechanisms down to the atomic scale.

The in-situ technique is advantageous to understand the fatigue behavior of 1D nanomaterials. One possible concern during in-situ fatigue testing in electron microscopes could be overexposure to the electron beam. The beam can cause localized damage, affecting the mechanical response. This problem can be overcome by using higher frequencies for fatigue loading to limit exposure time. Figure 5.11 shows a case study on fatigue testing of Ni nanowire (Zhang et al. 2017a). The experimental setup shown in Fig. 5.11a consists of a nanoindenter and a push-to-pull (PTP) micromechanical device. The nanowires suspended on the PTP device experience uniaxial tension when the device is subjected to compression by the indenter probe. The tests can be conducted in both displacement control and load control modes.

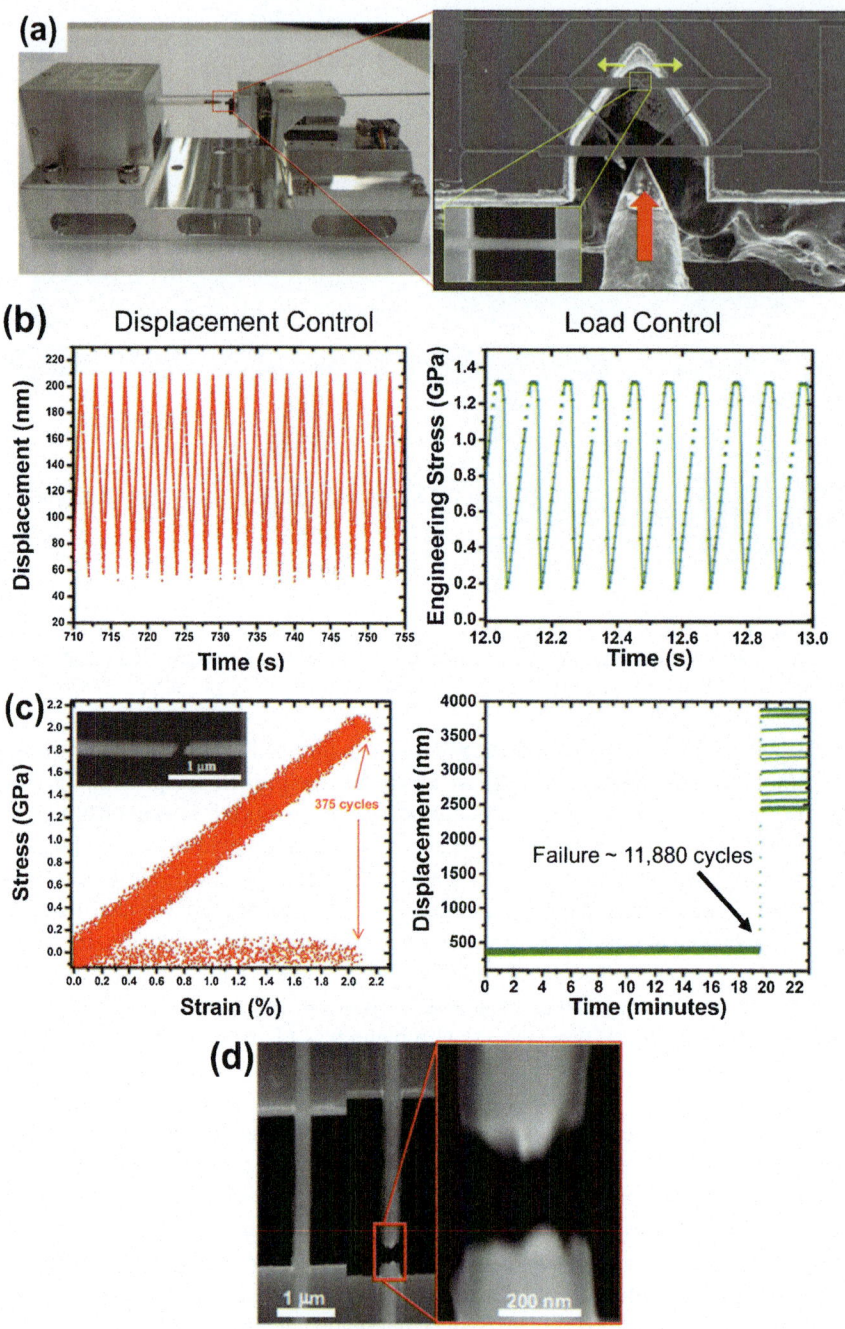

Fig. 5.11 In-situ fatigue testing of Ni nanowire: (**a**) test setup consisting of in-situ nanoindenter and a push-to-pull nanomechanical device, (**b**) programmed input for displacement control and load control fatigue tests, (**c**) fatigue response of the nanowire for 0.5 and 10 Hz frequency tests, and (**d**) SEM micrograph showing fracture morphology. (Reproduced/adapted with permission from Zhang et al. (2017a))

The plots in Fig. 5.11b show the programmed input for the two modes of fatigue testing. In displacement control, the peak displacement value or strain, frequency, and the number of cycles are the input parameters whereas the load or stress response is the output. In the load control mode, the load is input and the displacement response is the output parameter. Figure 5.11c shows the fatigue response of the nanowire for the tests conducted with different frequencies. The plot in the left corresponds to 0.5 Hz frequency and displacement control test. The nanowire failed after ~375 cycles, which manifested as a drop in the stress–response. The plot in the right shows fatigue response (displacement-time) for 10 Hz frequency and load control mode. The failure was seen to happen after ~11,880 loading cycles, as the displacement value jumps after failure when there is no resistance to deformation. High-resolution imaging allows close observation of failure mechanisms. The SEM micrograph in Fig. 5.11d revealed tapering on the fracture surface. Therefore, in-situ testing facilitates the quantitative and qualitative assessment of the fatigue performance of 1D nanomaterials.

1D nanomaterials are known to exhibit size effect in mechanical properties. As the diameter decreases, enhancement in stiffness, yield strength, and ultimate tensile strength are often observed. In-situ testing is an excellent approach to correlate the dimensions with the mechanical properties. A case study on Ag nanowire is shown in Fig. 5.12 (Zhu et al. 2012). The nanowires with the diameter ranging from 30 to 130 nm were subjected to tensile loading. The yield strength was recorded to be as high as 2.64 GPa for ~34 nm diameter and a low of 0.71 GPa for 130 nm diameter nanowire. The log-log plot of yield strength vs. diameter is shown in Fig. 5.12a. The curve revealed that the nanowire follows the power-law size effect. The peak yield strength recorded by the in-situ tensile test approached the theoretical tensile strength of Ag, which is around 3.5 GPa. Young's modulus also demonstrated prominent size effect, with remarkable stiffening as the diameter was reduced below 80 nm (Fig. 5.12b). The dashed line in Fig. 5.12b shows Young's modulus of "bulk" Ag. Unlike yield strength where diameter-dependence was observed starting from 130 nm, the size effect in Young's modulus is observed for much lower diameters. It is noteworthy that not all mechanical properties necessarily exhibit a strong size effect. For instance, yield strain for the same nanowire was seen to be almost independent of nanowire diameter. In-situ testing allows to quantify these nanoscale size effects otherwise difficult to establish using conventional test methods without high-resolution imaging.

The application of the in-situ technique has also been demonstrated for the characterization of "composite nanofibers." Figure 5.13 shows a case study on in-situ tensile testing of CNT/SiC composite nanofiber (Yang et al. 2016). The core-shell structure of the composite fiber is shown in Fig. 5.13a. Real-time SEM imaging revealed that SiC fractured first and CNTs acted as bridges (Fig. 5.13b). Eventually, the CNT core was pulled out and enabled crack deflection resulting in a stepped fracture surface. The quantitative measurements were also made and the stress–strain curve corresponding to the in-situ test is shown in the figure. Two-stage deformation behavior was observed. The initial linear regime corresponds to the elastic

Fig. 5.12 Size effect in mechanical properties of Ag nanowires: (**a**) yield strength, and (**b**) Young's modulus as a function of nanowire diameter. (Reproduced with permission from Zhu et al. (2012))

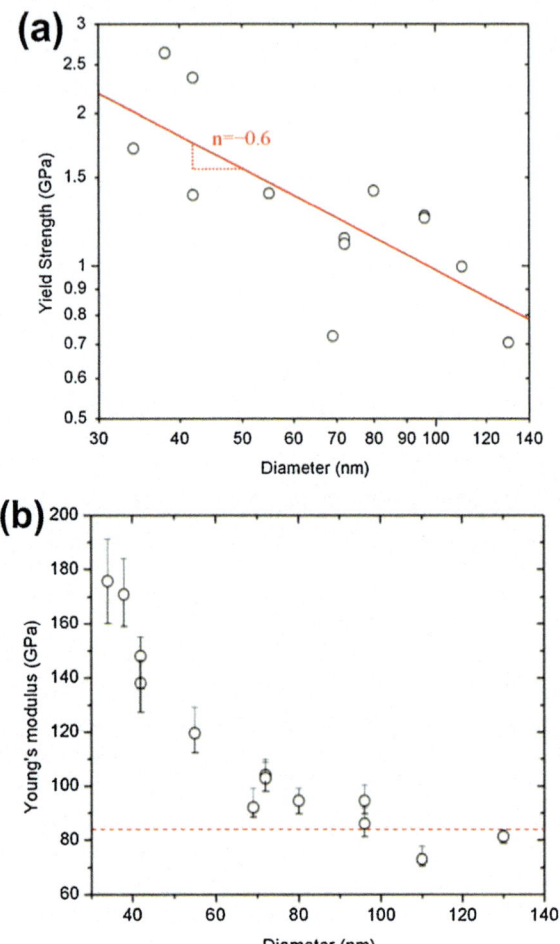

deformation. Thereafter, the nonlinear regime is characterized by load drops, changing slopes, plateauing of the stress, and final rupture. This non-linearity arises due to mechanisms such as CNT bridge formation, interfacial sliding, and nanotube pullout. Just like pristine nanomaterials, this composite nanofiber was also reported to exhibit size effect. As the CNT/SiC nanofiber diameter was reduced from over 400 nm to under 100 nm, the fracture strength increased from 3.8 GPa to 4.6 GPa. The elastic modulus calculated from the initial linear region also showed size effect, as the modulus increased from ~127.1 GPa to ~198.5 GPa. This example demonstrates that real-time imaging is advantageous to decipher the interplay of multiple mechanisms in composite 1D nano-systems.

In nanocomposite engineering, modification of the 1D nanofiller surface by chemical functionalization is a commonly adopted approach in order to achieve desired interface adherence. However, functionalization can influence the

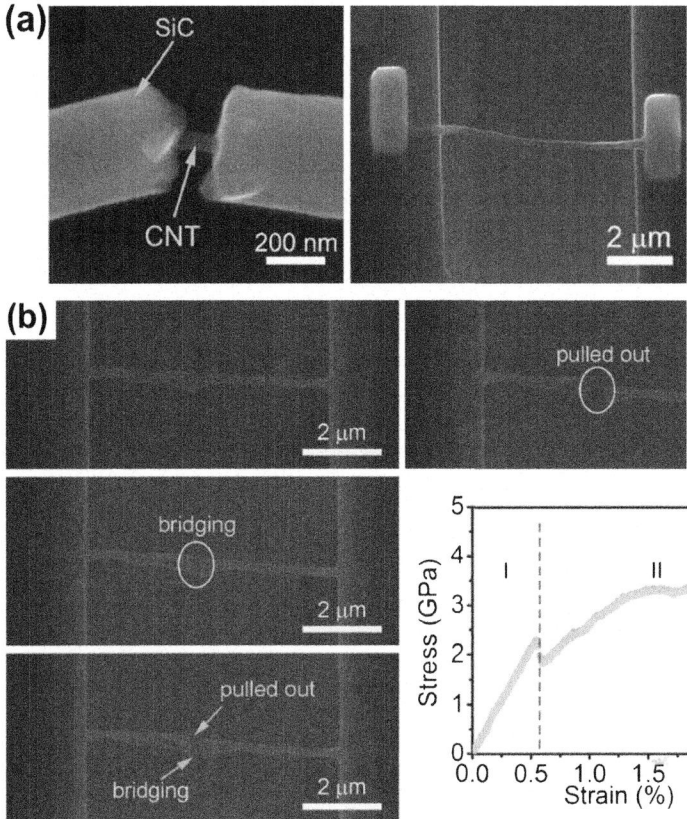

Fig. 5.13 (a) SEM micrograph of CNT/SiC composite nanofiber and in-situ SEM tensile set-up, (b) in-situ SEM testing of the composite nanofiber. (Reproduced/adapted with permissions from Yang et al. (2016))

mechanical properties of nanomaterials. The in-situ approach can be useful to probe the effect of chemical functionalization on the mechanics of nanomaterials. Figure 5.14 shows in-situ SEM tensile testing of carbon nanofibers with different surface modifications. This example demonstrates the characterization of pristine, fluorinated and amino-functionalized nanofibers. The fluorinated nanofibers consist of C-F and CF_2 groups, whereas amino-functionalized fibers are characterized by the presence of NH and NH_2 groups. In-situ tensile testing revealed significant strengthening of nanofibers because of fluorination (stress–strain plot in Fig. 5.14). The strength of fluorinated nanofibers was seen to be ~2.67 GPa, whereas pristine nanofibers were characterized by the average strength of ~1.89 GPa. There was a nominal drop in corresponding strain to failure: from ~6.3% for pristine nanofiber to ~5.7% for fluorinated fibers of comparable diameter. Also, the calculated elastic modulus was observed to increase from ~25.8 GPa (for pristine nanofiber) to ~45.8 GPa for fluorinated fiber. However, partial defluorination and grafting of

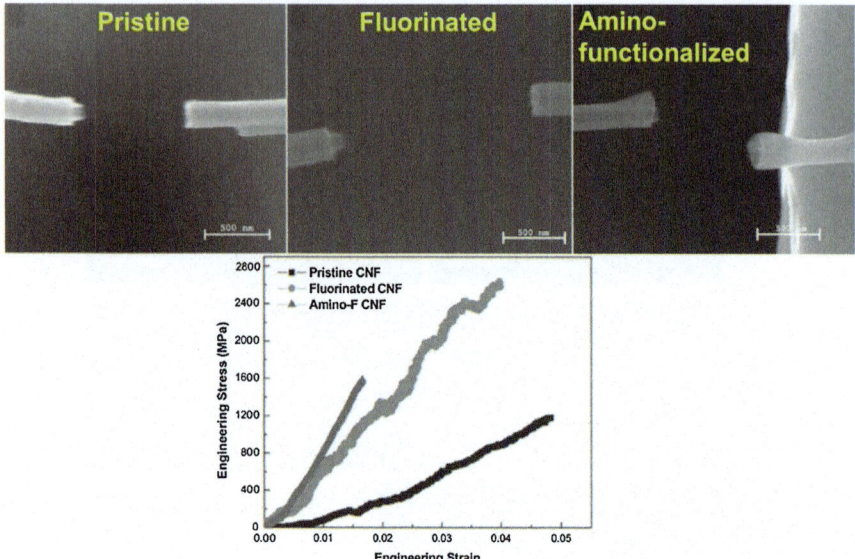

Fig. 5.14 In-situ tensile testing and comparison of mechanical response of chemically functionalized carbon nanofibers. (Reproduced/adapted with permission from Zhang et al. (2012))

ethylenediamino groups on these fluorinated fibers caused the strength to drop to ~1.75 GPa and the modulus to drop to ~31.6 GPa, which are in the same range as the pristine fibers. The most striking effect is on the strain, as the amino-functionalized fibers exhibit much lower strain to failure (~5.3%). The in-situ imaging revealed a cup-and-cone style failure of the nanofibers. This example highlights the importance of in-situ testing to interrogate the effect of surface-level structural modifications on the mechanical response of nanomaterials.

5.1.3 2D Nanomaterials

Nanomaterials which have two of the dimensions not confined to the nanoscale are referred to as 2D nanomaterials. Typically, 2D nanomaterials are single-layer materials consisting of atomic monolayer. Therefore, their thickness is extremely fine (Å-nm length scale). They are found in the form of sheets, flakes, membranes or films. Because of ultrafine thickness, it is rather difficult to examine their mechanical properties. In-situ testing in high-resolution microscopes is an effective approach to assess their mechanical response. Figure 5.15 demonstrates in-situ tensile testing of graphene, a 2D form of carbon with simultaneous real-time SEM imaging (Zhang et al. 2014). The experimental setup shown in Fig. 5.15a consists of an in-situ

5.1 In-Situ Characterization of Nanomaterials

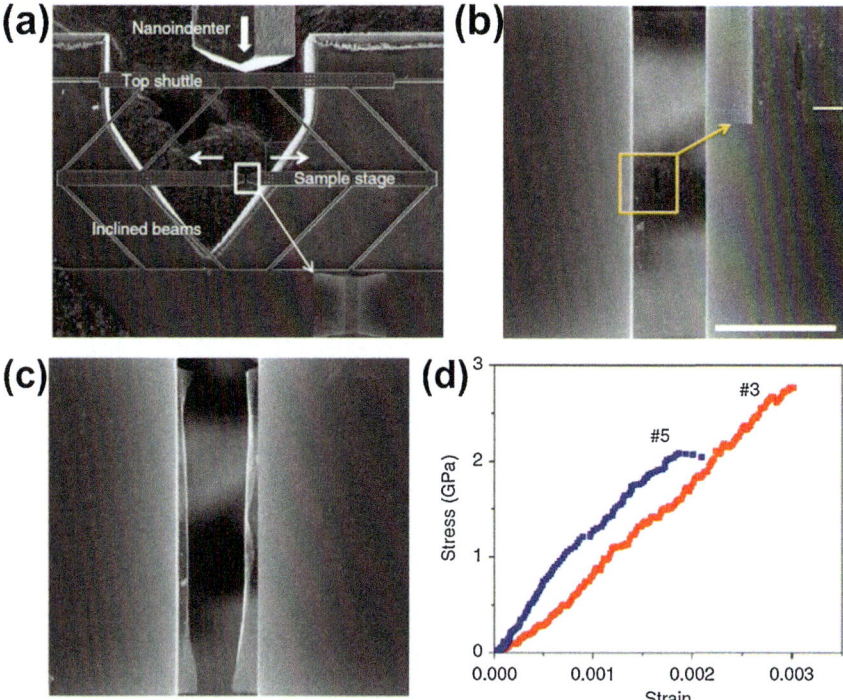

Fig. 5.15 In-situ tensile testing of 2D graphene: (**a**) experimental setup consisting of nanoindenter and push-to-pull micromechanical device, (**b**) graphene with a pre-crack prior to the test, (**c**) SEM micrograph of the specimen after splitting, and (**d**) stress–strain curve showing brittle failure. (Reproduced with permission from Zhang et al. (2014))

nanoindenter and a push-to-pull micromechanical device. A zoomed-in SEM micrograph of the graphene membrane subjected to tensile loading is shown in Fig. 5.15b. The inset to the figure shows a small pre-crack introduced in the graphene membrane to determine fracture toughness. Tensile loading resulted in the growth of the central crack, eventually leading to failure or splitting of graphene into two halves (shown in Fig. 5.15c). The quantitative stress–strain relationship was also recorded, and a linear response was noticed until the fracture point (Fig. 5.15d). Since the response is brittle in nature, Griffith's theory of brittle fracture can possibly be used to determine the critical stress (σ_c) for the onset of fast fracture (with a central pre-crack of length $2a_0$):

$$\sigma_c = \sqrt{\frac{2\gamma E}{\pi a_0}} \tag{5.11}$$

where E is Young's modulus and γ is the surface energy. Rearranging the expression, so as to club together the experimentally determined parameters (σ_c and a_0) together:

$$\sigma_c\sqrt{a_0} = \sqrt{\frac{2\gamma E}{\pi}} \tag{5.12}$$

The product in the left-hand side should be a constant for Griffith's theory to hold true. In-situ experiments on different graphene specimens with variable crack size (from 33 to 1256 nm) established this product indeed stays constant (around 2.25 MPa $\sqrt{m}$) within the margins of error (standard deviation of 0.35 MPa $\sqrt{m}$). Therefore, in-situ testing establishes that Griffith's theory is applicable for 2D nanomaterials. Quantitative in-situ testing is also useful to determine the fracture toughness of 2D nanomaterials using the relationship:

$$K_c = \sigma_c\sqrt{\pi a_0} \tag{5.13}$$

The fracture toughness of graphene in the case study shown in Fig. 5.15 was computed to be ~4 MPa $\sqrt{m}$. This example highlights the relevance of the in-situ mechanics approach to test and verify the applicability of classical theories in nanomaterials and nanoscale systems.

In-situ testing is used to probe the properties and response of "linked" nanomaterials. Figure 5.16 demonstrates in-situ TEM testing of carbon-linked graphene oxide nanosheets (Cao et al. 2016). The sample consists of a graphene oxide sheet with a pre-crack, which was linked by the deposition of amorphous carbon using a high energy electron beam. The linked region is clearly visible in the TEM micrograph before tensile loading (Fig. 5.16a). During mechanical loading, it was observed the failure initiated at the interface where amorphous carbon was deposited (Fig. 5.16b). Once initiated, the crack was seen to propagate and open further (Fig. 5.16c). The final TEM snapshot in Fig. 5.16d shows the failed nanosheet. The quantitative stress–strain measurements revealed a failure strain of ~4% and the failure strength to be ~4.4 GPa. This example illustrates the usefulness of a high-resolution in-situ approach to study the deformation of interfaces in nanomaterials.

In-situ imaging is applied to probe the mechanical characteristics of "patterned" nanomaterials. Figure 5.17 demonstrates the in-situ tensile stretching of graphene nanoribbons (Liao et al. 2017). The image of the push-to-pull device used for stretching the ribbons is shown in Fig. 5.17a. The nanoribbons had a width less than 100 nm, and length around ~380 nm. The testing was accompanied with TEM imaging. Figure 5.17b shows nanoribbons before loading, during stretching, and after fracture. It is noteworthy that the electron beam can influence the mechanical response. The beam effect is minimized by keeping the illumination step as low as possible (less than 3). Additionally, changing of illumination step, magnification, switching between imaging, spot and spectrum modes should be avoided, since strained nanoribbons are susceptible to failure. This is because changing lenses cause fluctuation of the magnetic field, which can possibly generate a current pulse in the indenter transducer. Since the indenter probe is electrically conductive, the current pulse may cause a jump of the tip. The quantitative measurements shown in Fig. 5.17c revealed the engineering strain (to failure) was ~3.18%. This value was

5.1 In-Situ Characterization of Nanomaterials

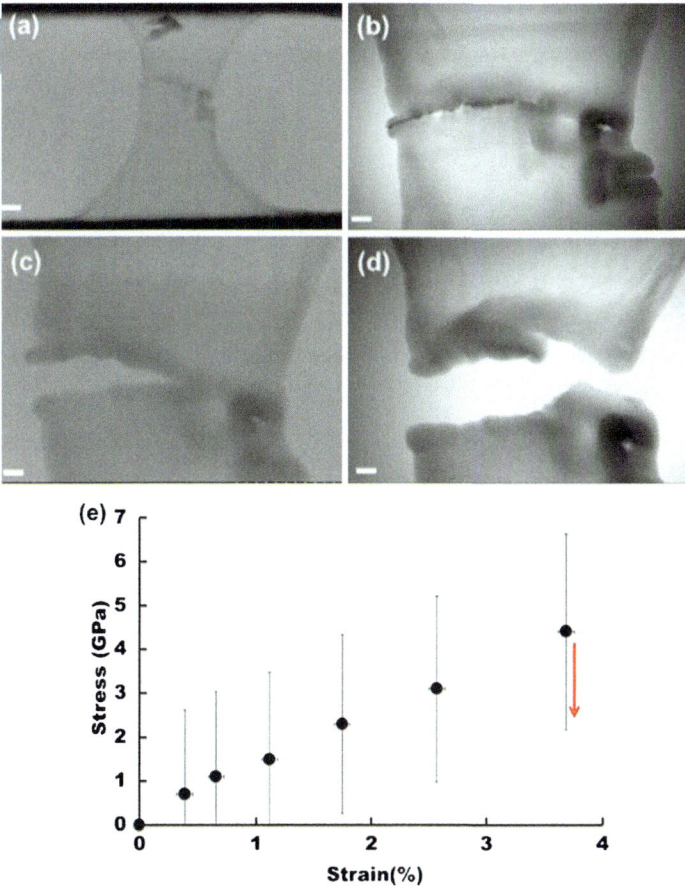

Fig. 5.16 Real-time TEM snapshots during tensile deformation of C-linked graphene oxide sheet: (**a**) before loading, (**b**) failure initiated near amorphous carbon deposit, (**c**) opening and propagation of the crack, and (**d**) failed/ruptured sheet. (**e**) Corresponding stress–strain response showing linear response and abrupt, catastrophic failure. (Reproduced/adapted with permission from Cao et al. (2016))

overestimated due to the presence of some unetched graphene membrane. The "true" failure strain was calculated to be around ~2.8–3%. This example demonstrates the usefulness of the in-situ technique for characterizing the mechanical properties of nanostructures with specific geometries and dimensions. This capability is particularly important for device applications of nanomaterials, where patterning is often carried out. As an example, the construction of graphene-based electronic devices requires specific patterning. Understanding of strain response of these nanostructures is vital for predictable performance of the devices.

The examples presented in this section showed the advantage of in-situ approach for probing the properties of nanomaterials, due to their ultralow dimensions. The next section address some of the key materials engineering aspects involving nanomaterials.

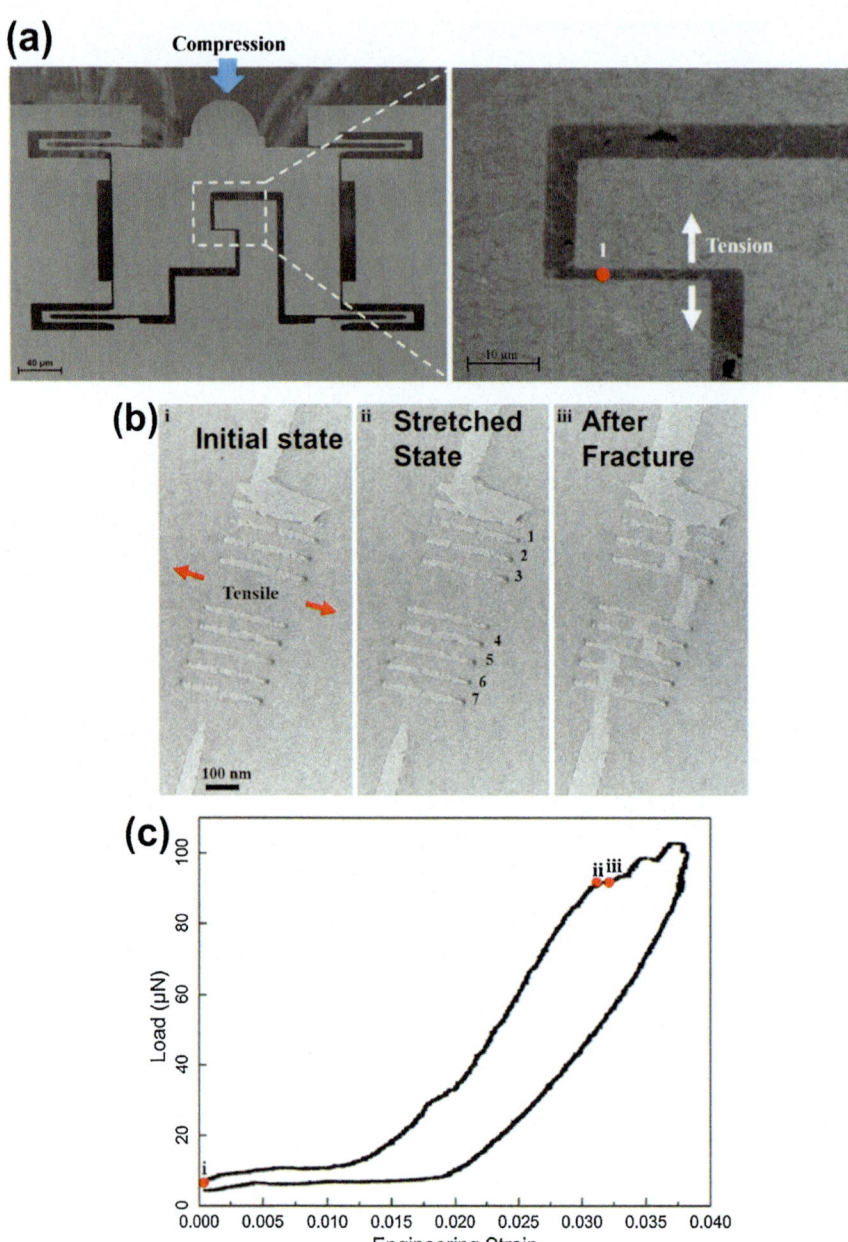

Fig. 5.17 (**a**) Experimental setup for in-situ tensile testing of graphene nanoribbons, (**b**) real-time TEM snapshots before, during and after tensile loading, and (**c**) corresponding load–strain response (point iii represents the point of fracture). (Reproduced/adapted from Liao et al. (2017) CC BY 4.0)

5.2 In-Situ Characterization of Composites

For most engineering applications, nanomaterials are used in conjunction with other bulk materials, such as metals, ceramics, or polymers. This section of the chapter discusses the suitability of in-situ characterization to probe the mechanics of advanced composites fabricated by introducing the nanofillers in the matrix material. The examination of interface mechanics is important as the adherence and debonding phenomena determine the performance of the composite material systems. This interfacial interaction zone is extremely localized, in Å-nm length scale. It is not possible to identify and probe the mechanics of interfaces by conventional characterization techniques, which are more geared toward large-scale sample testing. The in-situ approach is highly desirable for understanding the deformation and failure phenomena at the interfaces. High-resolution imaging allows identifying the regions of interest to perform mechanical tests. Real-time imaging throws light into complex mechanical interactions between multiple constituents of the composite system along the interface. The mechanical instrumentation for such investigations should be capable of sensing ultralow loads and displacements.

Figure 5.18 illustrates the in-situ pullout of CNTs embedded in Al, in order to determine interfacial shear strength (IFSS) in CNT-Al composites (Zhou et al. 2016; Yi et al. 2017). An AFM cantilever was used to pull the nanotubes protruding out from the fracture surface (test setup shown in Fig. 5.18a) (Zhou et al. 2016). The dangling nanotubes were clamped onto the cantilever tip by electron-beam-induced deposition (EBID), as shown in Fig. 5.18b. Once clamped, the nanotube is subjected to tensile force until it is pulled out of the Al matrix. The force (F) required to initiate pull out is determined based on the AFM cantilever deflection (δ) using the following equations:

$$\delta = \frac{2}{3}\theta L \quad (5.14)$$

$$F = k\delta \quad (5.15)$$

where k is the force constant of the cantilever, θ is the change of deflection angle of the cantilever before and after loading, and L is the cantilever length. The interfacial shear strength (τ_i) can then be determined using the equation:

$$\tau_i = \frac{F}{\pi d l_e} \quad (5.16)$$

where d is the average nanotube diameter and l_e is the effective length of the tube embedded in Al. The effective embedded length can be determined by imaging the nanotube before and after pull out (shown in Fig. 5.18c). A combination of strong and weak interfaces was observed: the tightly bonded nanotubes were characterized by IFSS ~25.8 MPa, whereas the weak interfaces exhibited a rather low IFSS

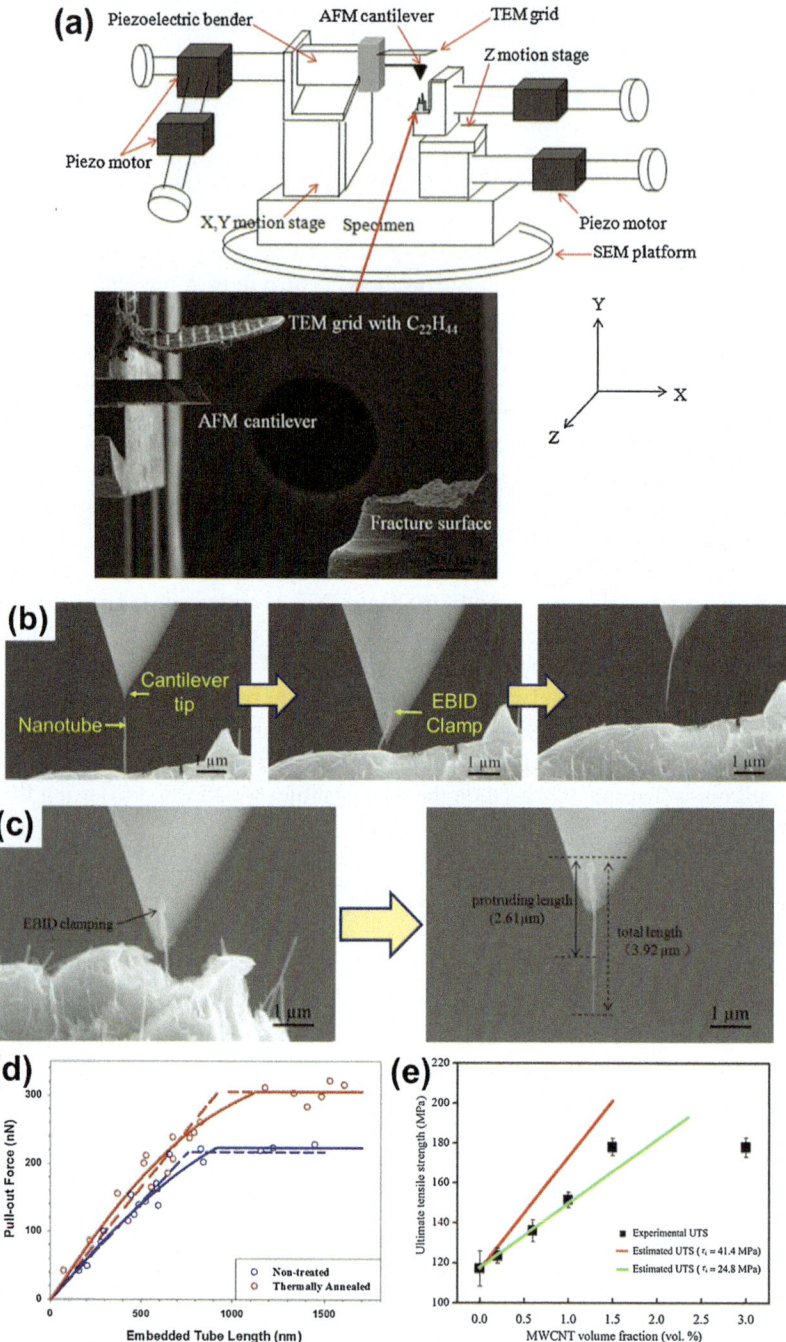

Fig. 5.18 In-situ pullout experiment to determine interfacial shear strength: (**a**) AFM-based test setup, (**b**) the process of clamping and pullout testing captured by SEM, (**c**) SEM images before and after to determine protruding and embedded lengths, (**d**) dependence of pullout force on embedded nanotube length and the effect of heat treatment, and (**e**) comparison of experimental UTS of Al-CNT with predicted UTS computed using the IFSS values. (Reproduced/adapted with permissions from Zhou et al. (2016), Yi et al. (2017))

of ~8.6 MPa (Zhou et al. 2016). The weak interfaces were more susceptible to slippage, and hence lower IFSS. The pullout force is seen to depend on the length of nanotube embedded in the matrix, as shown in Fig. 5.18d (Yi et al. 2017). When the embedded length exceeds a critical length, the required pullout force stays constant. Additionally, the thermal treatment affects the metal–nanotube interface. Due to temperature-induced diffusion, the interfacial bonding is improved which results in higher forces required for nanotube pullout. This is shown in Fig. 5.18d. The ability of the in-situ approach to quantify the effect of post-processing treatment on interface mechanics is promising from the standpoint of advanced material development.

Once IFSS is known, the tensile strength of the composite (Al-CNT) can be predicted using the shear lag model:

$$\sigma_C = V_f \frac{\tau_i l}{d} + \sigma_M (1 - V_f) \tag{5.17}$$

where V_f is the volume fraction of nanotube added to Al, l is the average length of nanotubes and σ_M is the UTS of the Al matrix. The equation above is for short fibers (with length < $l_{critical}$), since the study employed shorter nanotubes (Zhou et al. 2016). Figure 5.18e shows a comparison of modeled and experimentally determined values of the UTS, and a reasonable agreement is seen. This attests the importance of the in-situ approach for the quantitative prediction of complex material systems.

The in-situ single nanotube pullout technique is also used to characterize ceramic-nanotube bonding. A case study on interfacial mechanics in Boron Nitride Nanotube (BNNT)—Silica nanocomposite film is shown in Fig. 5.19 (Yi et al. 2019). However, not all the tests resulted in complete pullout: loading of nanotubes can also cause a fracture or telescopic sliding where the inner tube shells detach from the outer wall and slip out (Fig. 5.19b). These events happen when the nanotube is strongly adhered to the ceramic, making de-bonding and interface sliding rather difficult to achieve. It is not possible to calculate IFSS from these scenarios. Pullout of nanotubes is required to determine IFSS at the ceramic/BNNT interface. The pullout forces ranged from 57 to 190 nN for embedded tube lengths varying from 152 nm to 2 μm. The linearly increasing force-length curve in Fig. 5.19a confirms the composite system displays the shear lag effect. A force plateau is reached at 165 nN, with corresponding critical nanotube embedded length of ~560 nm. The average IFSS for the BNNT–silica interface is calculated to be 43.7 MPa. The IFSS value depends on the nanotube diameter. The avg. IFSS was seen to vary from a low of ~25.8 MPa for a nanotube with ~3.9 nm diameter, to a high of ~53.1 MPa when the nanotubes are very fine (less than 2 nm in diameter).

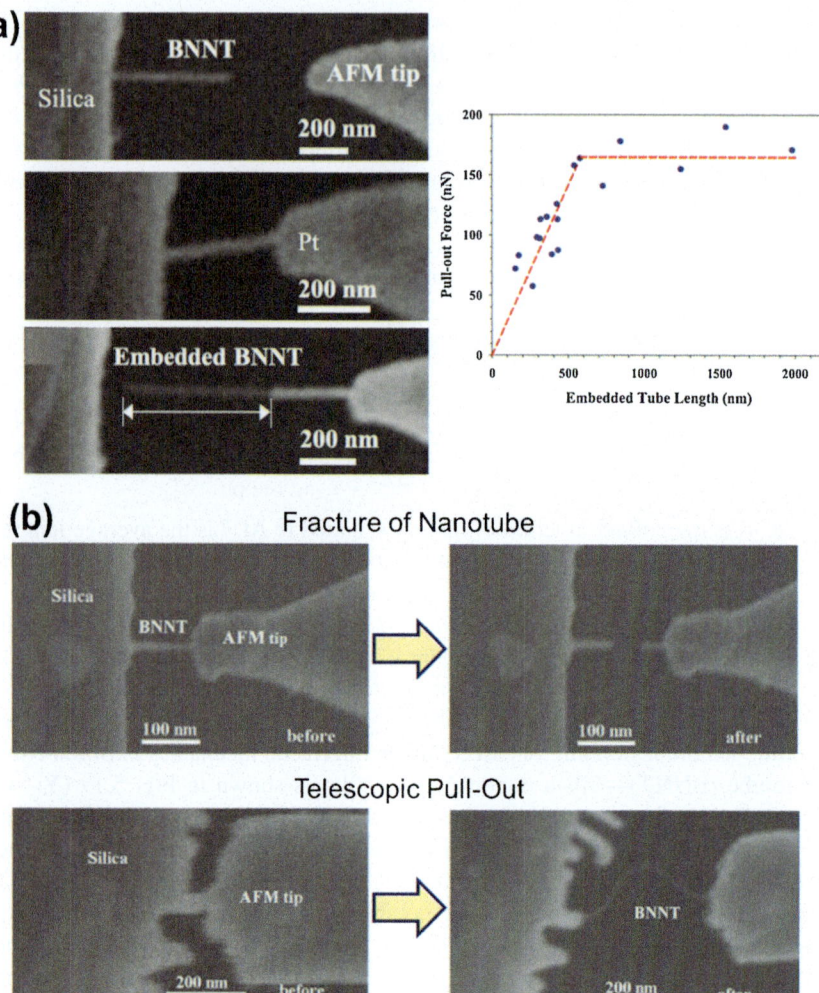

Fig. 5.19 In-situ pullout in SEM to study BNNT-silica interface: (**a**) real-time SEM images showing the progression of pullout and a plot of interfacial shear strength as a function of embedded tube length, and (**b**) instances where nanotube fractures or shows telescopic sliding instead of simple pullout. (Reproduced/adapted with permission from Yi et al. (2019))

The application of the pullout technique to characterize the polymer–nanofiller interface is shown in Fig. 5.20 (Tsuda et al. 2011). It is possible to compute the tensile stresses experienced by the nanotubes experiencing pullout forces using the expression:

$$\sigma_{max} = \frac{F_{pullout}}{\pi \left(D_{CNT}/2\right)^2} \tag{5.18}$$

5.2 In-Situ Characterization of Composites

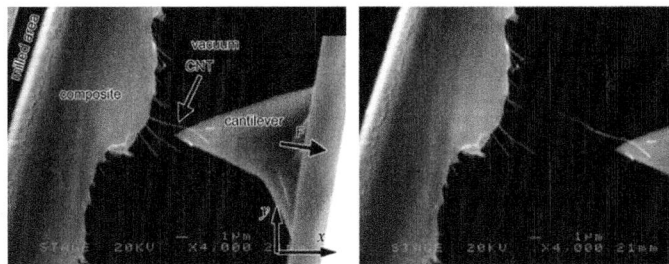

Fig. 5.20 In-situ pullout experiment to determine interfacial shear strength for PEEK-CNT nanocomposite. (Reproduced with permission from Tsuda et al. (2011))

The tensile stress was calculated to be around 3 GPa using the above expression. This value is much lower than the intrinsic tensile strength of the nanotubes, which verifies that there was no nanotube breakage during the pullout and the force readings indeed correspond to the interfacial resistance. The IFSS was found to vary in the range of 3.5–7 MPa for this case study. The effect of hot pressing was also examined, and IFSS was measured to be in the range of 6–14 MPa after the polymer composite system was hot-pressed at 573 K and 1 MPa pressure. This is a twofold jump, clearly showing that hot pressing impacts the polymer–nanotube interface. This example shows the importance of the in-situ measurement approach for establishing processing-property correlations by taking into account extremely localized nanoscale interactions, which are otherwise difficult to capture by conventional bulk-scale mechanical testing.

During composite synthesis, interface modification is a commonly adopted approach to achieve desired mechanical performance. An example of defect engineering on graphene for fabricating copper–graphene composites is shown in Fig. 5.21 (Chu et al. 2018). Plasma treatment of graphene was performed to introduce surface defects. It was observed that composites fabricated using plasma-treated graphene exhibit superior tensile strength, as shown in Fig. 5.21a. In-situ mechanical characterization can be useful to understand how the engineered interfaces influence the load-bearing capability. Real-time SEM snapshots showing tensile deformation of Cu reinforced with plasma-treated graphene are shown in Fig. 5.21b. It was observed that the crack initiates at the prefabricated notch, which propagates with multiple deflection events. Crack branching and formation of altogether new cracks during progressive loading was observed. These deflection events absorb energy and enhance the loads required for complete failure. The treated graphene sheets are partially pulled out along the fracture surface, as opposed to untreated graphene which can exhibit near-complete sliding or pullout. This is because the interface bonding was rather weak without plasma treatment. This example demonstrates the suitability of in-situ investigations to correlate interface engineering strategies with deformation mechanisms.

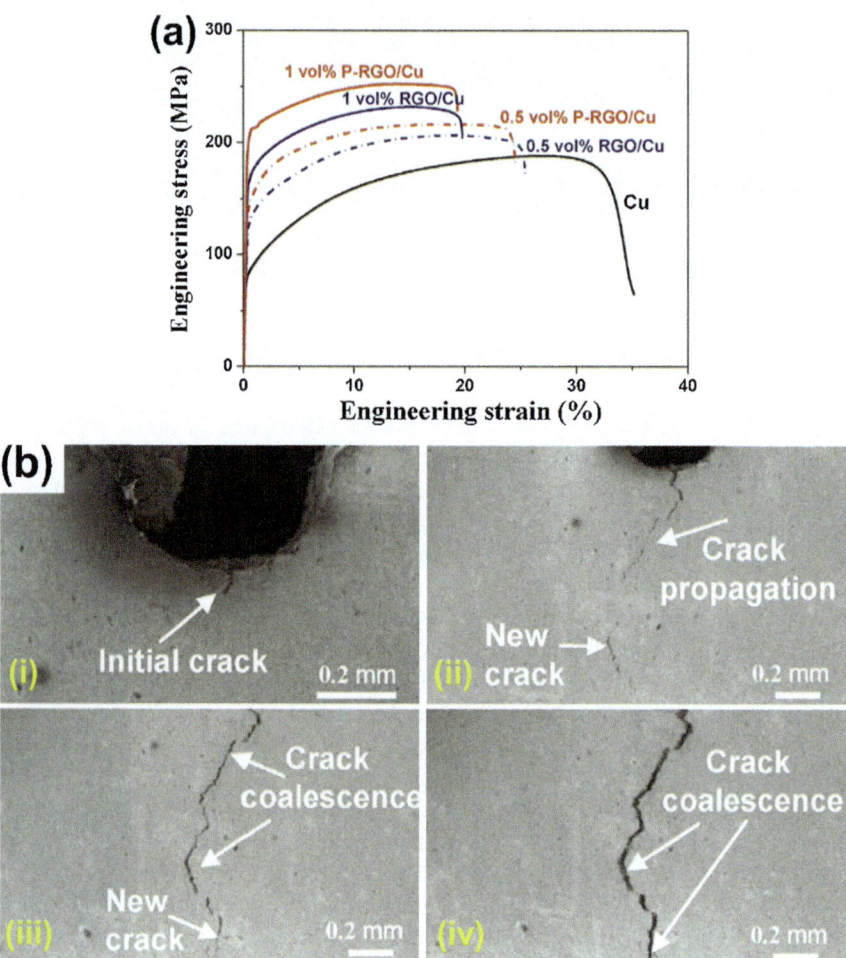

Fig. 5.21 (a) Stress–strain curves comparing the effect of defect engineering on the load-bearing ability of Cu-graphene composites, and (b) real-time SEM snapshots showing crack propagation, deflection, branching, and coalescence mechanisms when plasma-treated nanofiller is introduced. (Reproduced/adapted with permission from Chu et al. (2018))

The case studies discussed so far primarily focused on highly localized mechanical testing of nano-interfaces. In-situ approach is also useful to evaluate the mechanics of composite materials at much larger length scales (micro, meso, and even macroscale). An example of in-situ SEM tensile testing of the Al-CNT composite is shown in Fig. 5.22 (Boesl et al. 2014). A comparison of stress–strain curves shows much higher yield strength and UTS for Al reinforced with 1 vol.% CNT as compared to unreinforced Al. Real-time SEM images in the figure clearly show that the nanotubes bridge the microcracks during failure. The nanotubes were seen to be

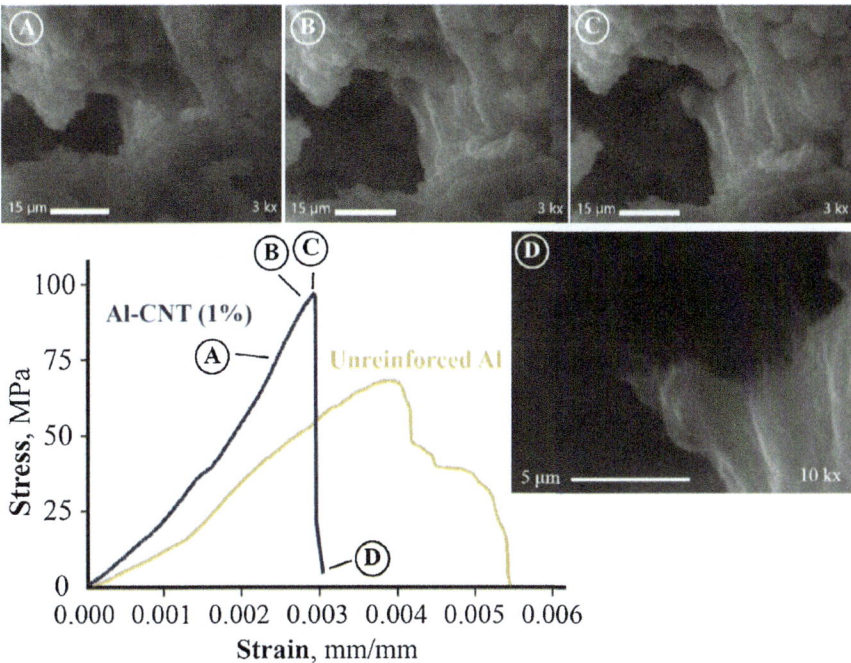

Fig. 5.22 Tensile testing and real-time imaging of Al-CNT composite. Comparative stress–strain curves show the strengthening of Al due to CNT addition. (Reproduced with permission from Boesl et al. (2014))

pulled out during tensile deformation. These nanotube bridges enhanced the resistance to failure, which explains the observed 40% increase in yield strength. The fracture strength of the composite (σ_C) can be predicted using the shear lag model for long fibers (when fiber length, $l > l_{critical}$):

$$\sigma_C = \sigma_{CNT} V_{CNT} \left(1 - \frac{l_{critical}}{2l}\right) + \sigma_M (1 - V_{CNT}) \quad (5.19)$$

where V_{CNT} is the volume fraction of CNT in the composite, σ_{CNT} is the tensile strength of CNT, l is the average nanotube length, and critical length ($l_{critical}$) is defined by the relation:

$$l_{critical} = \frac{\sigma_{CNT} D}{2\tau_m} \quad (5.20)$$

where D is the nanotube diameter and τ_m is the shear strength of the metal matrix. Based on these equations, the strength of the composite was calculated to be around ~90–110 MPa, which matches very well with the in-situ experimental value of ~95.5 MPa (Fig. 5.22). Therefore, the shear lag fiber strengthening model was established to be valid for the CNT-metal matrix composite system. After failure, the pulled-out protrusions as long as 7 μm were observed via real-time imaging, which attests load transfer from Al to CNT reinforcement during mechanical testing. The interfacial load transfer is essential for strengthening, and real-time imaging enables direct observation of the signs of stress transfer. The readers are referred to the Supplementary Video, Video 5.1 where carbon nanotube pullout is observed via real-time SEM imaging.

The addition of nanofillers is known to improve the fracture toughness of ceramics. Figure 5.23 illustrates the application of in-situ SEM imaging to decipher the mechanisms because of graphene addition to TaC (Zhang et al. 2016). The load–displacement curve shown in Fig. 5.23a corresponds to the indentation of pure TaC and TaC reinforced with 5 vol.% graphene nanoplatelets (GNP). It is noteworthy that this case study pertains to "high-load" indentation, with peak load in the order of hundreds of Newton. The readers are referred to Supplementary Videos, Videos 5.2 and 5.3, where deformation mechanisms for pure TaC and TaC-GNP captured via SEM are shown. Real-time imaging allows an understanding of the mechanisms activated in different load regimes. Pure TaC grains were susceptible to sliding along the porous interfaces. Long cracks, with an average length of ~610 μm were observed. The cracks were arrested after graphene addition, with ~450 μm average length after 5 vol.% GNP addition. This resulted in lower residual damage area post indentation, as observed from in-situ imaging. The total impact area went down from ~1.36 mm^2 (for TaC) to ~0.2 mm^2 (for TaC-5GNP). The number of major cracks due to indentation-induced stresses also went down from 6 to 3. A comparison of adding different weight fractions of graphene is shown in Fig. 5.23b. Graphene-rich samples demonstrated significant local chipping, but crack propagation was arrested. Real-time video (Video 5.3) showed crack deflection due to high graphene content, promoting energy dissipation and arresting the overall crack length. This example shows the usefulness of the in-situ approach to establish the qualitative and quantitative correlation between filler fraction and mechanical response of composites.

The use of multiple imaging techniques for studying the same material system can shed light on the mechanisms activated at different length scales. For instance, nanometer resolution is desirable to observe highly local events in a composite, such as interfacial sliding or de-bonding. On the other hand, meso-macro-scale imaging is desirable to observe stretching, necking and failure propagation. An example of multi-scale in-situ imaging for a 3D graphene foam-polymer composite is demonstrated in Fig. 5.24 (Nieto et al. 2015b). Figure 5.24a shows in-situ optical imaging of the porous composite under tension. The stress–strain curve revealed remarkable stretchability of the composite, with failure strains exceeding 270%. The optical snapshots show necking prior to failure. Higher-resolution SEM imaging allows observation of the microscopic origin of necking in the samples (Fig. 5.24b). PLC polymer bridges are formed during tension, which undergo extensive stretching. These bridges resist or delay fracture, evidenced by a significant jump in strain to failure from ~8.5% for pure graphene foam to ~276% for PLC-graphene foam. Eventually, the overstrained bridges were seen to snap one by one. The readers are referred to the Supplementary Videos 5.4 and 5.5, showing real-time SEM imaging of polymer bridging.

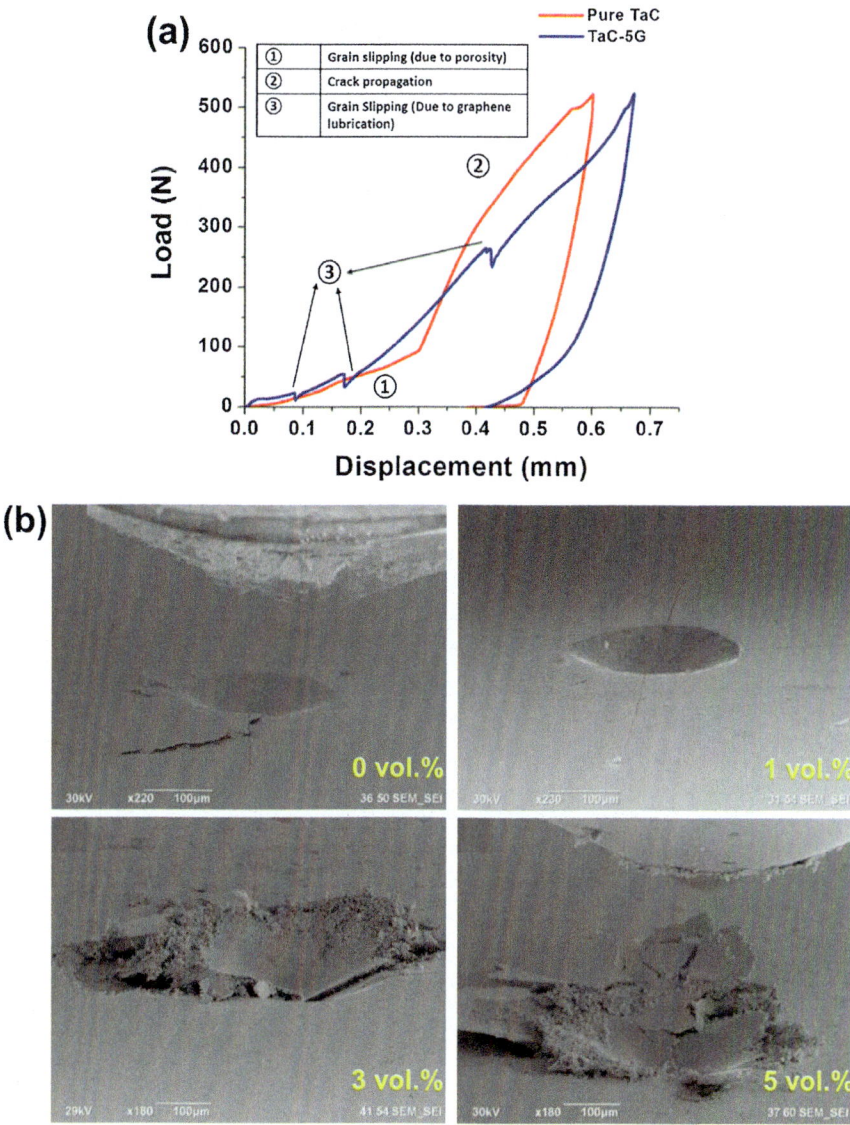

Fig. 5.23 In-situ high-load indentation of graphene-TaC composite: (**a**) load–displacement curve comparing the mechanical response of pure TaC and TaC-5 vol.% GNP, and (**b**) SEM micrographs comparing the indentation deformation mechanisms for different volume percentages of graphene filler in TaC. (Reproduced/adapted with permission from Zhang et al. (2016))

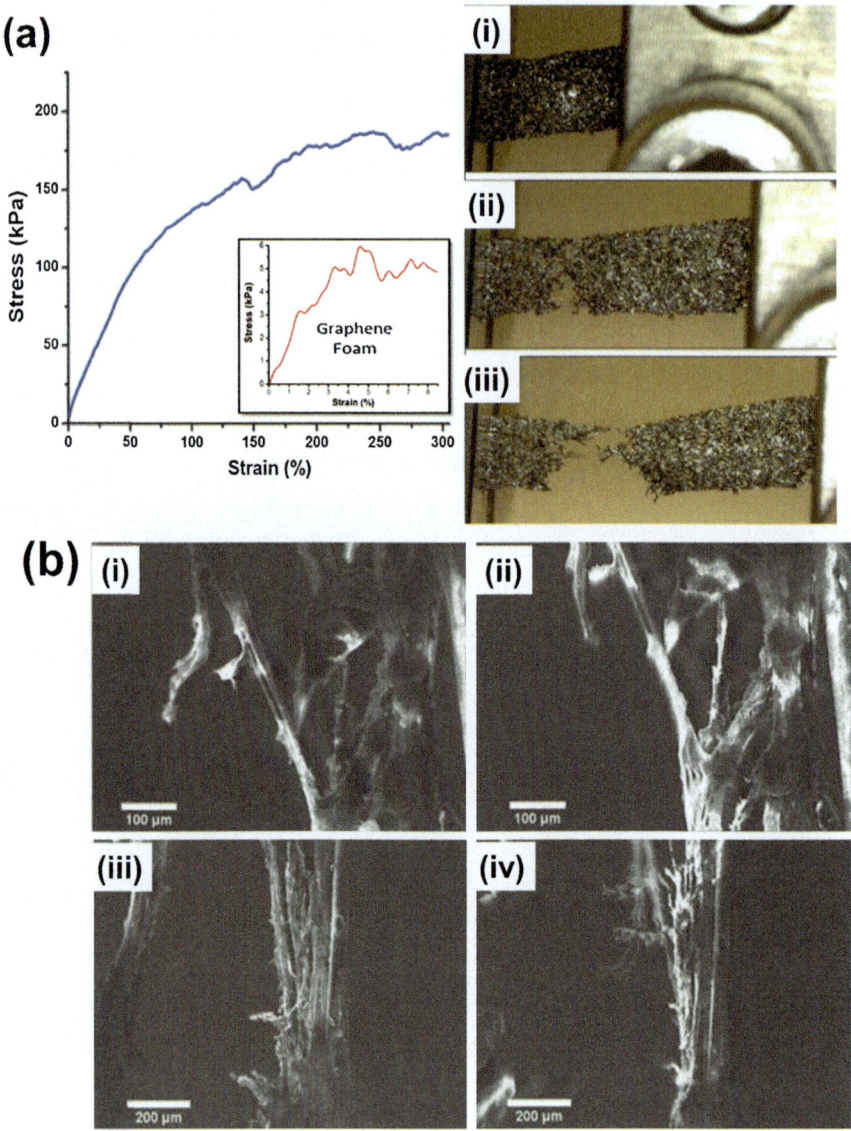

Fig. 5.24 Multi-scale imaging of a polymer-graphene foam composite under tension: (**a**) in-situ optical imaging (along with stress–strain plot) shows necking and failure, and (**b**) in-situ SEM imaging shows bridging, stretching, and snapping mechanisms. (Reproduced/adapted with permission from Nieto et al. (2015b))

In-situ testing is particularly advantageous to study the mechanics of "low-dimension" nanocomposite systems. These low-dimension systems could be ultra-thin films, membranes, fibers, foams, or mats, which have fine features. The case study in Fig. 5.25 shows in-situ testing of poly(methyl methacrylate)/poly(ethylene

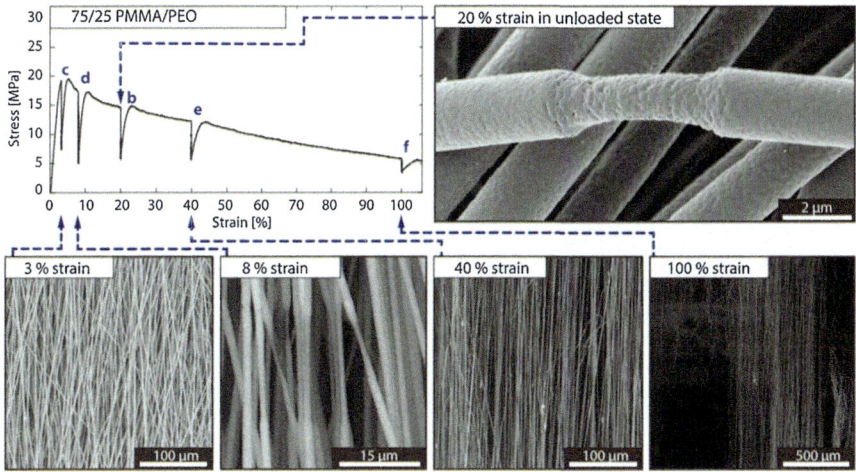

Fig. 5.25 Stress–strain curve and real-time SEM snapshots showing deformation of PMMA/PEO composite at different strains. (Reproduced from Andersson et al. (2014) CC-BY-NC-ND)

oxide) (abbreviated as PMMA/PEO) fiber mats (Andersson et al. 2014). The mats with 75% PMMA and 25% PEO composition were subjected to tension, and their deformation was observed via SEM. Real-time imaging showed thinning of the fibers as the sample was stretched. Necking was seen even at strains as low as ~8%. The necked regions then increased and eventually, the fibers constituting the mat started to fail. This behavior resulted in a progressively decreasing stress response (shown in Fig. 5.25). It is noteworthy that the stress relaxation seen in the stress–strain plot is due to pauses made during the test for superior quality imaging of the fibers at different strains. This is a challenge with in-situ imaging: pausing the test for imaging can affect the load readings due to material-dependent phenomena such as creep, strain hardening, or stress relaxation. It is advisable to perform a few tests without pauses to obtain a complete stress–strain response, which was reported by the authors in the original study (Andersson et al. 2014). The failure of mat occurred at strains exceeding 130%. Extensive necking resulted in a dramatic drop in the cross-section area of the mat, from the initial 3.8 to 0.5 μm^2. As a result, the true stress prior to failure was calculated to be much higher (~150 MPa) than the engineering stress (~14.8 MPa). Quantitative measurements allowed the determination of several mechanical properties. The composite mats displayed a toughness of 17.1 MJ/m^3, a tensile strength of 18.7 MPa, and an elastic modulus of 0.72 GPa. 75/25 PMMA/PEO composition was seen to display the best toughness value. Therefore, the in-situ mechanics approach is helpful to decipher the effect of different compositions on load-bearing mechanisms, enabling the engineering of composites with superior properties.

In-situ testing can be highly informative for designing composites with constituents that display a significant difference in deformation behavior. An example of a metal/ceramic multilayer system is shown in Fig. 5.26 (Li et al. 2015). In-situ

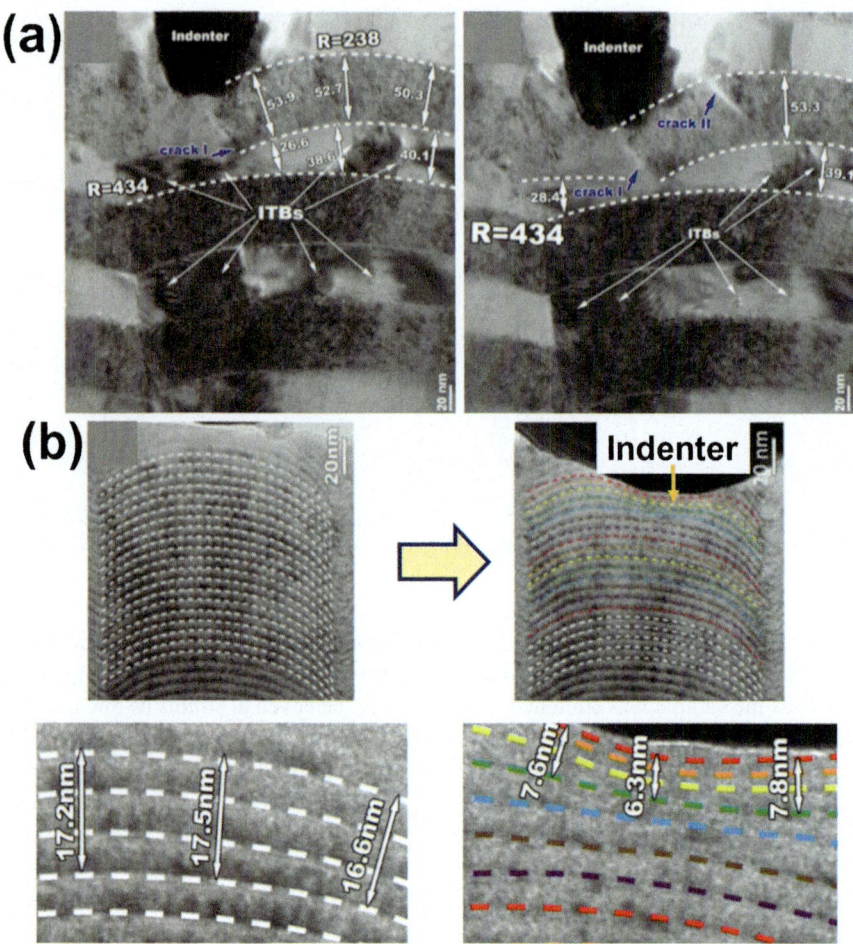

Fig. 5.26 In-situ indentation of multilayer Al-TiN composite system with real-time TEM imaging for variable layer thicknesses: (**a**) 50 nm Al-50 nm TiN composite, and (**b**) 2.7 nm Al-2.7 nm TiN composite. (Reproduced/adapted from Li et al. (2015) CC-BY-NC-ND)

indentation of Al/TiN multilayer composite was performed with real-time TEM imaging. Nanocomposites with different individual layer thicknesses were examined, such as 50 nm Al-50 nm TiN, 5 nm Al-5 nm TiN and 2.7 nm Al-2.7 nm AlN. The TEM micrographs in Fig. 5.26a correspond to the nanocomposite with 50 nm thick layers, revealing the Al layer deformed plastically and displayed thickness reduction. Contrary to this, no thickness reduction was seen for the TiN layer. Migration of incoherent twin boundaries (marked as ITBs in the micrographs) was observed due to indentation loading. Mode I cracks were observed to form in the TiN layer due to tensile stresses. However, no cracks were formed in the nanocomposite with 2.5 nm thick layers (shown in Fig. 5.26b). This difference is attributed to the size effect in mechanical properties, where plasticity is seen to be enhanced as the thickness goes down. The HRTEM images in Fig. 5.26b confirm the enhanced plasticity in the nanocomposite with ultrathin layers. The thickness of the first three bilayers decreased from 17.5 to 6.3 nm. The plastic deformability of TiN is because of mobile lattice dislocation, confirmed by in-situ TEM imaging in the study (Li et al. 2015). The ability to tailor the plasticity of nanocomposites is highly desirable for load-bearing applications, and the in-situ mechanics approach provides the toolsets required to establish structure–property correlations at nanometer length scale.

5.3 In-Situ Characterization of 3D Architectures and Metamaterials

Design and fabrication of 3D architectures, with well-defined unit cells, low mass densities, and superior load-bearing ability have gained traction in recent years. These 3D architectures, referred to as mechanical metamaterials, display excellent damage tolerance and recoverability upon mechanical loading. In-situ characterization is helpful to correlate the architecture/design with the mechanical response. The size of individual features in mechanical metamaterials can be extremely fine (nano-micrometer length scales), necessitating the use the high-resolution imaging to identify, probe, and observe the region(s)/feature(s) of interest. Figure 5.27 demonstrates in-situ compression of a ceramic sponge structure composed of TiO_2 nanofibers performed inside SEM (Wang et al. 2017). It was observed that the sponge

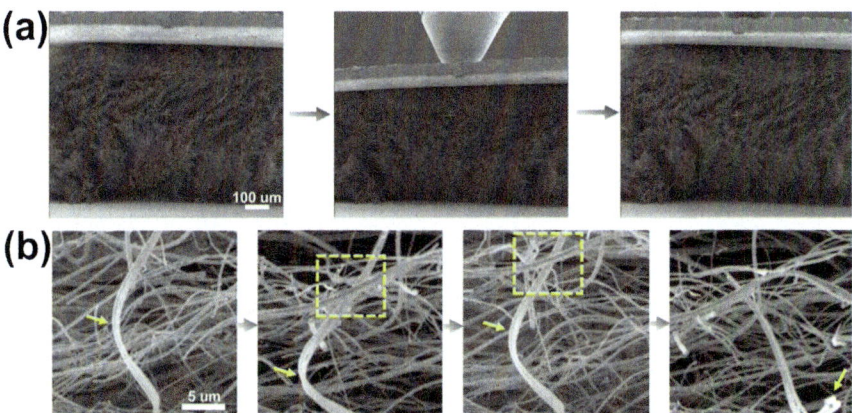

Fig. 5.27 In-situ SEM compression of a ceramic sponge showing: (**a**) ability of the sponge to revert to its original state after unloading (from 30% strain state), and (**b**) high-magnification view of bending, inter-tube sliding and recoiling shown by individual nanofibers constituting the foam. (Reprinted with permission of AAAS from Wang et al. (2017). © The Authors, some rights reserved; exclusive licensee American Association for the Advancement of Science. Distributed under a Creative Commons Attribution NonCommercial License 4.0 (CC BY-NC))

structure reverted back to its original shape after the compressive loads (~30% strain) are removed (Fig. 5.27a). Real-time imaging during the loading-unloading cycle showed the activation of mechanisms, such as narrowing and recovery of sponge pores, relative motion between the nanofibers in contact, and bending and springback exhibited by the nanofibers (high-magnification real-time SEM snapshots shown in Fig. 5.27b). As compressive strains were increased (up to 40–50%), irreversible deformation of the constituent nanofibers was observed. Densification of the foam structure at 50% strain resulted in a steep increase in the measured stress values. This example highlights the importance of 3D architectures because ceramics are typically prone to brittle failure. However, the porous, spongy structure allows for effective load redistribution and prevent local failure.

Section 5.1 delved into the mechanics of 2D nanomaterials, such as graphene. 3D architectures of graphene have gained popularity because of superior properties, ease of engineering composite materials, easy handling and multifunctional applications (Chen et al. 2011; Nardecchia et al. 2013; Shehzad et al. 2016; Bustillos et al. 2018; Idowu et al. 2018). In order to engineer free-standing forms of nanomaterials, it is vital to understand their mechanical stability. The case study in Fig. 5.28 demonstrates the in-situ mechanical investigation of a free-standing 3D graphene foam (Nieto et al. 2015a). The foam under tension was seen to fracture along the 45° plane (Fig. 5.28a). In-situ SEM imaging revealed the graphene foam branches tend to rotate and align along the loading axis, as shown in Fig. 5.28b. Readers are referred to the Supplementary Video 5.6, showing real-time SEM imaging of realignment of foam walls along the loading axis. The branches and the nodes exhibited different rates of rotation: while the nodes aligned rapidly with a rotation rate of ~3.08°/s, the alignment of the branches was much slower ~0.58°/s. Real-time imaging revealed necking exhibited by the aligned branches. As the branches were further strained, they started failing. These local branch failure events manifest as dips in the stress response shown in Fig. 5.28a. The real-time imaging of necking and fracture can be seen in the Supplementary Video 5.7.

The 3D architectures of graphene are gaining popularity to engineer composite metamaterials. An example of the in-situ investigation of a 3D graphene-ceramic metamaterial is demonstrated in Fig. 5.29 (Zhang et al. 2017b). The honeycomb structured graphene-Al_2O_3 metamaterial was subjected to compression cycles. The metamaterial demonstrated reversible compressibility up to 80% strains (Fig. 5.29a). This remarkable elasticity is rare for a ceramic composite. Real-time snapshots from in-situ SEM compression of the metamaterial are shown in Fig. 5.29b. For relatively lower strains (up to ~20%), elastic compression of the microcell was observed. Further increase in compressive strain induced out-of-plane bending of the cellular wall. Severe folding of the wall was seen for strains exceeding 50%, leading to elastic buckling. Coupling of graphene template and ceramic deposit produces a robust structure, which resists local stress concentration and premature failure. Superior load-transfer ability is at the core of ductile deformation behavior seen in this material system. High-resolution imaging provides the microscopic understanding of load-bearing by advanced composite metamaterials.

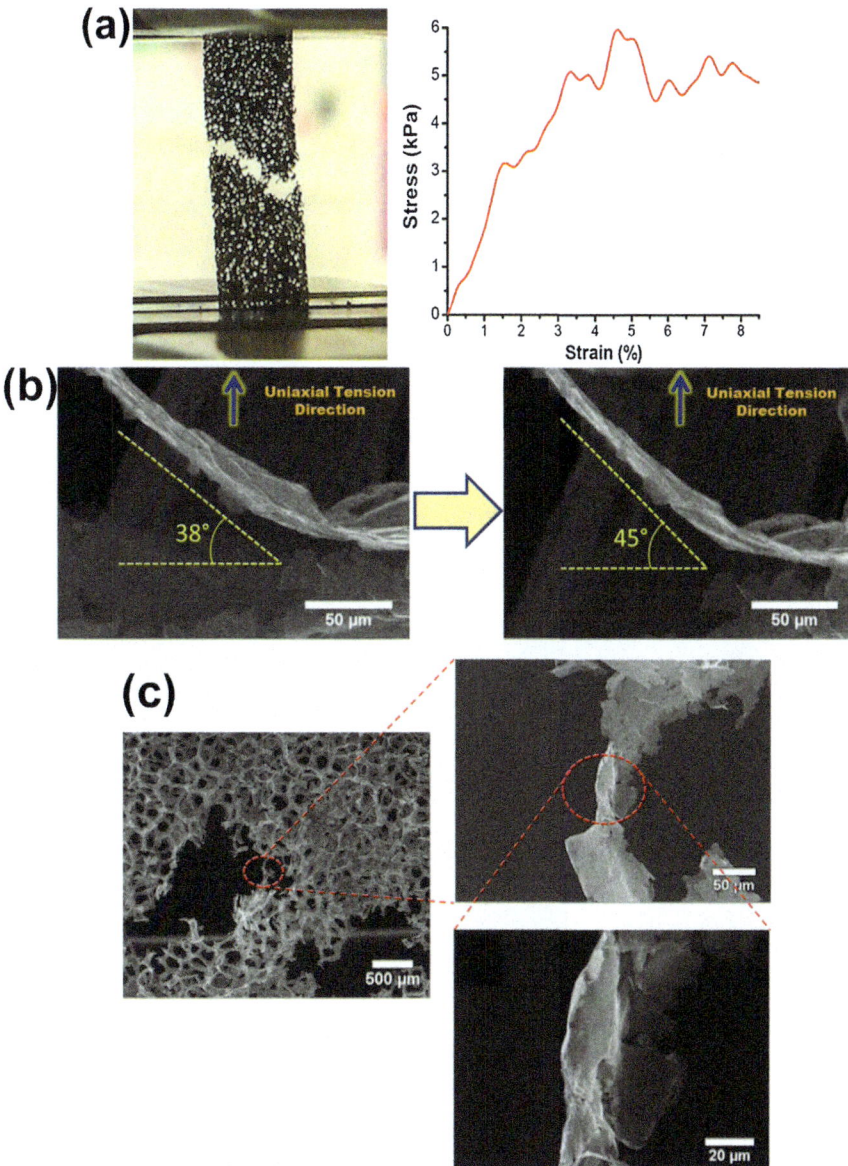

Fig. 5.28 In-situ testing of a 3D graphene foam: (**a**) stress–strain curve and the optical image captured during failure shows fracture along the 45°plane, (**b**) SEM imaging shows rotation and realignment of graphene foam walls, and (**c**) multi-scale real-time SEM images showing necking and bridging by strained foam branches. (Reproduced/adapted with permission from Nieto et al. (2015a))

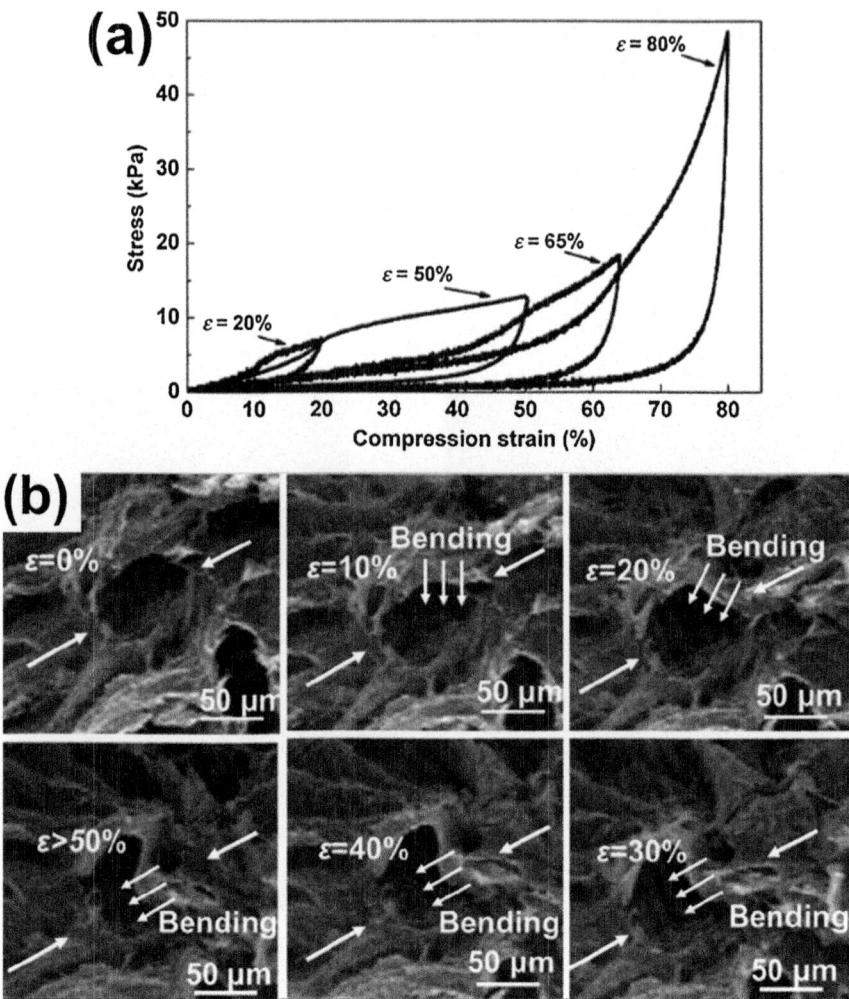

Fig. 5.29 Compression of a 3D graphene-ceramic metamaterial: (**a**) reversible deformation behavior demonstrated by the metamaterial for compressive strains up to 80%, and (**b**) real-time SEM snapshots of the microcell undergoing compression, bending, and folding at different strains. (Reproduced with permission from Zhang et al. (2017b))

The case studies discussed above pertain to the mechanical investigation of stochastic foams or 3D structures. Ordered cellular architectures, with well-defined geometries display superior mechanical properties, such as stiffness and yield strength. The predictable mechanical deformation of metamaterials can be exploited for applications in reconfigurable structures, auxetic stent, flexible electronics, robotics, smart actuators, and impact energy absorption to list a few (Wu et al. 2019). In-situ investigations are highly informative to study the effect of changing

5.3 In-Situ Characterization of 3D Architectures and Metamaterials

different geometric parameters on the load-bearing capability of the structures. Figure 5.30 demonstrates in-situ compression of octahedron metallic nanolattices in SEM (Montemayor and Greer 2015). The lattices consisted of *unit cells* about 5–20 μm size and hollow tubes with *wall thickness* ranging from 200–635 nm. The effect of changing the unit cell angle on the stress-strain response of the lattice is shown in Fig. 5.30a. Increasing the angle from 30° to 60° caused the yield stress to go up from 161.3 kPa to 858.2 kPa. An order of magnitude jump in stiffness was observed, from 8.42 MPa for 30° unit cell to 85.2 MPa for 60° unit cell. It was also observed that the onset of yielding happened at relatively lower strain for the 60° unit cell angle. The vertical members bear more load for 60° samples, whereas the horizontal elements of the lattice contribute more toward load-bearing for 30° specimens. In addition to the unit cell angle, the thickness of the hollow walls also influenced the mechanical response of the nanolattices. Figure 5.30b shows comparative stress–strain curves and in-situ SEM images for 45° lattices with three different wall thicknesses: 200, 327, and 635 nm. About ~120% enhancement in yield stress was noticed when wall thickness was increased from 200 to 327 nm. Real-time SEM imaging revealed deformation initiated at the base of the structures. The lattices are expected to demonstrate either Euler buckling or yielding, as the tubes experience compressive loads. The critical stresses for the onset of the two mechanisms can be obtained by the following expressions:

$$\sigma_{buckling} = 8\pi^3 E \cos^2\theta \sin\theta \left(\frac{D_{min}}{l}\right)^2 \left(\frac{D_{min} + 3D_{max}}{l}\right)\left(\frac{t}{l}\right) \quad (5.21)$$

$$\sigma_{yield} = 4\pi \sin\theta \left(\frac{t}{l}\right)\left(\frac{D_{min} + D_{max}}{l}\right)\sigma_{y(Au)} \quad (5.22)$$

where D_{max} is the major axis, D_{min} is the minor axis, l is the tube length, t is the thickness, and θ is the angle. These parameters are marked in the drawing shown in Fig. 5.30c. The stresses required to initiate buckling were calculated to be extremely high, in the range of ~2.4 GPa for a 60° lattice to ~16.4 GPa for a 45° lattice. The threshold stresses for yielding were an order of magnitude lower: ~164–290 MPa. Therefore, all the nanolattice tubes should fail by yielding, which is in confirmation with the SEM imaging (shown in Fig. 5.30a, b). Yielding was seen to initiate at the hollow nodes, because of local stress concentration. This case study highlights the importance of the in-situ characterization approach for assessing the validity of mechanics theories for metamaterials.

The superior mechanical performance of architected lattices makes them suitable candidates for engineering nanocomposites. Figure 5.31 demonstrates a case study where the in-situ approach was used to assess the deformation characteristics of a polymer-high entropy alloy composite (Zhang et al. 2018a). The structure was composed of polymer struts, which were coated with alloy film (with thickness varying from ~14 to 126 nm). The lattices, with unit cell sizes varying from

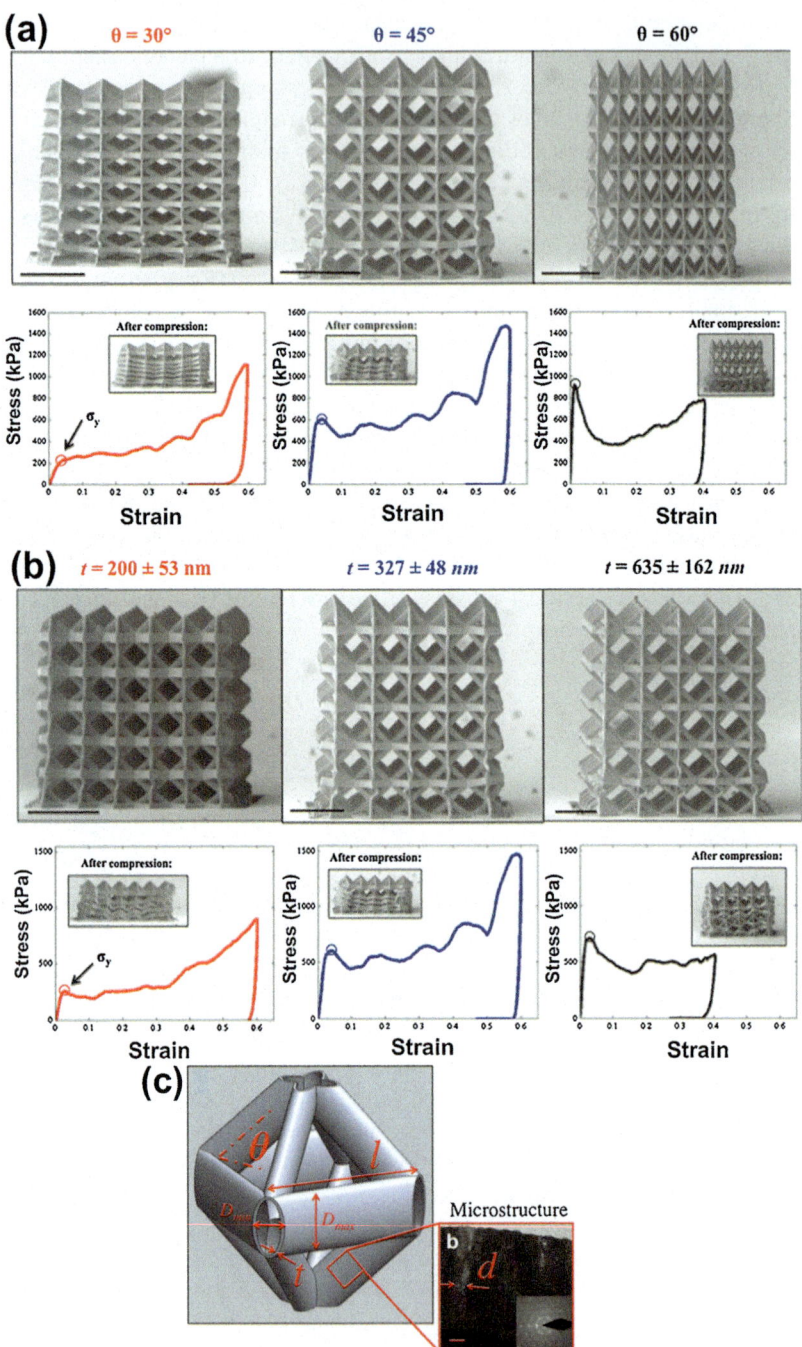

Fig. 5.30 In-situ SEM compression of octahedron nanolattices: (**a**) comparison of deformation behavior of lattices with angles varying from 30° to 60°, (**b**) comparison of deformation behavior of lattices with tube wall thickness varying from 200 to 635 nm, and (**c**) unit cell schematic showing key dimensions and the TEM micrograph of the metallic wall with grain size (d) marked. (Reproduced with permission from Montemayor and Greer (2015) The American Society of Mechanical Engineers)

5.3 In-Situ Characterization of 3D Architectures and Metamaterials

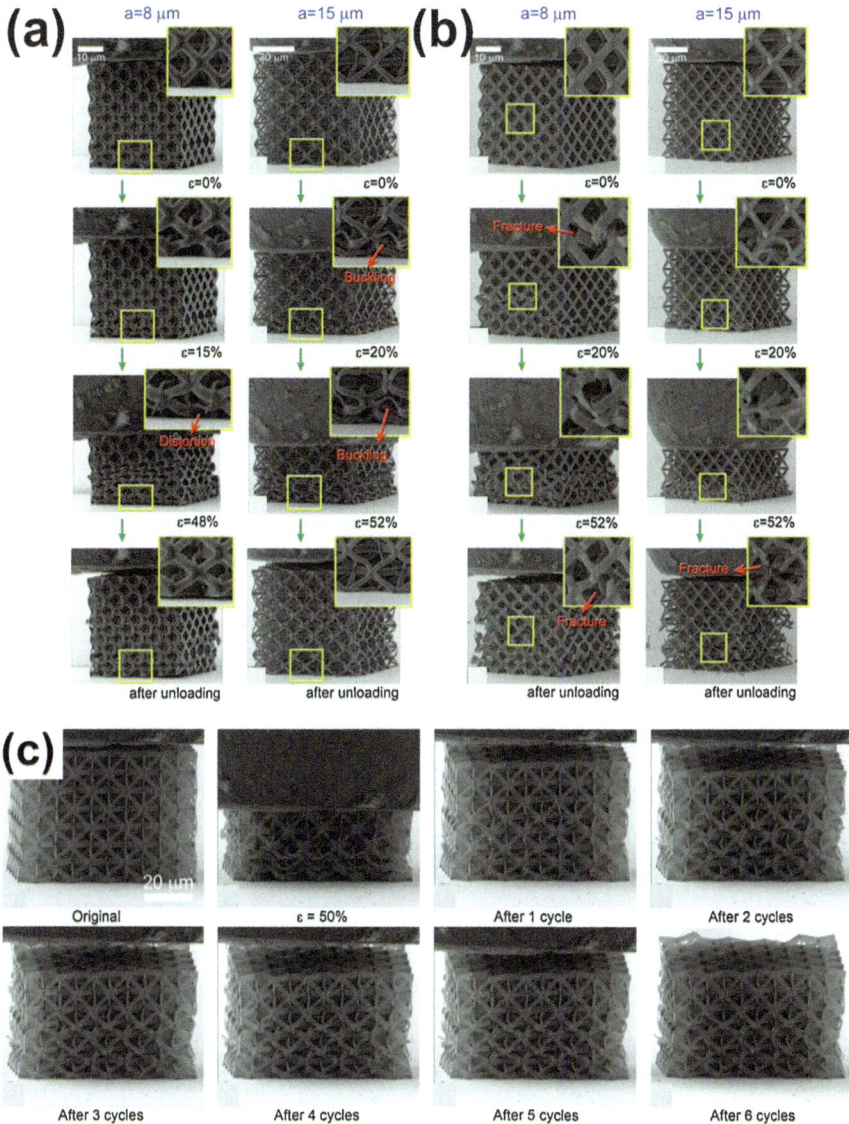

Fig. 5.31 In-situ SEM compression of polymer-high entropy alloy composite lattice for: (**a**) lattice with thinner alloy coating ($t = 14.2$ nm), and (**b**) thicker coatings ($t = 94.3$ nm). (**c**) Real-time imaging of the composite lattice subjected to cyclic compression (up to 6 cycles). (Reproduced with permission from Zhang et al. (2018a))

~8–15 µm were investigated. In-situ imaging revealed that buckling propagated from the bottom of the lattice to the top. Buckling resulted in extensive rotation of the struts. The thinner struts ($t = 14.2$ nm) show near-complete recovery after being subjected to strains exceeding 50% (Fig. 5.31a). However, thicker struts ($t = 94.3$ nm) experienced fracture, causing irreversible damage (Fig. 5.31b). A combination of fracture and buckling mechanisms was observed for the lattices with thicker struts. In-situ testing is useful to evaluate the long term structural stability of the 3D structures. 93% recovery was observed after the first compression cycle up to 50% strain (Fig. 5.31c). After multiple loading cycles, damage accumulation resulted in enhanced residual strain and the recoverable lattice height dropped to ~85% after six cycles. The real-time snapshots in Fig. 5.31c show excellent damage tolerance of the structure over six loading-unloading-reloading cycles, as the structure did not fail catastrophically. These in-situ observations were helpful to determine optimal alloy coating thickness for achieving desired mechanical properties, such as specific modulus, specific strength, and energy absorption per unit volume. Alloy coating thickness in the range of 14–50 nm was found to be desirable, and higher thicknesses resulted in brittle behavior. Therefore, real-time imaging during mechanical testing can provide critical insights for engineering composite lattices with intended mechanical characteristics.

In-situ characterization of 3D architectures can be performed at elevated temperatures as well, to understand load-bearing capability as a function of temperature. The case study shown in Fig. 5.32a illustrates high-temperature compression of 3D CNT structures in SEM at temperatures ranging from 25 to 750 °C (Bhowmick et al. 2018). The CNT structure in this study had evenly distributed micro-porosities. The peak stress value was seen to decrease with increasing temperatures. The slope of the initial linear region also decreased at elevated temperatures, indicating a drop in elastic modulus. The modulus of the 3D structure dropped from 640 MPa at room temperature to 150 MPa at 750 °C. The effective elastic modulus of a parallel array of nanotubes can be modeled by the relationship:

$$E' = \frac{E_{0\varphi}}{1+\tan^2(\theta)\lambda'^2/3}\left(1-\frac{T}{T^*}\right)^\alpha \tag{5.23}$$

where E_0 is the intrinsic modulus of CNT, φ is the nominal density factor, θ is the average inclination angle, T^* and α are the critical parameters, and λ' is expressed as:

$$\lambda' = \lambda_1(1-i) + i\lambda_g \tag{5.24}$$

where λ_1 and λ_g are local and global slenderness., and i is the interaction factor. Plateau or peak strength is modeled as:

$$\sigma_p = \frac{\pi\left(1+\tan^2(\theta)\lambda'^2/3\right)}{4\lambda'^2}E' \tag{5.25}$$

Based on the above equations, the coefficients T^*, α, θ and i can be determined by fitting together with the experimental data. Once these parameters are computed, the modulus can be predicted at any temperature using Eq. 5.23. Therefore, in-situ

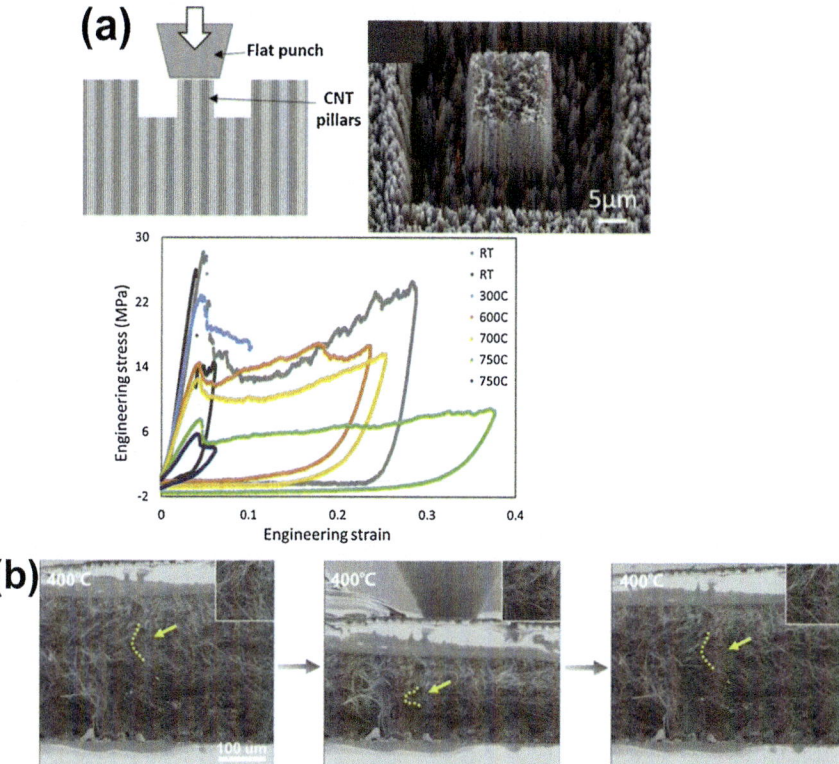

Fig. 5.32 High-temperature in-situ investigation of 3D architectures: (**a**) compression of a 3D CNT structure consisting of parallel nanotube arrays, and (**b**) high-temperature compression of ceramic nanofiber sponge. (Reproduced with permission from Bhowmick et al. (2018), Wang et al. (2017) CC BY-NC 4.0)

investigations of nanostructured architectures can be advantageous to develop models for predicting mechanical properties.

In-situ investigations as a function of temperature are vital for high-temperature engineering applications of 3D architectures. A case study on real-time imaging of compression of ceramic nanofibers sponge at 400 °C is shown in Fig. 5.32b (Wang et al. 2017). The TiO_2 sponge was compressed inside SEM for multiple loading-unloading-reloading cycles. Real-time imaging revealed the stability of the sponge, as the compressive strain of 23% induced a low ~5% residual strain after 10 cycles. The nanofibers constituting the foam were seen to experience bending upon mechanical loading and springback after unloading. The stability of nanofibers at elevated temperatures makes the ceramic sponges promising for high-temperature applications. The sponge demonstrated energy dissipation ability (~1.52 mJ/cm^3 in the first cycle), which is close to its room temperature dissipation capability. This case study highlights the suitability of the in-situ approach to probe the mechanics of 3D architectures across the length and temperature scales.

5.4 In-Situ Mechanics of Smart Materials

Smart materials or intelligent materials are responsive to external stimuli. One such material class is shape memory material system, which can mechanically respond to temperature changes. Despite mechanical deformation, these materials can recover or revert to their original shape by exposure to transition temperature. This behavior opens doors to multiple applications, such as sensing, actuation, device construction, foldable structures, self-healing, or damage recovery. In-situ deformation, along with high-temperature exposure can provide a mechanistic understanding of transformation phenomena. Figure 5.33 demonstrates the application micropillar compression technique at variable temperatures to study high-temperature shape memory effect in a ceramic (Zeng et al. 2017). These $2Y_2O_3$-$5TiO_2$-ZrO_2 specimens had the austenite start temperature of 476 °C, below which compression loading induces shape change because of tetragonal to monoclinic transformation. The austenite finish temperature of the specimen was 510 °C, and heating above this temperature enables shape recovery. Therefore, in-situ compression at 400 °C was accompanied with a prominent displacement plateau in the load–displacement curve (Fig. 5.33). Additionally, the formation of striped patterns was observed via SEM imaging on the pillar wall. This observation is attributed to the de-twinning or rotation of monoclinic variants (Zeng et al. 2017). Complete shape recovery and

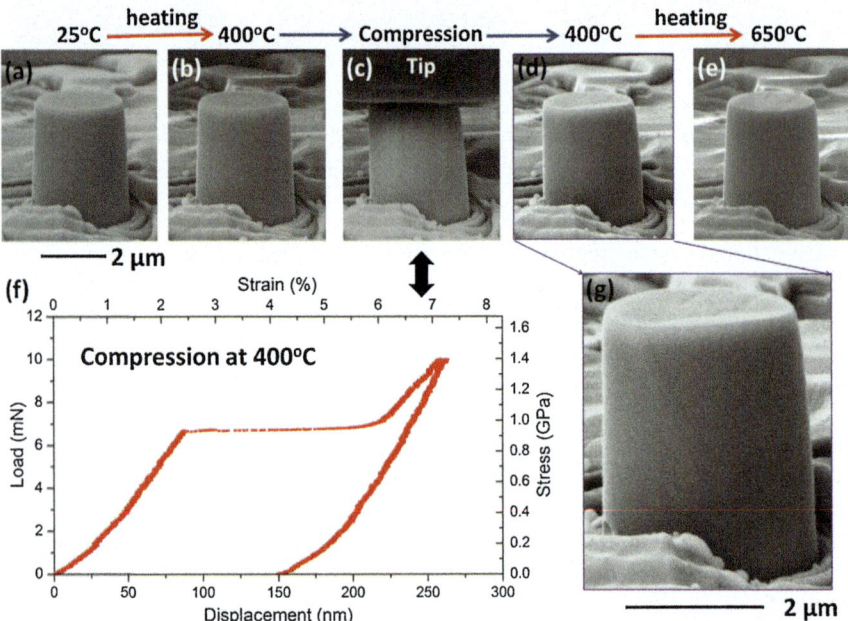

Fig. 5.33 In-situ SEM investigation of shape memory effect in zirconia investigated by high-temperature micropillar compression technique. (Reproduced with permission from Zeng et al. (2017))

disappearance of striped patterns were observed by real-time imaging after exposing the compressed pillar to 650 °C.

In-situ deformation in TEM provides insights into stress-induced phase transformations in shape memory alloys. A case study on in-situ TEM indentation of single-crystalline $Ni_{54}Fe_{19}Ga_{27}$ is shown in Fig. 5.34 (Liu et al. 2014). A stress plateau was seen at 180 MPa. Real-time imaging in diffraction mode revealed the emergence of 10 M martensite at 43 s, followed by 14 M at 70 s (Fig. 5.34a). During unloading, 10 M/14 M reverted to 10 M at 135 s, which disappeared after further unloading (at 178 s). Bright field TEM imaging during indentation at different locations in the shape memory alloy revealed first and second types of martensitic transformation: (a) $L2_1$-to-10 M/14 M at relatively lower stresses, and (b) $L2_1$-to-$L1_0$ transformation at much higher stresses (Fig. 5.34b). During first transformation, a dark ring band was formed, and the 10 M phase fronts were seen to propagate in the direction away from the tip (toward the lower left). During second type of martensitic transformation, stress contours emanated from the lower right side of the indenter tip. Further indentation loading triggered the emission of multiple groups of $L1_0$. During the unloading cycle, the reverse martensitic transformation was observed via TEM imaging, characterized by the disappearance of dark bands and de-twinning. 10 M was seen to revert to $L2_1$ austenite, albeit with some residual twins. This example

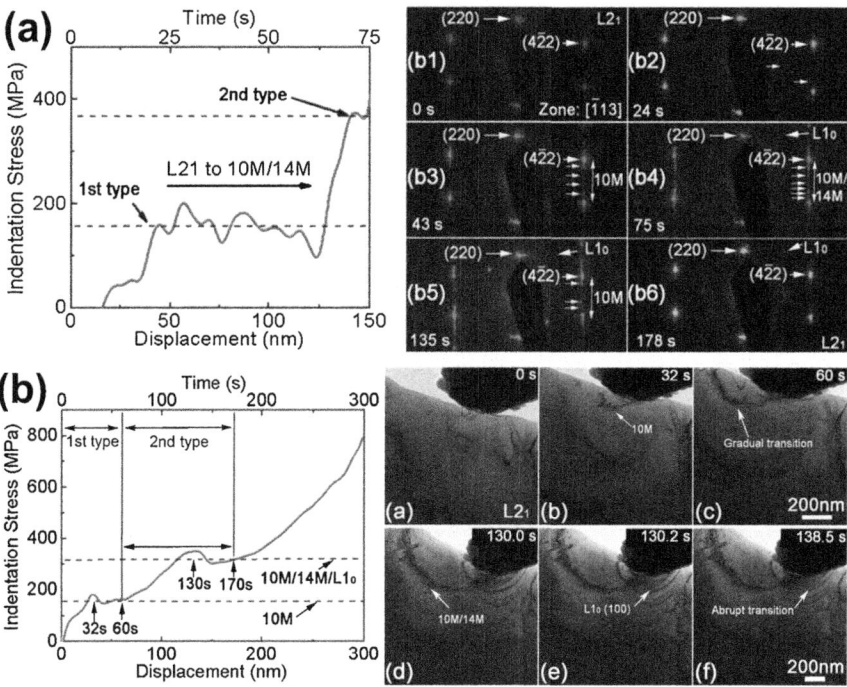

Fig. 5.34 In-situ TEM investigation of phase transformations in shape memory magnetic alloy, captured in: (**a**) diffraction mode, and (**b**) bright field imaging mode, with corresponding stress-displacement plots. (Reproduced with permission from Liu et al. (2014))

illustrates the significance of high-resolution real-time imaging to develop a fundamental mechanistic understanding of phase transformations in smart materials. These insights can be beneficial to engineer materials with tailorable transformation characteristics.

Coupling mechanical deformation with electrical measurements during in-situ tests can be useful to probe piezoelectric materials. Piezoelectric materials display interdependence between mechanical stresses and electrical response. Figure 5.35 illustrates a case study, where a 3D-architected ZnO was subjected to mechanical compression and the corresponding open-circuit voltage response was measured (Yee et al. 2019). ZnO is a piezoelectric material. It was seen that the application of ~200 nm displacement on ZnO tetrakaidecahedron resulted in a ~0.52 mV voltage drop. This experimentation confirmed the electromechanical coupling in the 3D piezoelectric architecture. Therefore, the in-situ measurement approach is instrumental in establishing the correlation between structure-mechanics-electrical characteristics. These in-situ investigations can be beneficial for engineering "smart material systems" for energy harvesting and device applications. Due to ultrafine feature sizes (as seen in the SEM micrographs in Fig. 5.35), it is rather difficult to perform such investigations using conventional ex-situ approaches and real-time imaging is highly desirable.

5.5 In-Situ Mechanics of Biological Materials

In-situ investigations play an important role in deciphering the mechanics of biological materials because of their complex, hierarchical structure. The resultant structure–property correlations are helpful in understanding biological phenomena in nature, aid in medical diagnosis, inform therapeutic strategies or provide inspiration to engineer bioinspired/biomimetic materials and structures for multifarious applications. Conventional, bulk-scale ex-situ mechanical testing approach does not capture the deformation characteristics of different constituents or the role of

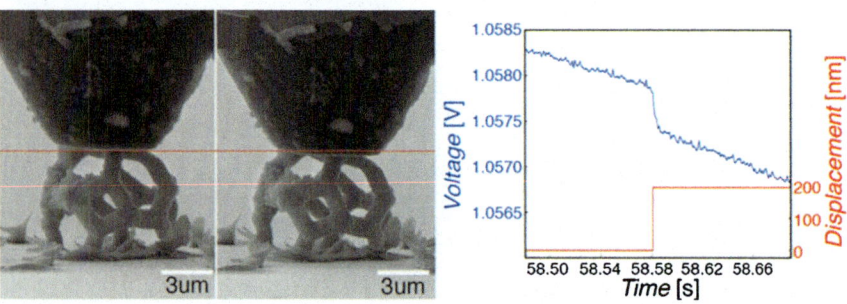

Fig. 5.35 Investigation of electro-mechanical coupling in ZnO tetrakaidecahedron architecture by performing open-circuit voltage measurement during in-situ compression. (Reproduced with permission from Yee et al. (2019))

5.5 In-Situ Mechanics of Biological Materials

interfaces. However, there are several challenges associated with the in-situ testing of biological samples. Often, the samples are time and environment sensitive. For instance, the mechanical properties of biological tissues are influenced by the presence or absence of liquid media. Drying of tissue samples can cause undesirable hardening and stiffening (Ebenstein and Pruitt 2006). Some biological specimens are sensitive to the electron beam, and prolonged exposure may cause sample degradation. Sample preparation for mechanical testing can also be tricky. Many materials are difficult to machine, and precautions need to be taken to avoid damaging the samples during the preparation step. Biological samples might also require customized fixtures for performing mechanical tests.

A case study on micropillar compression of lamellar bone is shown in Fig. 5.36 (Schwiedrzik et al. 2014). The micropillars machined out of dry ovine osteonal bone were subjected to uniaxial loading along with real-time SEM imaging. Nanoindentation, a common approach to extract mechanical properties of bones,

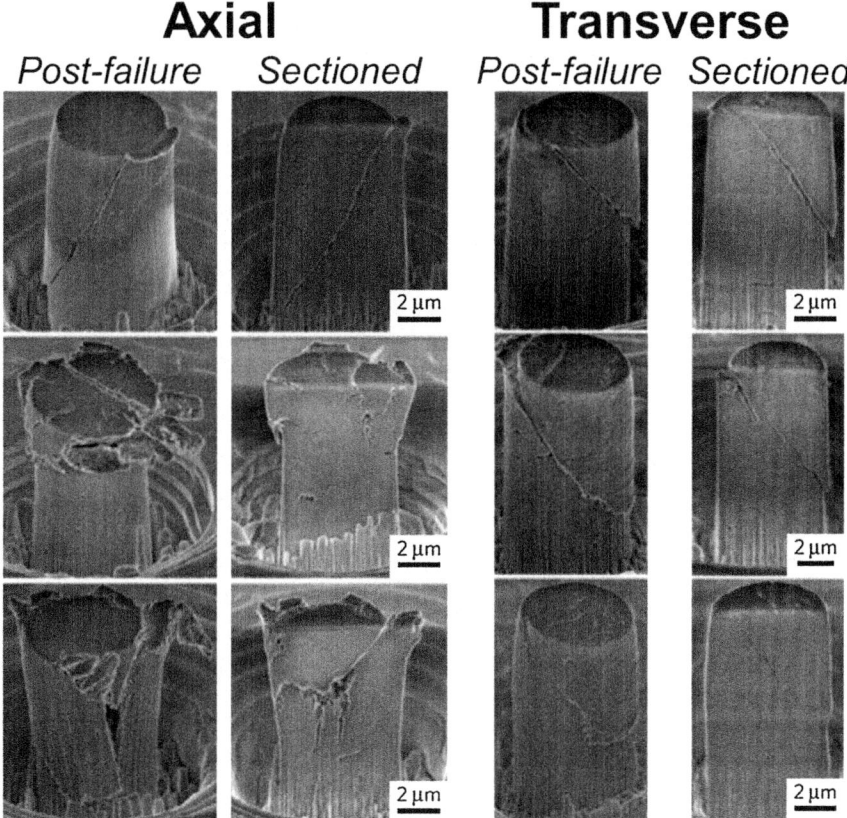

Fig. 5.36 SEM images of the bone micropillars and FIB cross sections showing deformation/ failure mechanisms activated for axial and transverse orientation loading. (Reproduced/adapted with permission from Schwiedrzik et al. (2014))

often ignores the effect of defects because of small sample sizes and interaction volumes. This results in an overestimation of stresses and other mechanical properties. The advantage of a micropillar compression is that it captures the effect of the defects and mechanisms arising due to the interactions between multiple constituents in the microstructure. Real-time SEM imaging revealed the activation of shearing, mushrooming and axial splitting mechanisms. For axial orientation, it was observed that 55% of the pillars failed by the development of shear planes, 25% demonstrated failure at the top of the pillar causing local cracking/delamination, while only 15% of the tested pillars exhibited axial splitting. On the other hand, compression along the transverse orientations was characterized by only shearing failure. These observations are shown in Fig. 5.36. The SEM imaging revealed isolated fibril bridging, ligament bridging, and crack deflection were the key toughening mechanisms. In addition to the observed mechanisms, the anisotropy in the mechanics of bone was also evident from the modulus values. The micropillars compressed in the axial direction were characterized by relatively higher modulus (~31.16 GPa), as compared to the transverse direction (~16.5 GPa). The post yielding mechanical response depended on the kind of mechanism activated during compression: activation of mushrooming and axial splitting induced strain hardening and softening, whereas the pillars that exhibited shearing displayed uninterrupted hardening until the failure point. In-situ investigations provide important information about the micromechanical behavior of bone, with implications in medical treatment and care.

Another related area where the mechanics of bone is important is bone-cutting tool interaction, because of relevance for orthopedic surgery and bone implants. The penetration of a sharp tool or a foreign implant on the bone may induce cracks and fracture. In-situ investigation of penetration deformation of bone can provide important mechanistic insights, which can, in turn, inform clinicians about safe practices and enable engineers to design implants that do not cause unintended damage to the bone. Figure 5.37 demonstrates a case study on the in-situ penetrative cutting of cortical bone with real-time micro-lens high-speed camera imaging (Li et al. 2014). The penetration investigations were performed parallel and perpendicular to the bone axis. Penetration along the longitudinal axis was responsible for material separation and crack propagation along the osteon direction (Fig. 5.37a). The deformation was brittle in nature, with a rather low fracture resistance. Contrary to this, penetration perpendicular to osteons demanded enhanced forces. This is because of higher stiffness and fracture toughness along the bone's longitudinal direction (as compared to circumferential and radial directions). The tool induced damage was brittle at the plexiform bone region (Fig. 5.37b). Fragmentation and peeling off of material were prominently observed. However, the osteonal structure was characterized by ductile damage (Fig. 5.37c). In-situ imaging provides insights into the local underlying microstructure-deformation correlation for bone, which can be helpful to develop advanced models, enhance the understanding of tool–bone interactions and facilitate designing of surgical tools.

Some of the lightweight and high-strength biological materials can offer cues to engineer materials with superior mechanical properties. In-situ investigations are

5.5 In-Situ Mechanics of Biological Materials

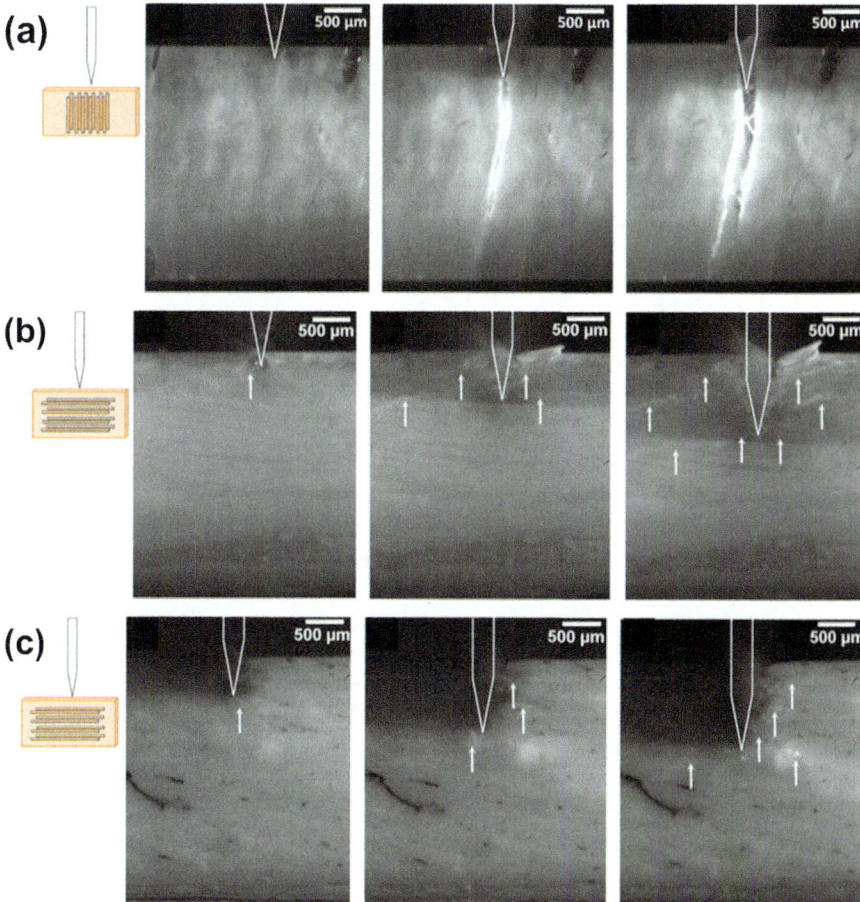

Fig. 5.37 In-situ high-speed camera videos showing cutting induced damage for: (**a**) cutting parallel to osteons, (**b**) cutting perpendicular to osteons at the plexiform bone region, and (**c**) ductile damage of the osteonal structure. (Reproduced from Li et al. (2014) CC BY 3.0)

useful to discern the load-transfer mechanisms in such materials. Figure 5.38 shows an example of mechanical testing of dragonfly wing with real-time imaging. The morphology, structure, and composition of insect wings are known to provide crack arrest ability along with resistance to airflow and alternating load application (Zhang et al. 2018b). Miniature samples can be prepared and tested from different regions of the wing. High frame rate imaging is required for imaging cracking behavior of the wings. Figure 5.38 shows real-time high-speed camera snapshots as the wing was subjected to tensile loading. In-situ imaging captured crack initiation, crack propagation and eventual fracture of the miniature wing specimen. The wing consisted of rigid veins and a flexible membrane. It was observed that the interfaces between the two were sites of stress concentration, creating small cracks. Further loading resulted in the appearance of cracks at the edges of the wing.

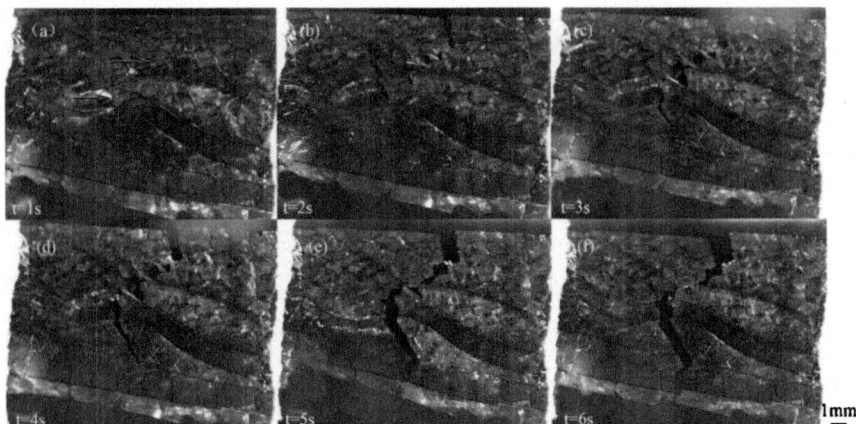

Fig. 5.38 Real-time imaging of crack initiation and propagation in dragonfly wing captured by a high-speed camera. (Reproduced with permission from Zhang et al. (2018b))

Electron microscope imaging is infeasible for soft samples, like live cells or tissues, because of the presence of liquid media. Inverted optical microscopy (OM) is better suited to image and test live biological samples. Figure 5.39a shows an in-situ experimental setup, where an indentation transducer is mounted on an inverted OM. Unlike conventional nanoindenters, the biological indenter shown in the figure is capable of applying larger displacements (>100 μm) while being able to resolve low loads (resolution <1 μN). This is useful for probing ultrasoft samples, which tend to deform significantly with minimal forces. OM imaging allows the identification of the regions of interest in the sample. An example of a cluster of cardiomyocytes is shown in the figure. Once the region is identified, the microscope arm on which the indenter is mounted can be lowered and the tip is made to approach the sample. The inset in Fig. 5.39b shows the real-time optical image of the approaching indenter tip, very close to the sample surface. The load–displacement response captured from a cluster of cardiomyocytes is shown in Fig. 5.39b. The tip used in this case study was 50 μm in diameter, which is 1–2 orders of magnitude larger than the typical nanoindentation probes (~100–1000 nm). Larger sized probes prevent excess stress concentration, making them suitable for testing soft samples. Typically, fluid cell probes with longer shafts (>1 mm) are used for indenting live samples. Longer tip shaft is required to be able to approach the samples submerged in a liquid medium and offset the effect of meniscus forces. An important feature in the force curve is the initial approach segment before indentation loading begins (encircled in Fig. 5.39b). This controlled initial approach ensures that the sample is not significantly indented/deformed during the initial approach, which is a major concern with ultrasoft samples and can often result in overestimation of stiffness. Recording the force–displacement response during approach is also useful to measure and correct the effect of the buoyancy forces associated with the liquid medium. The elastic

5.5 In-Situ Mechanics of Biological Materials

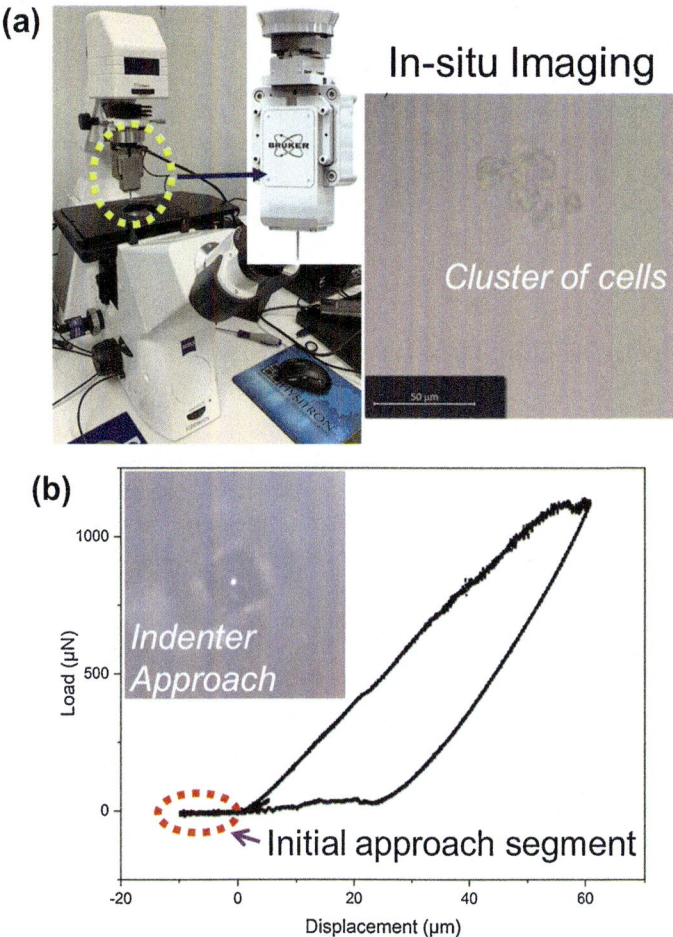

Fig. 5.39 (a) Inverted optical microscope based in-situ indenter (an OM image of a cluster of cardiomyocytes prior to indentation is also shown), and (b) load–displacement response captured from the cluster of cells (inset: OM image of the indenter tip approaching the sample surface). (Courtesy: Agarwal group (FIU), Darryl Dickerson (FIU) and CELL-MET NSF Engineering Research Center (Nanosystems ERC for Directed Multiscale Assembly of Cellular Metamaterials with Nanoscale Precision))

modulus was determined to be ~180 kPa by fitting the loading segment of the curve (Hertz model):

$$P = \frac{4E}{3(1-\upsilon^2)}\sqrt{R}\delta^{3/2} \tag{5.26}$$

where R is the tip radius, δ is tip displacement, P is indentation force, υ is Poisson's ratio of the sample, and E is the elastic modulus of the sample. If the samples are transparent/translucent, inverted OM can be used to record the deformation video. A key limitation of the inverted OM-based in-situ testing is the inability to image opaque and thick samples.

These examples illustrate the suitability of the in-situ approach to develop mechanistic insights into the deformation of biological materials. Biological samples are trickier to test than conventional engineering materials. Advancements in instrumentation to make mechanical testers compatible with biological microscopes will lead to wider acceptance of the in-situ measurement approach.

5.6 In-Situ Mechanics of Irradiated Materials

Mechanical properties of materials are influenced by radiation exposure. Understanding the mechanics of irradiated materials is critical for applications, such as nuclear energy, where materials should demonstrate structural integrity and mechanical stability in radiation environments. In-situ mechanical investigations at multiple length scales can be insightful to decipher how radiation exposure influences the mechanical response of materials. Figure 5.40 demonstrates a case study on in-situ micromechanics of a Pile Grade-A (PGA) nuclear graphite containing elongated filler particles (Liu and Flewitt 2017). The PGA graphite was exposed to neutron radiation and CO_2 radiolytic oxidation. Figure 5.40a compares electron micrographs of PGA before and after irradiation (in matrix and filler regions). Circular pores were seen in the matrix after irradiation. The filler regions, on the other hand, showed pore reduction post-irradiation. This difference in microstructure transformation resulted in different mechanical responses from the matrix and the filler regions, observed via in-situ SEM tests. Two PGA matrix cantilever samples demonstrated brittle fracture and lower fracture strengths of 154 and 357 MPa (Fig. 5.40b). This is much lower than the fracture strength of unirradiated PGA (measured to be 264 and 335 MPa for two cantilever specimens). Contrary to this, the cantilevers tested in the filler regions of the irradiated sample demonstrated much higher fracture strengths (exceeding 652 GPa). This strength is higher than the unirradiated sample, which is ascribed to neutron irradiation hardening. The crack opening was observed via in-situ imaging, as shown in Fig. 5.40c. This case study highlights the importance of the in-situ mechanics approach to evaluate the effect of neutron irradiation on mechanical properties of specific microstructure features/components.

The application of in-situ TEM investigation has also been demonstrated for examining the effect of ion beam irradiation on mechanical deformation. Figure 5.41 shows a case study on nanocompression of Cu single crystal irradiated by a proton beam (Kiener et al. 2011). The irradiated pillar was characterized by multiple defects, shown in Fig. 5.41a. Pillar compression caused bowing and escape of short dislocation segments (shown in Fig. 5.41b–d). Mechanical loading up to 9.4% strain resulted in a localized slip step (Fig. 5.41f). This large slip step was seen only for irradiated Cu. The irradiated pillars also deviated from the unirradiated specimens

5.6 In-Situ Mechanics of Irradiated Materials

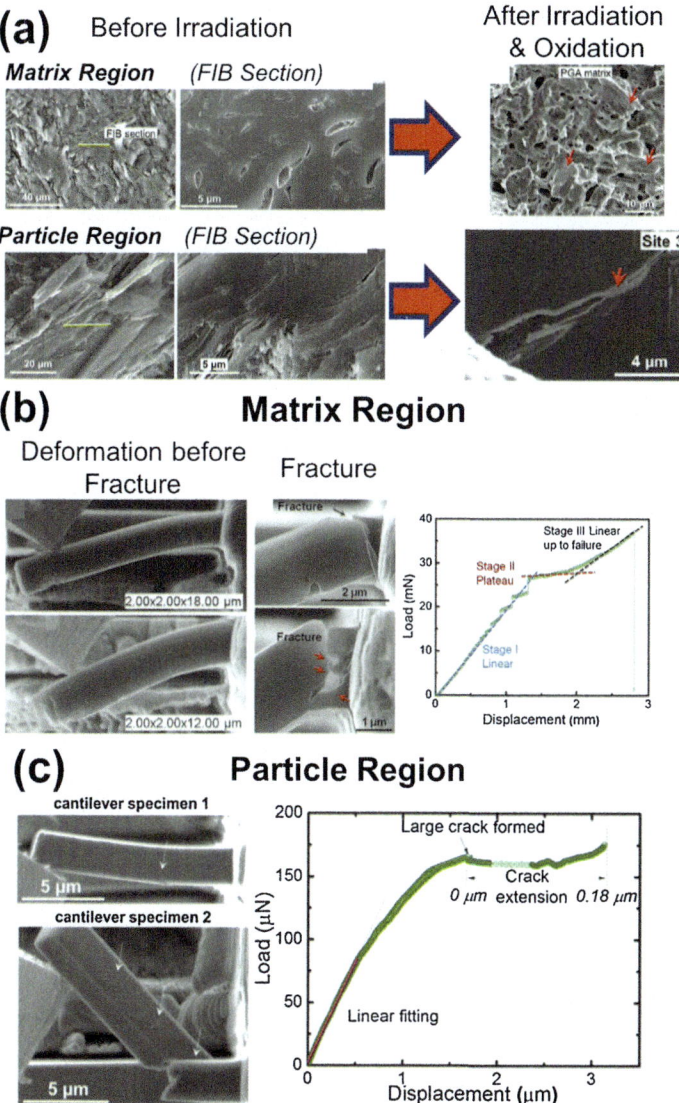

Fig. 5.40 (a) SEM micrographs comparing the effect of neutron irradiation and oxidation on the microstructure of PGA, (b) in-situ micro-cantilever deflection of irradiated PGA matrix, and (c) in-situ micro-cantilever deflection of irradiated particle-reinforcement region. (Reproduced/adapted with permission from Liu and Flewitt (2017))

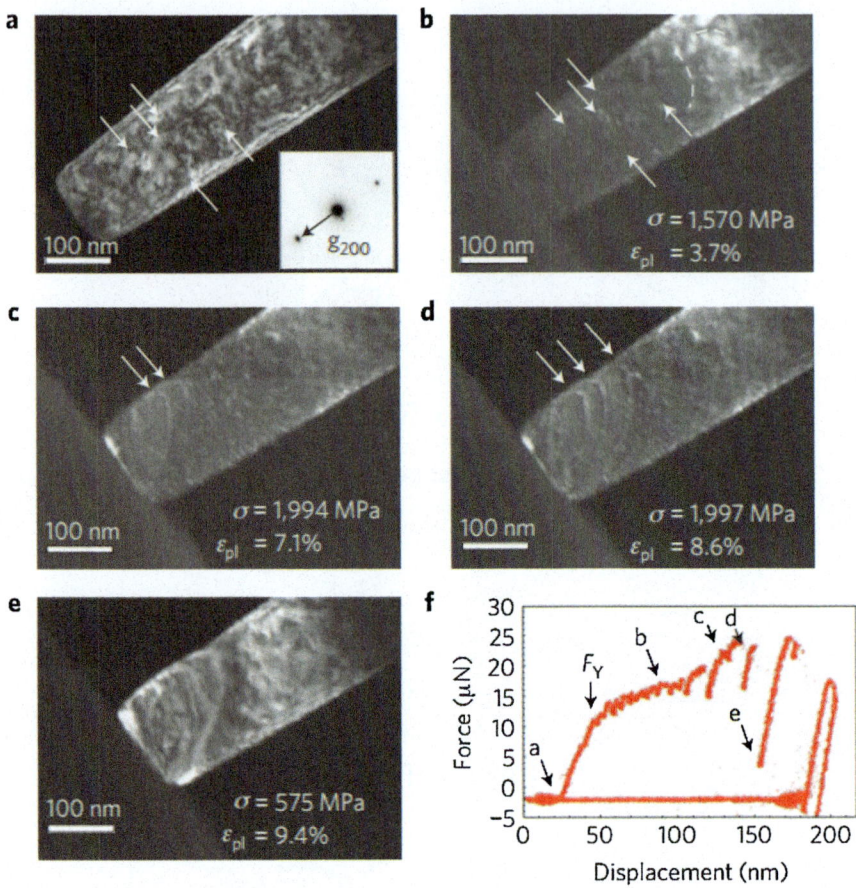

Fig. 5.41 Real-time TEM images and force–displacement plot for compression of an ion beam irradiated Cu pillar. (Reproduced with permission from Kiener et al. (2011))

in terms of the nature of size effect in yield stress: while the unirradiated pillars displayed size-dependent yield stress throughout the size range (100–1000 nm), the irradiated specimens displayed size effect only up to 400 nm due to dislocation source limitation. The size independence above 400 nm was ascribed to interactions between the moving dislocations and the irradiation defects in the pillar. These interactions determine the plasticity mechanisms in Cu.

These examples illustrate the application of in-situ imaging approach at multiple length scales to decipher the mechanics of irradiated materials. The fundamental understanding of deformation mechanisms and their correlation with the microstructure of materials exposed to radiations is important for engineering advanced radiation-resilient materials.

5.7 Summary

The application of in-situ measurement techniques for studying different material systems was presented in this chapter. High-resolution imaging is useful for probing the mechanical response of nanomaterials, because of their ultrafine dimensions. Different case studies on in-situ testing of 0D, 1D and 2D nanomaterials were discussed. Real-time imaging provides mechanistic insights into the fracture, dislocation activities, the role of inherent defects, effect of structural hierarchy, and the influence of tweaking dimensions on their mechanical response. Examples of different loading scenarios, such as tension, compression, buckling, fatigue, and cantilever deflection of nanomaterials were highlighted in the chapter. The application of the in-situ mechanics approaches to determine size effects on different mechanical properties was discussed. The case studies on real-time imaging to visualize strengthening mechanisms in different composites were presented in the chapter. Multi-scale imaging, in optical, scanning electron, and transmission electron microscopes provide information about deformation mechanisms activated at multiple length scales. The in-situ approach is well suited to decipher load-transfer/stress-distribution mechanisms in 3D architectures and mechanical metamaterials. Tweaking the geometry or dimensions of unit cells can dramatically alter the load-bearing capability of 3D architectures. Coupling in-situ mechanical measurements with heating or electrical measurements open the avenues to probe smart materials, such as shape memory and piezoelectric materials. These materials display coupling between mechanical-thermal or mechanical-electrical phenomena. Real-time imaging provides important insights into these correlations. The in-situ approach is promising to examine the mechanics of biological materials. Such investigations are useful to study anisotropy and heterogeneity of mechanical response. Understanding of deformation mechanisms is advantageous for engineering biomimetic materials, developing tools for orthopedic surgery, and designing implants with superior mechanical compatibility. Toward the end, the application of the in-situ approach for probing the mechanical response of materials exposed to radiations is discussed. Understanding the deformation characteristics of irradiated materials is useful for assessing their usability in radiation environments. These applications show the versatility of in-situ mechanical characterization and its relevance to a wide diversity of materials science problems.

Questions and Assignments

1. Which of the following in-situ imaging techniques is suitable for observing dislocation events in real time during the deformation of nanoparticles?

    (a) Optical imaging.
    (b) Scanning electron microscope.
    (c) Transmission electron microscope.
    (d) All of the above.

2. What are the benefits of probing local interfacial mechanical properties, such as interfacial shear strength, for the matrix/nanofiller interface? What are the qualitative and quantitative information one can derive from the in-situ assessment of the interfaces?

3. The examples shown in Figs. 5.18, 5.19 and 5.20 pertain to interfacial shear strength measurement of nanotube–matrix interfaces for metal, ceramic, and polymer matrices, respectively. Can you adopt the same AFM pullout approach to measure interfacial shear strength for a 2D nanofiller-reinforced composite? List the potential challenges, if any, related to pullout testing of a 2D nanofiller. Write the equations for computing the interfacial shear strength and tensile stresses for a 2D nanofiller. (Hint: Modify the equations for the nanotubes mentioned in Sect. 5.2. You may assume the 2D plane of the nanofiller to be a perfect rectangle with length, l and breadth, b. Assume the thickness of the nanofiller to be "t").

4. What is the advantage of having high-temperature in-situ mechanical testing capability for studying shape memory effect? Frame your response by taking into consideration the following three factors:

   (a) Importance of temperature stimulus.
   (b) Role of mechanical stresses to induce transformations.
   (c) Information acquired by real-time imaging.

   You may use a published scientific article from literature (other than the case studies discussed in this chapter) to elaborate your response.

5. What are the challenges associated with in-situ mechanical testing of biological materials? Explain and elaborate with relevant real-world examples. Consider the following aspects for framing your response:

   (a) Environment sensitivity.
   (b) The complexity of sample preparation.
   (c) Real-time imaging issues.

6. Biomimetics is the practice of seeking inspiration from nature to engineer/create materials, systems, processes, or models with superior performance. Bioinspiration can be useful to develop materials with superior mechanical properties, such as strength, toughness, and stiffness. How can the in-situ mechanics approach be helpful to engineer and optimize the properties of biomimetic materials? Explain with an example.

   Note: Consider both quantitative (stress–strain response) and qualitative (real-time imaging) aspects of in-situ investigations.

7. A table summarizing indentation-derived mechanical properties of osteonal bone is reproduced below. It can be seen that the modulus values are enhanced for dry samples. This poses a challenge for in-situ testing of bone samples in typical SEMs since the vacuum environment requires the samples to be dried

before. What could be a possible approach to test moist bone specimens in SEM?

(Hint: Think in terms of imaging instrument design considerations).

Table 1 | Mean±s.d. of the indentation modulus E^*, indentation hardness H_{IT}, elastic work W_{el} and total work W_{tot} as well as the number of experiments N for wet and dry indentations in the axial and transverse directions.

Hydr.	Direct.	E^* (GPa)	H_{IT} (GPa)	W_{el} (pJ)	W_{tot} (pJ)	N
Dry	Axial	27.5±2.2	1.01±0.13	1,837±258	8,069±1,012	50
Dry	Transv.	19.0±1.8	0.67±0.08	1,636±214	6,482±745	72
Wet	Axial	22.8±1.6	0.60±0.11	1,083±249	4,849±907	67
Wet	Transv.	14.5±1.6	0.51±0.08	1,313±204	4,125±756	83

Reference: Schwiedrzik et al. *Nature Materials* **13**, 740–747. (Reproduced with permission)

8. Low-density mechanical metamaterials display enhanced load-bearing ability (with impressive strength-to-weight ratio). However, in-situ imaging during compression has revealed failure tends to initiate at the nodes or point of intersection for multiple struts (see Sect. 5.4). What could be the factors responsible for failure initiation at these points? What general precautions or modifications would you recommend while designing 3D architectures, so as to avoid premature failure?

9. An SEM micrograph of a polymer microbeam subjected to mechanical loading is shown in the figure below.

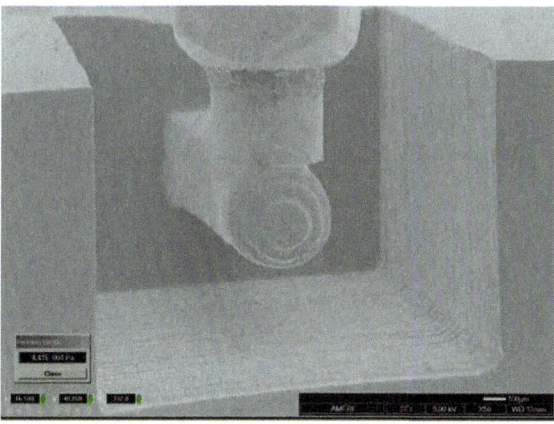

Figure: SEM micrograph of a polymer beam subjected to mechanical loading by a flat-ended punch. (The beam is 1.5 mm long and has a diameter of 300 μm). (Courtesy: Agarwal group (FIU), CELL-MET NSF Engineering Research Center)

(a) Polymers are known to display time-dependent or viscoelastic mechanical behavior. Propose a detailed experimental scheme to investigate the viscoelastic response of this polymer specimen by "in-situ" testing (inside SEM). Your experimental scheme should be comprehensive and provide information pertaining to instrumentation, test parameters, real-time imaging, output, and analysis of the resultant data.

(b) What could be the benefits of real-time imaging as compared to the conventional ex-situ approach? What are the challenges associated with in-situ measurement?

(Note: The chapter covered some examples of in-situ fatigue investigations. Seek inspiration from those examples to propose your experimental scheme.)

10. The figure below (left) shows the SEM micrograph of a 3D printed polymeric honeycomb lattice. A flat-ended punch can be used to extract the overall mechanical response of the architecture. How can you measure the modulus of individual arms or nodes in the microstructure? Suggest a suitable in-situ (SEM) measurement technique and instrumentation.
Note: The individual arms have a width of ~2 μm.

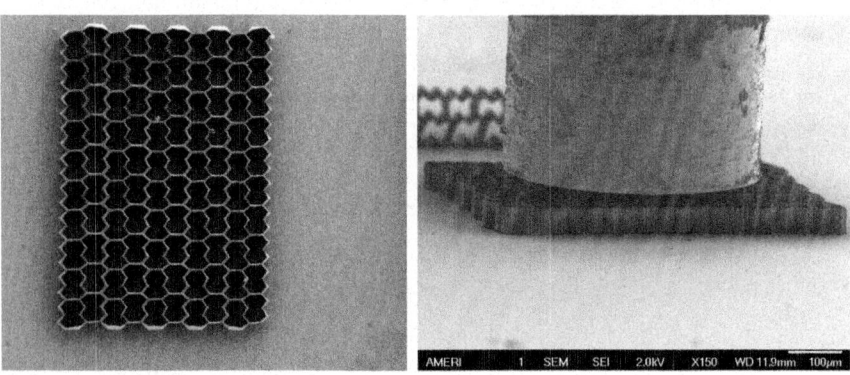

Figure: SEM micrograph of 3D printed honeycomb structure (left) and use of a flat-ended punch for in-situ compression inside SEM (right). (Courtesy: Agarwal group (FIU) and CELL-MET NSF Engineering Research Center)

References

Alducin D, Borja R, Ortega E et al (2016) In situ transmission electron microscopy mechanical deformation and fracture of a silver nanowire. Scr Mater 113:63–67. https://doi.org/10.1016/j.scriptamat.2015.10.011

Andersson RL, Ström V, Gedde UW et al (2014) Micromechanics of ultra-toughened electrospun PMMA/PEO fibres as revealed by in-situ tensile testing in an electron microscope. Sci Rep 4:6335. https://doi.org/10.1038/srep06335

Bhowmick S, Ozden S, Bizão RA et al (2018) High temperature quasistatic and dynamic mechanical behavior of interconnected 3D carbon nanotube structures. Carbon N Y 142:291–299. https://doi.org/10.1016/j.carbon.2018.09.075

Boesl B, Lahiri D, Behdad S, Agarwal A (2014) Direct observation of carbon nanotube induced strengthening in aluminum composite via in situ tensile tests. Carbon N Y 69:79–85. https://doi.org/10.1016/j.carbon.2013.11.061

Bustillos J, Zhang C, Boesl B, Agarwal A (2018) Three-dimensional graphene foam-polymer composite with superior deicing efficiency and strength. ACS Appl Mater Interfaces 10:5022–5029. https://doi.org/10.1021/acsami.7b18346

Cao C, Howe JY, Perovic D et al (2016) In situ TEM tensile testing of carbon-linked graphene oxide nanosheets using a MEMS device. Nanotechnology 27:28LT01

Casillas G, Palomares-Báez J, Rodríguez-López J et al (2012) In situ TEM study of mechanical behaviour of twinned nanoparticles. Philos Mag 92:4437–4453. https://doi.org/10.1080/14786435.2012.709951

Chen H, Gao Y, Zhang H et al (2004) Transmission-electron-microscopy study on fivefold twinned silver nanorods. J Phys Chem B 108:12038–12043. https://doi.org/10.1021/jp048023d

Chen Z, Ren W, Gao L et al (2011) Three-dimensional flexible and conductive interconnected graphene networks grown by chemical vapour deposition. Nat Mater 10:424–428. https://doi.org/10.1038/nmat3001

Chu K, Wang J, Liu Y, Geng Z (2018) Graphene defect engineering for optimizing the interface and mechanical properties of graphene/copper composites. Carbon N Y 140:112–123. https://doi.org/10.1016/j.carbon.2018.08.004

Deneen Nowak J, Mook WM, Minor AM et al (2007) Fracturing a nanoparticle. Philos Mag 87:29–37. https://doi.org/10.1080/14786430600876585

Ebenstein DM, Pruitt LA (2006) Nanoindentation of biological materials. Nano Today 1:26–33. https://doi.org/10.1016/S1748-0132(06)70077-9

Huang JY, Zheng H, Mao SX et al (2011) In situ nanomechanics of GaN nanowires. Nano Lett 11:1618–1622. https://doi.org/10.1021/nl200002x

Idowu A, Boesl B, Agarwal A (2018) 3D graphene foam-reinforced polymer composites – a review. Carbon N Y 135:52–71. https://doi.org/10.1016/j.carbon.2018.04.024

Issa I, Amodeo J, Réthoré J et al (2015) In situ investigation of MgO nanocube deformation at room temperature. Acta Mater 86:295–304. https://doi.org/10.1016/j.actamat.2014.12.001

Kiener D, Hosemann P, Maloy SA, Minor AM (2011) In situ nanocompression testing of irradiated copper. Nat Mater 10:608–613. https://doi.org/10.1038/nmat3055

Li S, Abdel-Wahab A, Demirci E, Silberschmidt VV (2014) Penetration of cutting tool into cortical bone: experimental and numerical investigation of anisotropic mechanical behaviour. J Biomech 47:1117–1126. https://doi.org/10.1016/j.jbiomech.2013.12.019

Li N, Wang H, Misra A, Wang J (2015) In situ nanoindentation study of plastic co-deformation in Al-TiN nanocomposites. Sci Rep 4:1–6. https://doi.org/10.1038/srep06633

Liao Z, Sandonas LM, Zhang T et al (2017) In-situ stretching patterned graphene nanoribbons in the transmission electron microscope. Sci Rep 7:211. https://doi.org/10.1038/s41598-017-00227-3

Liu D, Flewitt PEJ (2017) Deformation and fracture of carbonaceous materials using in situ micromechanical testing. Carbon N Y 114:261–274. https://doi.org/10.1016/j.carbon.2016.11.084

Liu F, Tang D-M, Gan H et al (2013) Individual boron nanowire has ultra-high specific young's modulus and fracture strength as revealed by in situ transmission electron microscopy. ACS Nano 7:10112–10120. https://doi.org/10.1021/nn404316a

Liu Y, Karaman I, Wang H, Zhang X (2014) Two types of martensitic phase transformations in magnetic shape memory alloys by in-situ nanoindentation studies. Adv Mater 26:3893–3898. https://doi.org/10.1002/adma.201400217

Montemayor LC, Greer JR (2015) Mechanical response of hollow metallic nanolattices: combining structural and material size effects. J Appl Mech 82:071012. https://doi.org/10.1115/1.4030361

Nardecchia S, Carriazo D, Ferrer ML et al (2013) Three dimensional macroporous architectures and aerogels built of carbon nanotubes and/or graphene: synthesis and applications. Chem Soc Rev 42:794–830. https://doi.org/10.1039/c2cs35353a

Nieto A, Boesl B, Agarwal A (2015a) Multi-scale intrinsic deformation mechanisms of 3D graphene foam. Carbon N Y 85:299–308. https://doi.org/10.1016/j.carbon.2015.01.003

Nieto A, Dua R, Zhang C et al (2015b) Three dimensional graphene foam/polymer hybrid as a high strength biocompatible scaffold. Adv Funct Mater 25:3916–3924. https://doi.org/10.1002/adfm.201500876

Schwiedrzik J, Raghavan R, Bürki A et al (2014) In situ micropillar compression reveals superior strength and ductility but an absence of damage in lamellar bone. Nat Mater 13:740–747. https://doi.org/10.1038/nmat3959

Shan ZW, Adesso G, Cabot A et al (2008) Ultrahigh stress and strain in hierarchically structured hollow nanoparticles. Nat Mater 7:947–952. https://doi.org/10.1038/nmat2295

Shehzad K, Xu Y, Gao C, Duan X (2016) Three-dimensional macro-structures of two-dimensional nanomaterials. Chem Soc Rev 45:5541–5588. https://doi.org/10.1039/c6cs00218h

Tang D-M, Ren C-L, Wei X et al (2011) Mechanical properties of bamboo-like boron nitride nanotubes by in situ TEM and MD simulations: strengthening effect of interlocked joint interfaces. ACS Nano 5:7362–7368. https://doi.org/10.1021/nn202283a

Tsuda T, Ogasawara T, Deng F, Takeda N (2011) Direct measurements of interfacial shear strength of multi-walled carbon nanotube/PEEK composite using a nano-pullout method. Compos Sci Technol 71:1295–1300. https://doi.org/10.1016/j.compscitech.2011.04.014

Vlassov S, Polyakov B, Dorogin LM et al (2014) Elasticity and yield strength of pentagonal silver nanowires: in situ bending tests. Mater Chem Phys 143:1026–1031. https://doi.org/10.1016/j.matchemphys.2013.10.042

Wang H, Zhang X, Wang N et al (2017) Ultralight, scalable, and high-temperature-resilient ceramic nanofiber sponges. Sci Adv 3:e1603170. https://doi.org/10.1126/sciadv.1603170

Wei X, Wang M-S, Bando Y, Golberg D (2010) Tensile tests on individual multi-walled boron nitride nanotubes. Adv Mater 22:4895–4899. https://doi.org/10.1002/adma.201001829

Wu W, Hu W, Qian G et al (2019) Mechanical design and multifunctional applications of chiral mechanical metamaterials: a review. Mater Des 180:107950. https://doi.org/10.1016/j.matdes.2019.107950

Xu F, Qin Q, Mishra A et al (2010) Mechanical properties of Zno nanowires under different loading modes. Nano Res 3:271–280. https://doi.org/10.1007/s12274-010-1030-4

Yang Y, Chen W, Hacopian E et al (2016) Unveil the size-dependent mechanical behaviors of individual CNT/SiC composite nanofibers by in situ tensile tests in SEM. Small 12:4486–4491. https://doi.org/10.1002/smll.201601113

Yang W, Yang J, Dong Y et al (2018) Probing buckling and post-buckling deformation of hollow amorphous carbon nanospheres: in-situ experiment and theoretical analysis. Carbon N Y 137:411–418. https://doi.org/10.1016/j.carbon.2018.05.047

Yee DW, Lifson ML, Edwards BW, Greer JR (2019) Additive manufacturing of 3D-architected multifunctional metal oxides. Adv Mater 31:1901345. https://doi.org/10.1002/adma.201901345

Yi C, Chen X, Gou F et al (2017) Direct measurements of the mechanical strength of carbon nanotube - aluminum interfaces. Carbon N Y 125:93–102. https://doi.org/10.1016/j.carbon.2017.09.020

Yi C, Bagchi S, Gou F et al (2019) Direct nanomechanical measurements of boron nitride nanotube — ceramic interfaces. Nanotechnology 30:025706

Zeng XM, Du Z, Tamura N et al (2017) In-situ studies on martensitic transformation and high-temperature shape memory in small volume zirconia. Acta Mater 134:257–266. https://doi.org/10.1016/j.actamat.2017.06.006

Zhang J, Loya P, Peng C et al (2012) Quantitative in situ mechanical characterization of the effects of chemical functionalization on individual carbon nanofibers. Adv Funct Mater 22:4070–4077. https://doi.org/10.1002/adfm.201200593

Zhang P, Ma L, Fan F et al (2014) Fracture toughness of graphene. Nat Commun 5:3782. https://doi.org/10.1038/ncomms4782

Zhang C, Boesl B, Silvestroni L et al (2016) Deformation mechanism in graphene nanoplatelet reinforced tantalum carbide using high load in situ indentation. Mater Sci Eng A 674:270–275. https://doi.org/10.1016/j.msea.2016.07.110

References

Zhang H, Jiang C, Lu Y (2017a) Low-cycle fatigue testing of Ni nanowires based on a micromechanical device. Exp Mech 57:495–500. https://doi.org/10.1007/s11340-016-0199-1

Zhang Q, Lin D, Deng B et al (2017b) Flyweight, superelastic, electrically conductive, and flame-retardant 3D multi-nanolayer graphene/ceramic metamaterial. Adv Mater 29:1605506. https://doi.org/10.1002/adma.201605506

Zhang X, Yao J, Liu B et al (2018a) Three-dimensional high-entropy alloy – polymer composite nanolattices that overcome the strength – recoverability trade-off. Nano Lett 18:4247–4256. https://doi.org/10.1021/acs.nanolett.8b01241

Zhang Z, Zhang L, Yu Z et al (2018b) In-situ mechanical test of dragonfly wing veins and their crack arrest behavior. Micron 110:67–72. https://doi.org/10.1016/j.micron.2018.05.003

Zhou W, Yamamoto G, Fan Y et al (2016) In-situ characterization of interfacial shear strength in multi-walled carbon nanotube reinforced aluminum matrix composites. Carbon N Y 106:37–47. https://doi.org/10.1016/j.carbon.2016.05.015

Zhu Y, Qin Q, Xu F et al (2012) Size effects on elasticity, yielding, and fracture of silver nanowires: in situ experiments. Phys Rev B Condens Matter Mater Phys 85:045443. https://doi.org/10.1103/PhysRevB.85.045443

Chapter 6
Interfacing In-Situ Mechanics with Image Correlation and Simulations

In-situ characterization in conjunction with analytical and computational tools can be useful to extract qualitative and quantitative information pertaining to the deformation of materials. Digital image correlation (DIC) is a powerful image analysis technique to quantify local as well as global strains experienced by a sample under mechanical loading. Determination of feature-specific strains provides insights into the deformation of multicomponent materials with complex microstructures. DIC analysis enables an understanding of stress transfer characteristics in 3D architectures. Local strains can be computed along different orientations, providing information about the degree of isotropy/anisotropy in microstructure response. DIC mapping is employed to examine crystallographic slip and lattice rotation during mechanical deformation. The chapter introduces and presents case studies on molecular dynamic (MD) simulations to complement real-time imaging. MD simulations provide insights into defect interactions, the role of flaws on deformation, work hardening, cyclic deformation as well as high-temperature mechanics of materials. In-situ imaging along with MD simulations reveal underlying atomic-scale phenomena during deformation. The validation of simulation by in-situ imaging is vital for developing reliable models capable of predicting mechanical phenomena in a wide diversity of materials. Table 6.1 below summarizes the application of analytical and computational tools/techniques along with in-situ experimental characterization to probe mechanical phenomena in different classes of materials.

Electronic supplementary material The online version of this chapter (https://doi.org/10.1007/978-3-030-43320-8_6) contains supplementary material, which is available to authorized users.

Table 6.1 Application of analytical and computational investigations to probe mechanical phenomena in a wide variety of material systems

Mechanical phenomena	Analytical/computational examination	Examples
Failure strains	Microstructure strain mapping	Local strains associated with crack initiation/propagation
Strain localization	Site-specific microstructure strains	Strains in a multicomponent microstructure, such as a sandwich composite
Stress transfer	Evolution of strains in regions away from the point of loading	3D architectures, Metamaterials
Degree of isotropy/anisotropy	Strain contours along different orientations	Materials processed by/subjected to severe deformation, such as hot or cold working
Stress concentration	Atomistic simulations investigation near the flaw	Notched specimens
Work hardening	Simulation of dislocation pileup	Cyclic mechanical loading of metals
Reversible cyclic deformation	Simulation of atomic-sale reversible mechanisms	Twinning-Detwinning during tension-compression cycles

(continued)

6.1 Digital Image Correlation Analysis

Table 6.1 (continued)

Mechanical phenomena	Analytical/computational examination	Examples
High-temperature mechanics	Simulation of bond breakage/reconstruction, thermal vibrations	Mechanical properties as a function of test temperature

Images reproduced with permissions from Embrey et al. (2017), Huang et al. (2018), Nautiyal et al. (2018a), Nautiyal et al. (2018b), Gu et al. (2013), Bufford et al. (2014), Lee et al. (2014), Bhowmick et al. (2018)

6.1 Digital Image Correlation Analysis

The previous chapter on the application of the in-situ approach for probing different material systems presented case studies showing the advantage of real-time imaging to understand the elastic, plastic, and failure mechanisms. In addition to qualitative assessment, in-situ videos can provide quantitative information, such as strain evolution, strain distribution, the extent of deformation in different regions and overall displacements. The determination of strains or displacements from deformation videos is accomplished by the digital image correlation (DIC) analysis technique. During DIC analysis, the real-time snapshots of samples subjected to mechanical loading are compared with respect to the original starting image/snapshot (before any external mechanical loads are applied). This comparative image analysis allows the quantification of strain evolution in real time. The most common application of this technique is to compute strains during a tensile test, precluding the need for using strain gauges. The determination of strains during the tensile test can be challenging for slender, miniature or fragile samples, where attaching an external strain gauge is not possible. DIC analysis of the tensile test video enables the computation of strains in such specimens. Apart from calculating net strains, the DIC technique is also used to obtain strain maps characterizing the magnitude of strain experienced by different regions in a sample under deformation. The resolution of strain maps is dependent on the imaging resolution. For instance, DIC analysis of real-time SEM micrographs can enable estimation of strains in specific microstructure features. The strain contours superimposed on the micrographs allow characterization of strain localization in the microstructure, determination of "critical" failure strains and degree of isotropy/anisotropy in microstructure deformation.

Figure 6.1a shows a case study on DIC analysis for the tensile test of a polymer composite reinforced with macroporous graphene foam (Embrey et al. 2017). In-situ optical imaging of composite stretching and failure are shown in Supplementary Videos, Videos 6.1 and 6.2. These in-situ videos were analyzed to

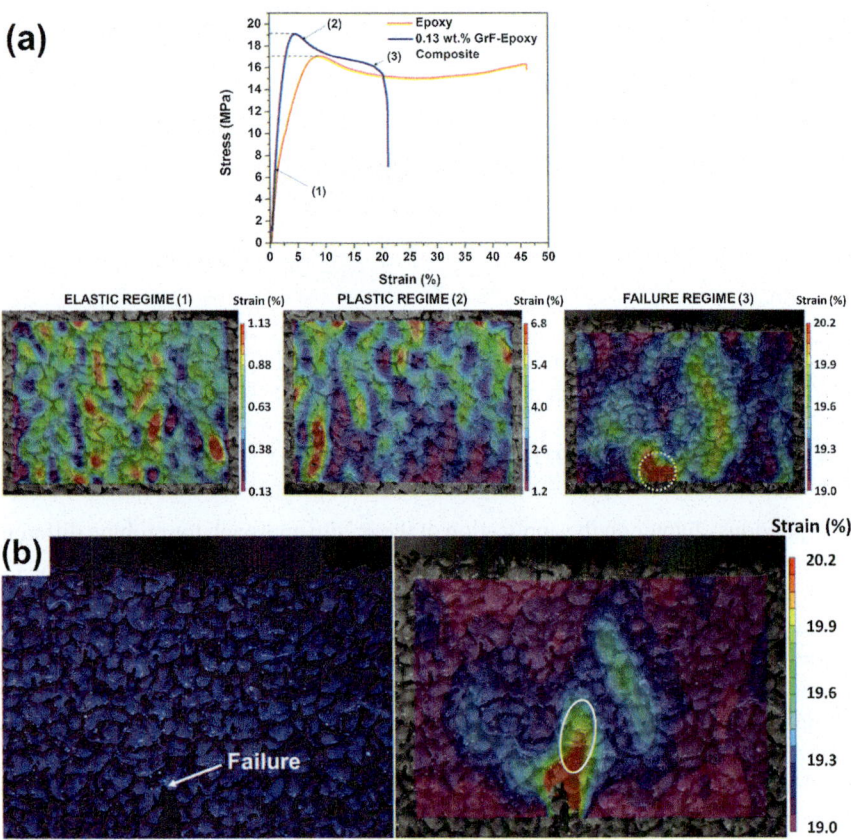

Fig. 6.1 DIC analysis of a polymer-graphene foam composite under tension: (**a**) stress–strain curve and strain maps in the elastic, plastic, and failure regimes, and (**b**) optical snapshot and the corresponding strain map capturing failure/crack propagation in the composite. (Reproduced/adapted with permission from Embrey et al. (2017))

determine overall strains experienced by the sample during mechanical loading, and the strain values were used to plot stress–strain curves. The distribution of local strains in the sample is also shown in the strain maps during different instants of the tensile test. Three maps, captured during elastic, plastic, and failure are reproduced in Fig. 6.1a. These strain maps consist of contours with different magnitudes of local strains. For instance, it was seen that the local strains at point (1) in the elastic regime can vary from a low of ~0.13% up to 1.13%. The strain distribution is a function of local microstructure features, such as the size and morphology of graphene foam cells. Strain mapping is a useful approach to determine failure strains. Figure 6.1b shows an optical snapshot of the specimen undergoing tensile failure and the corresponding strain map. The local strain contour in the failure region (crack tip) showed the failure strain to be ~20%. The red-colored strain contour shows strain localization at the crack tip and also indicates the crack propagation

6.1 Digital Image Correlation Analysis

pathway. Therefore, the DIC technique provides critical quantitative information pertaining to the failure of materials.

DIC analysis can be useful to measure strains along different orientations in a material with respect to the loading direction. This is demonstrated in Fig. 6.2a, where strain maps along two different orientations are shown for the in-situ SEM tensile loading of Ti (Huang et al. 2018). X-direction corresponds to the loading/tensile axis, and Y is perpendicular to the loading direction. The maps show local strain values in the sample at different macro-strains (ε varying from 0.5% to 4.5%). Strain localization was observed as the sample was deformed, evident from maps captured at $\varepsilon = 4.5\%$. However, a layered composite of Al and Ti behaved differently

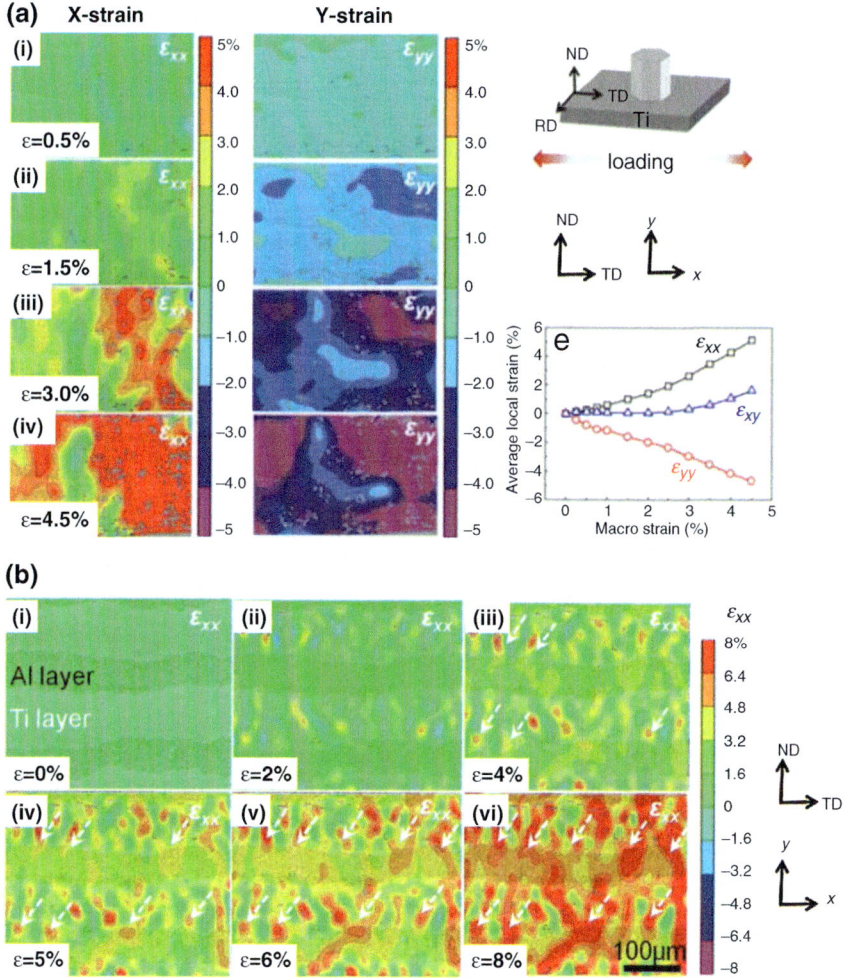

Fig. 6.2 High-resolution DIC analysis of in-situ tensile testing for: (**a**) pure Ti, and (**b**) Ti-Al layered composite. (Reproduced with permission from Huang et al. (2018))

to tensile loading (Fig. 6.2b). There was some degree localization in the Ti layer for strains up to ~2%. With further loading, strain localization was seen at the Al/Ti interface and eventually strain transfer to the softer Al phase occurred (at higher macro-strains of ~6% and 8%). DIC analysis revealed that the Ti/Al interface is critical for strain transfer, evident from the maps showing localization (high strain value contours) along the interfaces. This strain transfer phenomenon weakens strain localization at Ti. Effective strain transfer was beneficial as it enabled the accommodation of higher strains by the sample, enhancing the overall ductility of the sample in comparison to pure Ti. This example shows the significance of high-resolution DIC to obtain insights into deformation characteristics at microstructural length scales, including the interactions between the constituents and their overall effect on the net mechanical response.

The application of in-situ measurement for 3D architectures/metamaterials was discussed in Chap. 5. Real-time imaging helps to observe load transfer in 3D architectures. DIC analysis enables the quantification of strain distribution as the structures are loaded. Figure 6.3a shows an example of the indentation loading of graphene foam (Nautiyal et al. 2018a). In-situ SEM imaging revealed excellent damage tolerance of the foam, as the structure demonstrated reversible, recoverable deformation. The in-situ SEM imaging of the flexible foam under compression loading is shown in Supplementary Video, Video 6.3. The strain maps were obtained by DIC analysis at different instants during mechanical loading. Figure 6.3b shows the maps at 100, 600, 1200 and 1700 µN loads. The peak strain values steadily increased from 0.18% at 100 µN to 2.18% at 1700 µN. Interestingly, the DIC maps revealed the foam branches which were several hundreds of micrometers away from the point of loading (tip-sample interaction zone) experienced strains. This attests long-distance stress transfer in the 3D structure. Strain redistribution prevented local stress concentration and enhanced the damage tolerance of the structure. Therefore, image correlation is a powerful technique to decipher the structural performance of 3D architectures.

DIC technique is particularly useful to evaluate deformation in materials with hierarchical microstructures. A case study on in-situ mechanics of cold-sprayed aluminum alloy is shown in Fig. 6.4 (Nautiyal et al. 2018b). Cold-spray is an additive manufacturing process, where powder particles are accelerated and deposited with supersonic velocities. These severely deformed particles, known as splats, are the building blocks of cold-sprayed microstructures (shown in the SEM micrograph in Fig. 6.4a). The overall deformation behavior of cold-sprayed materials is greatly influenced by the deformation characteristics of splats and mechanical interactions between them. Real-time imaging has been demonstrated to be useful to probe the local deformation of splats and correlate it with the overall bulk response of the material. Figure 6.4b demonstrates the experimental scheme, where the cyclic bending of a cold-sprayed 6061 Al alloy coating (deposited on 6061 Al substrate) was imaged in SEM in real-time. DIC analysis of the high-resolution SEM images revealed that the microstructure did not deform as a single rigid unit; instead, individual splats were seen to experience different localized strains. This is shown in the strain maps in Fig. 6.4c. The differences in the magnitude, as well as the direction

6.1 Digital Image Correlation Analysis

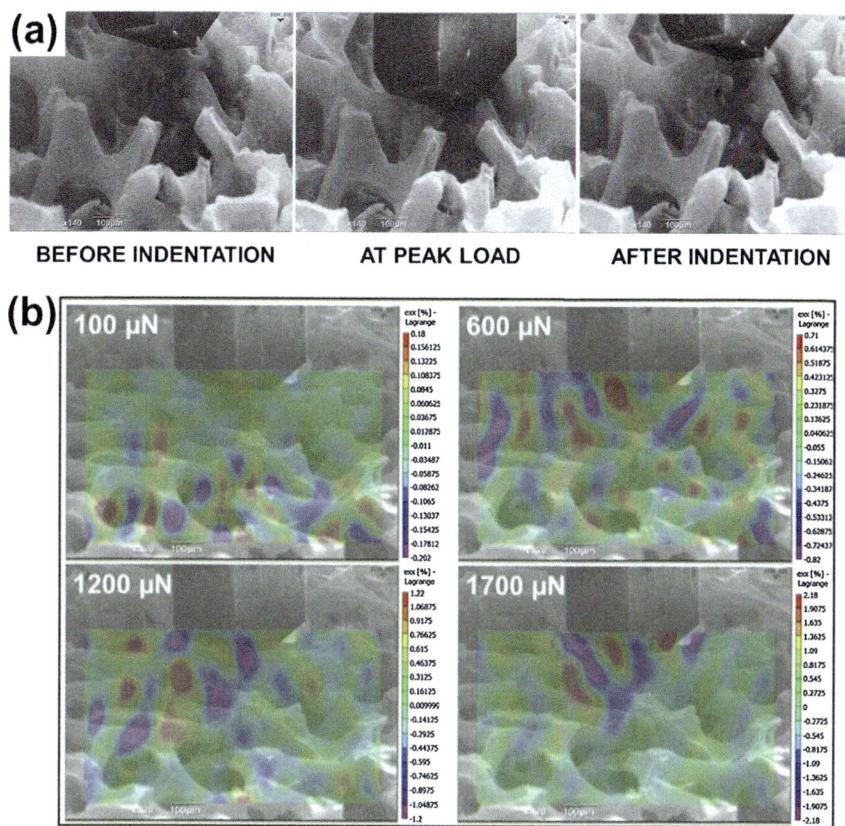

Fig. 6.3 (**a**) In-situ indentation of 3D graphene foam in SEM, and (**b**) DIC strain maps of the foam under compression at different mechanical loads. (Reproduced with permission from Nautiyal et al. (2018a))

of strains experienced by the splats, suggests inter-splat sliding is a prominent mechanism in cold-sprayed materials, as opposed to their bulk counterparts which do not comprise of splats. This splat sliding mechanism can be observed in Supplementary Video, Video 6.4, which shows the evolution of strain contours in real time as the coating is subjected to mechanical loading. The splat sliding behavior was strongly influenced by the processing conditions. For instance, the air-sprayed coating showed an isotropic response, evident from ε_x and ε_y values in the strain–time plot (Fig. 6.4c). However, the response of Helium-sprayed coating was anisotropic, as the strains in X and Y directions were significantly different (Fig. 6.4d). The deformation was arrested in Y direction. The X and Y directions correspond to the orientations perpendicular and parallel to the spray direction, respectively. Therefore, the DIC technique can be a useful tool for correlating manufacturing process parameters with the resultant mechanical response at microstructural length scales. It is well known that post-manufacturing treatment of materials,

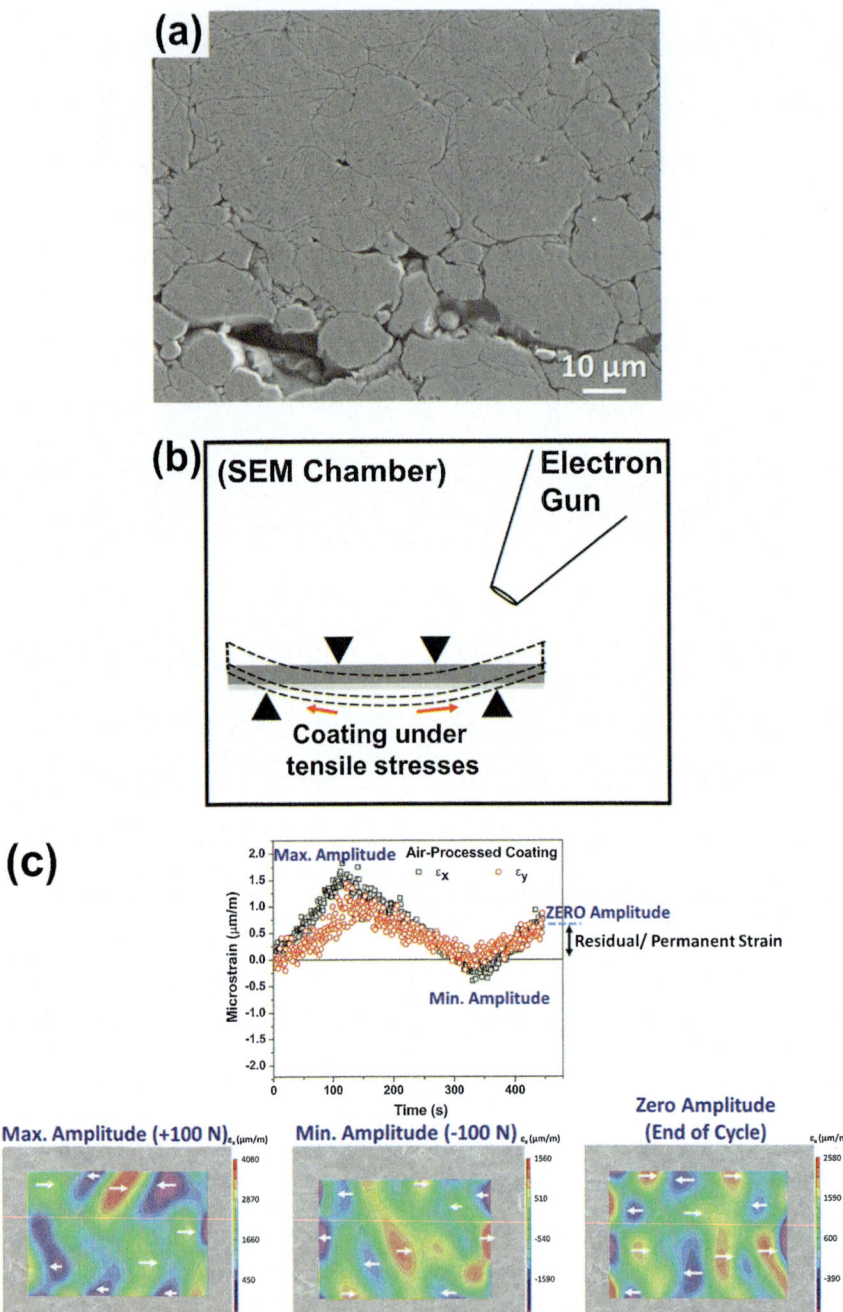

Fig. 6.4 (**a**) SEM micrograph of a cold-sprayed microstructure, (**b**) in-situ experiment scheme to study deformation of cold-sprayed 6061 Al alloy, (**c**) strain–time plots and strain maps obtained by DIC analysis of cold-sprayed 6061 Al, (**d**) strain–time plot and strain maps showing anisotropy in mechanical response, and (**e**) strain–time plot obtained by DIC analysis comparing the effect of heat treatment on mechanical response. (Reproduced with permission from Nautiyal et al. (2018b))

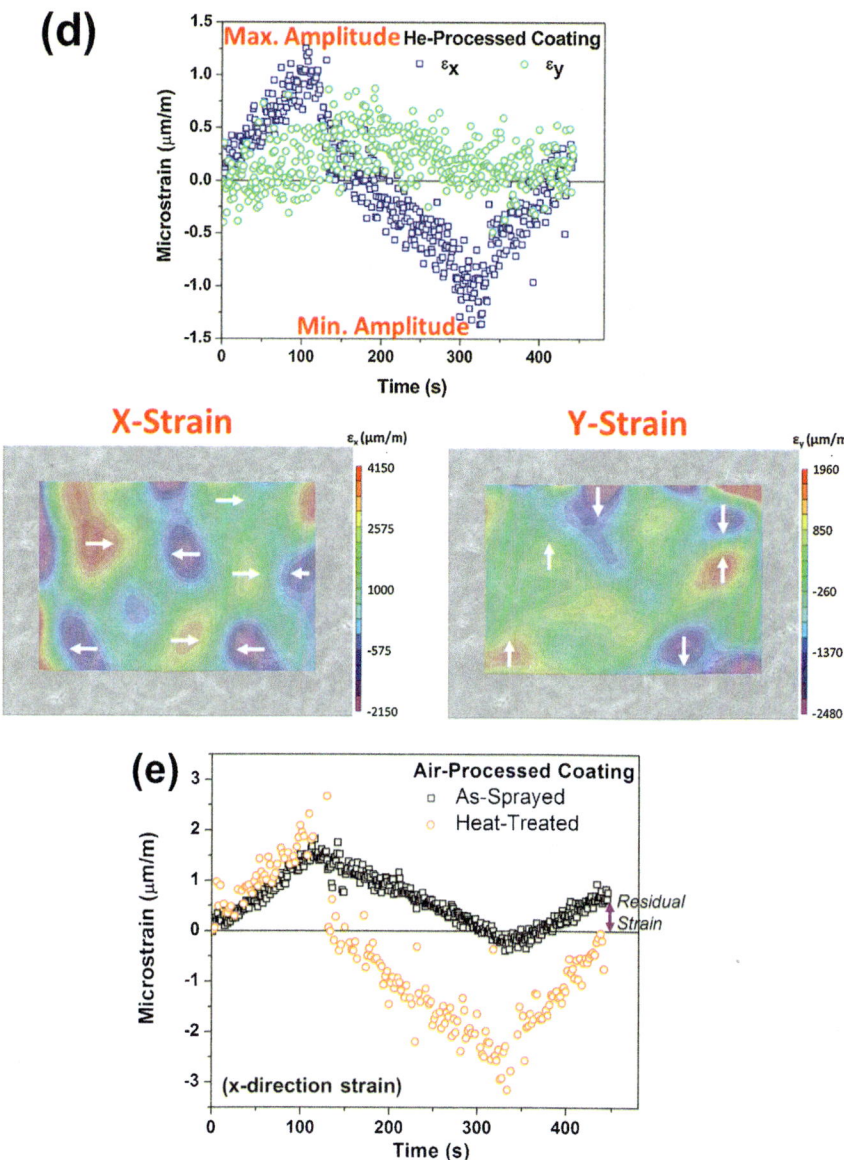

Fig. 6..4 (continued)

such as heat treatment or mechanical working, alters their properties. DIC analysis of in-situ videos can be performed to assess the effect of these treatments on the mechanical response of the microstructure. As an example, the strains computed by DIC analysis of cold-sprayed 6061 Al before and after heat treatment are compared in Fig. 6.4e. It was observed that the final, residual strain (after complete unloading) was reduced from ~0.22 μm/m to 0.03 μm/m due to heat treatment. This observation was ascribed to improved splat-splat bonding due to thermal diffusion. Considering the recent upsurge of interest in additive manufacturing processes, in-situ mechanics in conjunction with DIC analysis can be highly informative to establish processing-microstructure-mechanics correlations for different material systems.

DIC mapping of deformation in small-scale samples is useful to examine crystallographic slip and lattice rotation. Figure 6.5 shows a case study on in-situ compression of Cu micropillar in SEM (Di Gioacchino and Clegg 2014). Speckle pattern was applied on the pillar surface by FIB—these speckles enable identification of pixel subsets in in-situ videos as they move during deformation. The tracking of the subset movement is used to calculate the displacement field. The deformation tensor, **F**, is expressed by the relationship:

$$\mathbf{F} = \mathbf{R}^e \mathbf{F}^p \quad (6.1)$$

where $\mathbf{R}^e$ is lattice rotation and $\mathbf{F}^p$ corresponds to the elements deforming due to moving dislocations through the lattice. Considering a Cartesian coordinate system (X_1, X_2, X_3) with $X_1 X_2$ plane on the sample surface, F_{11} and F_{21} can be determined by the equations:

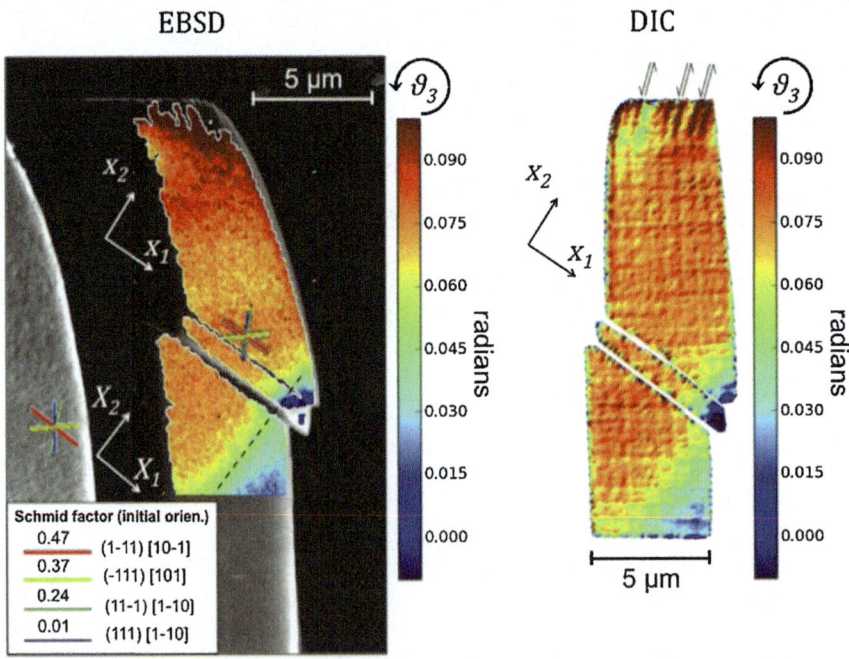

Fig. 6.5 Comparison of in-plane rotation in Cu micropillar measured by EBSD and DIC analysis techniques. (Reproduced with permission from Di Gioacchino and Clegg (2014))

$$F_{11} = R_{11}^e F_{11}^p + R_{12}^e F_{21}^p \tag{6.2}$$

$$F_{21} = R_{21}^e F_{11}^p + R_{22}^e F_{21}^p \tag{6.3}$$

In this equation, $F_{21}^p = 0$. If $\mathbf{R}^e$ (ϑ_3, X_3) is small, lattice rotation ϑ_3 can be expressed as:

$$\vartheta_3 = \frac{F_{21}}{F_{11}} \tag{6.4}$$

Determination of ϑ_3 is useful to derive the rotational matrices, $\mathbf{R}^e(\vartheta_3, X_3)$, which can then enable calculation of $\mathbf{F}^p$:

$$\mathbf{F}^p = \mathbf{R}^{eT} \mathbf{F} \tag{6.5}$$

where $\mathbf{R}^{eT}$ is the transpose of $\mathbf{R}^e(\vartheta_3, X_3)$.

ϑ_3 values were obtained by DIC and observed to be location-dependent (as shown in Fig. 6.5). ϑ_3 was close to zero or slightly negative at the right-hand side of the pillar base. On moving away from this region along the direction X_1 marked in the figure, the υ_3 was seen to increase steadily up to 0.07 radian. X_1 axis corresponds to the slip direction. The rotation values were observed to stay steady along X_2 axis, which is perpendicular to the slip direction. The rotation angles obtained by DIC were found to be in agreement with the values calculated from EBSD. The comparative side-by-side maps superimposed on the SEM micrograph of the pillar are shown in Fig. 6.5. The determination of ϑ_3 with certainty using the in-situ measurement approach allows the calculation of $\mathbf{F}^p$ based on Eq. (6.5). This example highlights the suitability of the DIC technique to extract deformation characteristics of materials during small-scale in-situ testing.

The image correlation approach can be applied to volumetric 3D image data acquired by computed tomography (CT). This 3D correlation, known as digital volume correlation (DVC), is useful to measure displacements and deformation in the whole volume of the material under in-situ loading (Forsberg et al. 2010; Brault et al. 2013). DVC technique can be applied to compute critical strains leading to failure initiation in materials. Figure 6.6a shows the nanoindentation of elephant dentin, imaged in-situ by microtomography (Patterson et al. 2016). Real-time imaging captured crack initiation and crack growth beneath the indenter tip. DVC analysis of the 3D-datasets allowed the determination of the strain field during nanoindentation, as shown in the figure. Dentin has a heterogeneous microstructure, composed of hydroxyapatite and collagen. DVC approach is highly informative for such complex systems since the strain contours can be superimposed on the internal microstructure image, providing insights into the fracture mechanics of the material. DVC analysis can map strain evolution along the microcrack path. This is shown in Fig. 6.6b for a cortical bone under compression loading (Christen et al. 2012). A key observation from the profile was that the local strains in the bone are

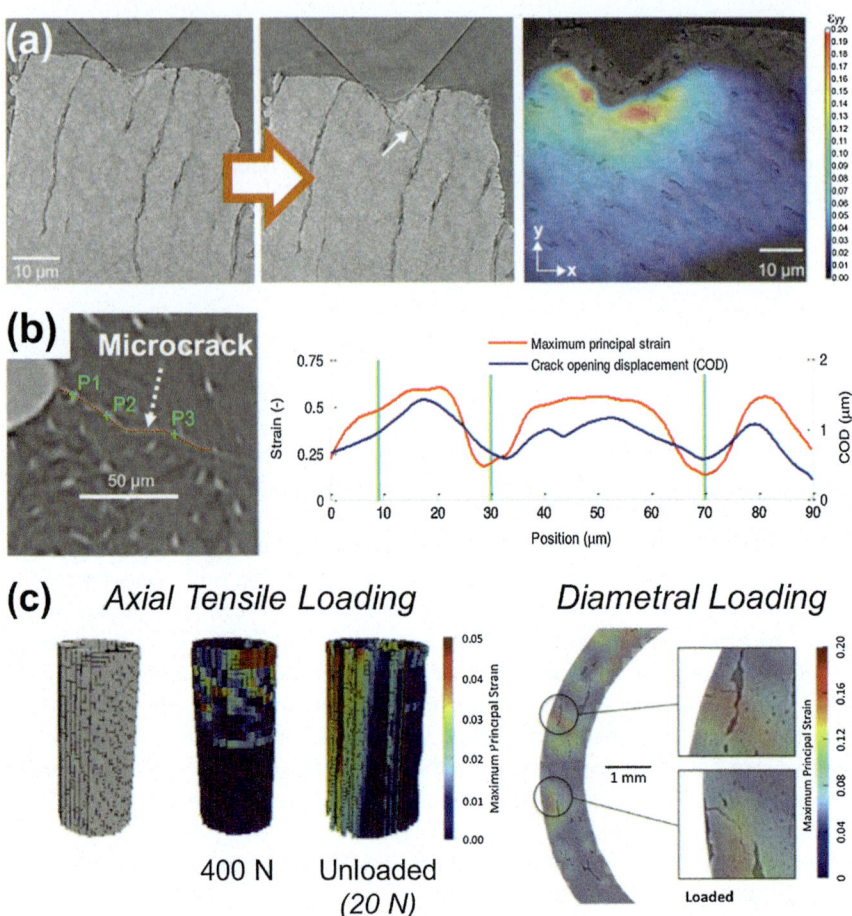

Fig. 6.6 (a) Indentation loading of elephant dentin causing microcrack growth and DVC analysis to determine microstructure strain, (b) local strain along the crack pathway in a cortical bone, and (c) strain evolution in a SiC/SiC$_{fiber}$ under tensile and diametral loading. (Reproduced/adapted with permissions from Patterson et al. (2016) Christen et al. (2012), Saucedo-Mora et al. (2016))

nonhomogeneous and sensitive to the surrounding microstructure, such as canals and osteocyte lacunae. Figure 6.6c illustrates the application of DVC to examine failure in a composite material (Saucedo-Mora et al. 2016). The figure shows strain distribution in a SiC/SiC$_{fibre}$ composite tube subjected to uniaxial tensile and diametral loading. DVC revealed nonhomogeneous strains in the tube, and there were residual strains even after unloading. Diametral loading of the tube induced some local cracking. DVC strain maps show the areas of high tensile strains are more susceptible to failure.

DIC technique discussed in this section is a promising approach to gain insights into the deformation behavior of materials. DIC enables the quantification of microstructure response to mechanical loading by estimation of global and local strains.

Analysis of the real-time videos along with the DIC strain maps can provide a host of information—failure strains, degree of isotropy/anisotropy, stress transfer, crystallographic slip, and lattice rotation. However, the DIC technique relies on in-situ videos as input for obtaining this information. The next section discusses the computational mechanical investigation approach, independent of in-situ results, to probe underlying mechanisms activated in response to external mechanical loading. The computational investigations in conjunction with in-situ testing can help decipher complex mechanical phenomena in materials at the atomic scale, discussed in detail in the next section.

6.2 Interfacing with Molecular Dynamic Simulations

Atomistic molecular dynamic (MD) simulations are performed to develop insights into deformation mechanisms and understand the in-situ observations captured by real-time videos. Validation of MD simulations by high-resolution in-situ videos is also vital to develop reliable models that can accurately predict the deformation characteristics of materials. MD simulations capture the response and interactions between multiple microstructural features, which play a crucial role in the mechanics of materials. Computational models can provide guidance for engineering material microstructures to achieve desired mechanical performance and properties. This section presents case studies on MD simulations performed on different classes of materials and for different mechanics problems.

A case study on indentation-compression of twinned nanoparticles is shown in Fig. 6.7 (Casillas et al. 2012). Figure 6.7a shows in-situ TEM as well as MD simulations for the mechanical test. MD simulations revealed strain localization at the tip-particle contact point. TEM images showed the bending of contour fringes in tetrahedra due to strain accumulation. Initially, the strain was observed to accumulate in tetrahedral 1 and 2. Eventually, the strain build-up was seen in tetrahedral 3 and 5 as well. The real-time imaging indicates the role of twin boundaries as strain filters. MD simulations supported the observations. After applying strain up to 20%, nucleation and movement of partial dislocations along the twin boundaries were observed in the simulations. Multiplication of partial dislocations was seen for further compressive loading, causing plastic deformation. New stacking faults are formed parallel to the twin boundaries. The stress maps for the nanoparticle under compression can be generated using the expression:

$$\sigma_i^{\alpha\beta} = \frac{1}{\omega_i \left\{ \frac{1}{2} m_i v_i^\alpha v_i^\beta + \frac{1}{2} \sum_i \left(-\frac{1}{r}\frac{\delta\varphi}{\delta r} \right) r^\alpha r^\beta |r=r_{ij} \right\}} \quad (6.6)$$

where α and β are the Cartesian coordinates, r_{ij} is the distance between the ith and jth atmos, φ is the interatomic potential, and v_i, ω_i and m_i are the effective velocities,

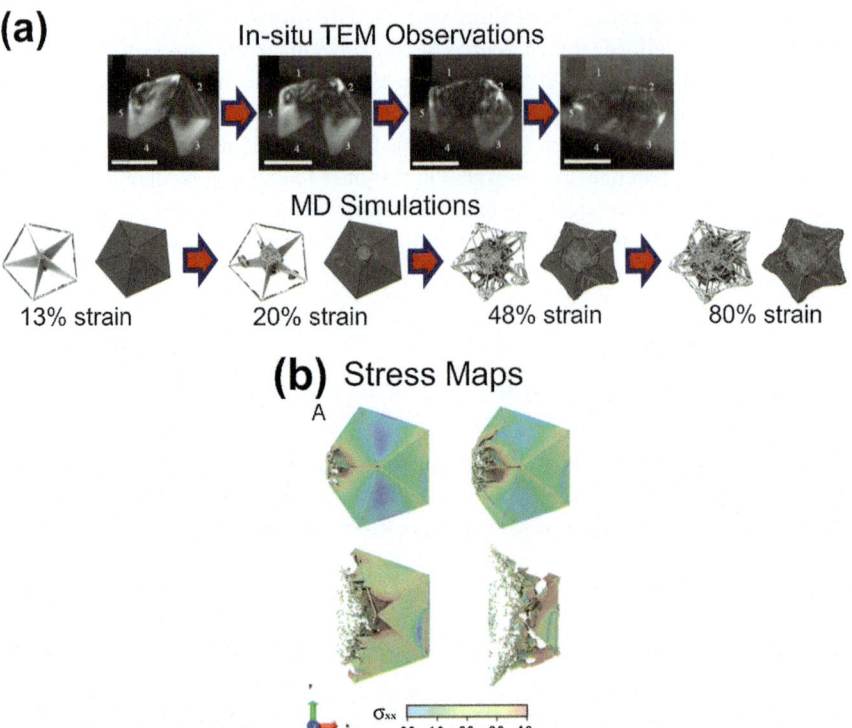

Fig. 6.7 (a) In-situ TEM images and MD simulation showing deformation of tetrahedral constituting the twinned nanoparticle, and (b) stress maps confirming the role of twin boundaries as stress filters. (Reproduced/adapted with permission from Casillas et al. (2012))

volume, and mass of the ith atom, respectively. Figure 6.7b demonstrates the magnitude of the stress per atom. The map revealed that while the left-hand region of the particle experiences extreme deformation (red-colored atoms), the tetrahedron away from the deformation region and close to the substrate (right-hand side) barely experienced stress buildup. The MD simulation findings are in agreement with the in-situ experimental observation that the twin boundaries are stress-filters in the nanoparticle.

Microstructure flaws influence the mechanical response of materials. The effect of flaws can be more complicated in nanoscale or nanostructured materials due to multiple competing phenomena. MD simulations can be helpful to examine the effect of flaws on the deformation behavior. An example of the tensile fracture of Pt nanocylinders is highlighted in Fig. 6.8 (Gu et al. 2013). The choice of the sample size to grain size ratio (D/d) is important for the mechanical characterization of nanosized samples, as extremely small values of D/d (below 5) can result in sample size dependent weakening. The nanocylinders in the study had a D/d ratio of 20 to extract size-independent mechanical characteristics. In-situ SEM testing revealed the failure did not necessarily happen at the flaw (Fig. 6.8a), and the ultimate tensile strength was independent of failure location. MD simulations were performed to understand the flaw insensitivity of sample strength (Fig. 6.8b). Grain boundary sliding and nucleation of partial dislocations were observed at the notch. This resulted in plastic deformation, causing reduction of local stresses. Therefore, the notch was no longer a major stress concentrator in the microstructure. The reduced stress levels were similar to typical intrinsic stress concentrators, such as grain boundary triple junctions. MD simulations indicate the competition between the external flaw and the intrinsic microstructure features plays an important role in determining if the failure of nanocrystalline materials would be sensitive to the flaw. This example shows the significance of MD simulations to understand in-situ observations by probing the interplay of local plasticity mechanisms in the specimens.

MD simulations in conjunction with in-situ mechanical deformation are helpful to understand work hardening in metals. In-situ TEM nanoindentation of aluminum showed migration and pileup of dislocations at the incoherent twin boundaries (Fig. 6.9a) (Bufford et al. 2014). Continuous obstruction and the pileup of dislocations caused distortion of the twin boundary. MD simulations also revealed nucleation and migration of lattice dislocations toward the twin boundaries during indentation (Fig. 6.9b). The lattice dislocations were seen to dissociate into interface disconnections upon meeting the twin boundaries. The interface disconnection was characterized by a Burgers vector of $\frac{1}{3}[111]$. The simulations also showed kink formation on the deformed twin boundary. The confinement of dislocations by the twin boundaries was observed from MD simulations, which leads to work hardening and high strength. This finding is in confirmation with the in-situ TEM images, which showed there was no dislocation transmission during work hardening. This example highlights the ability of MD simulations to capture defect interactions in the microstructure during deformation.

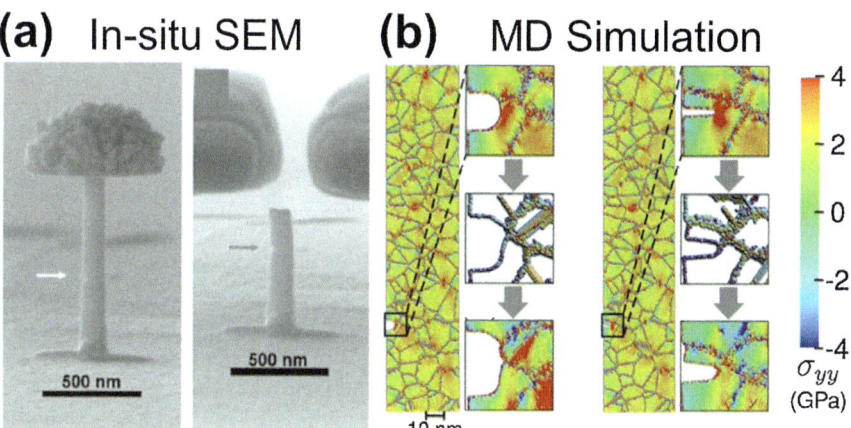

Fig. 6.8 Tensile deformation and failure of a Pt nanocylinder: (**a**) in-situ SEM images of notched cylinder before and after failure, and (**b**) MD simulations reveal plasticity mechanisms at notch. (Reproduced/adapted with permission from Gu et al. (2013))

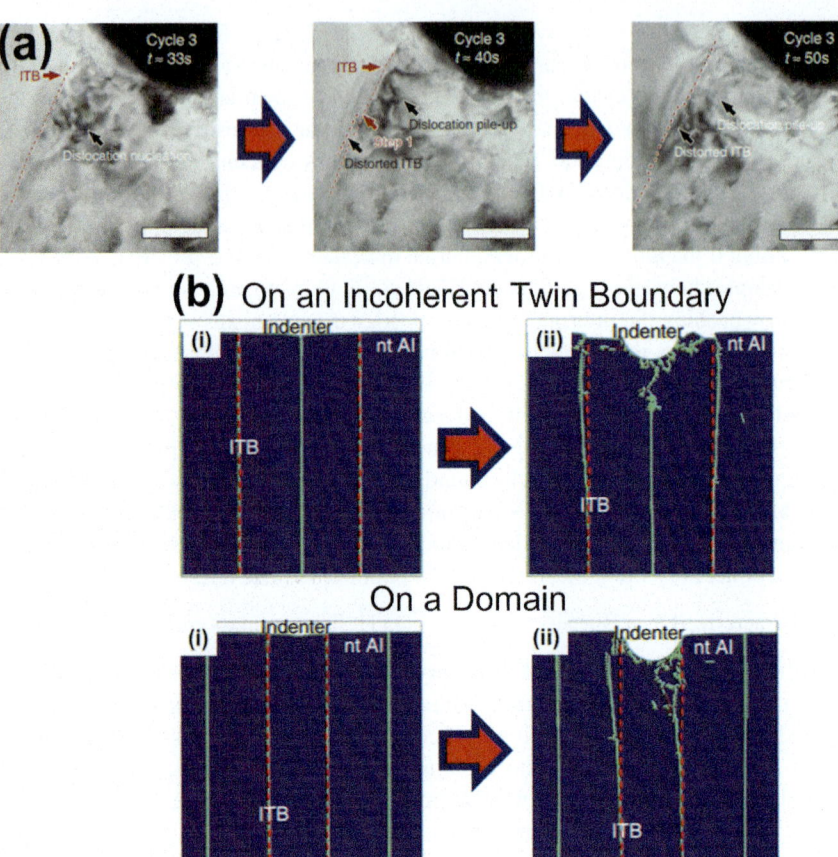

Fig. 6.9 In-situ imaging and MD simulations to study work hardening mechanisms: (**a**) in-situ TEM nanoindentation, and (**b**) MD simulations showing migration, obstruction, and pileup of dislocations. (Reproduced/adapted with permission from Bufford et al. (2014))

Cyclic loading-unloading can result in reversible activation of deformation mechanisms in materials. MD simulations are insightful to decipher the underlying atomic-scale mechanisms that govern the cyclic mechanical deformation behavior. Figure 6.10a shows real-time TEM images as a gold nanowire was subjected to uniaxial tension and compression cycles (Lee et al. 2014). Tensile deformation caused twinning in the nanowire. As the twinned nanowire was subjected to compression, shrinking and disappearance of the twins were seen. On reloading the same nanowire (under tension), twinning was observed again (Fig. 6.10a). The interplay of deformation mechanisms is associated with the interchange in Schmidt factor for the leading and trailing partial dislocations. MD investigations show the activation of reversible twinning-detwinning mechanism in the nanowire (Fig. 6.10b). The simulations revealed that the glide of partial dislocations at the twin boundaries is responsible for this observation (Lee et al. 2014). Considering

6.2 Interfacing with Molecular Dynamic Simulations

the importance of the twin boundaries, atomic-scale MD investigations focused on the boundaries were also performed. Layer-by-layer growth of the twin was seen as a partial dislocation with a/6$\left[11\bar{2}\right]$ Burgers vector (white-colored atoms) glided downward on the twin boundary (Fig. 6.10c). During detwinning (under compression), the layer-by-layer removal of the twin was observed due to partial dislocation (with the same Burgers vector) gliding in the opposite direction. This case study demonstrates the advantage of performing multi-scale computational investigations in conjunction with real-time mechanical testing to decode the deformation mechanisms in materials.

MD simulations also provide insights into the effect of temperature on the mechanical response of materials. Figure 6.11 shows a case study on in-situ SEM compression of CNT pillars at elevated temperatures (Bhowmick et al. 2018). MD simulations provide information on the change in bond number during deformation. Room temperature compression resulted in a decrease in n/n_0 ratio (where n_0 is the initial bond number and n is the current bond number). However, compression at 1000 K was characterized by bond reconstruction as n/n_0 ratio first decreased and then increased (Fig. 6.11b). Despite bond formation at elevated temperature, the overall network was weakened due to higher thermal vibrations. This manifested as buckling of the CNT pillar structure at elevated temperature. The role of covalently bonded interconnects on the structural stability of the pillars was probed by MD simulations A comparative stress–strain plot for compression of CNT pillars with and without covalent interconnects is shown in Fig. 6.11c. Significant enhancement in load-bearing capability was noticed due to interconnects. Pillars without interconnections were susceptible to localized bending at relatively lower strains, which manifested as multiple drops in the stress response.

This section highlights the application of MD simulations to decode deformation mechanisms in different materials. The simulations provide additional information, which can complement in-situ imaging to obtain a holistic understanding of the microstructure-mechanics relationship. The computational investigations are useful to probe multiple competing phenomena active in a material system, and their net effect on the mechanical response. In-situ testing with high-resolution imaging is useful to assess the validity of the simulations.

6.3 Summary

The application of analytical and computational techniques in conjunction with in-situ measurement was discussed in this chapter. DIC analysis of high-resolution real-time videos provides quantitative information of microstructure deformation in materials. Global strains during mechanical tests can be computed by DIC analysis of the specimens under mechanical loading, which is useful to extract stress-strain relationships. Strain maps superimposed on real-time micrographs provide information about local, site-specific strain evolution in the microstructure. High-resolution DIC to estimate local strains is useful to decipher the deformation characteristics of

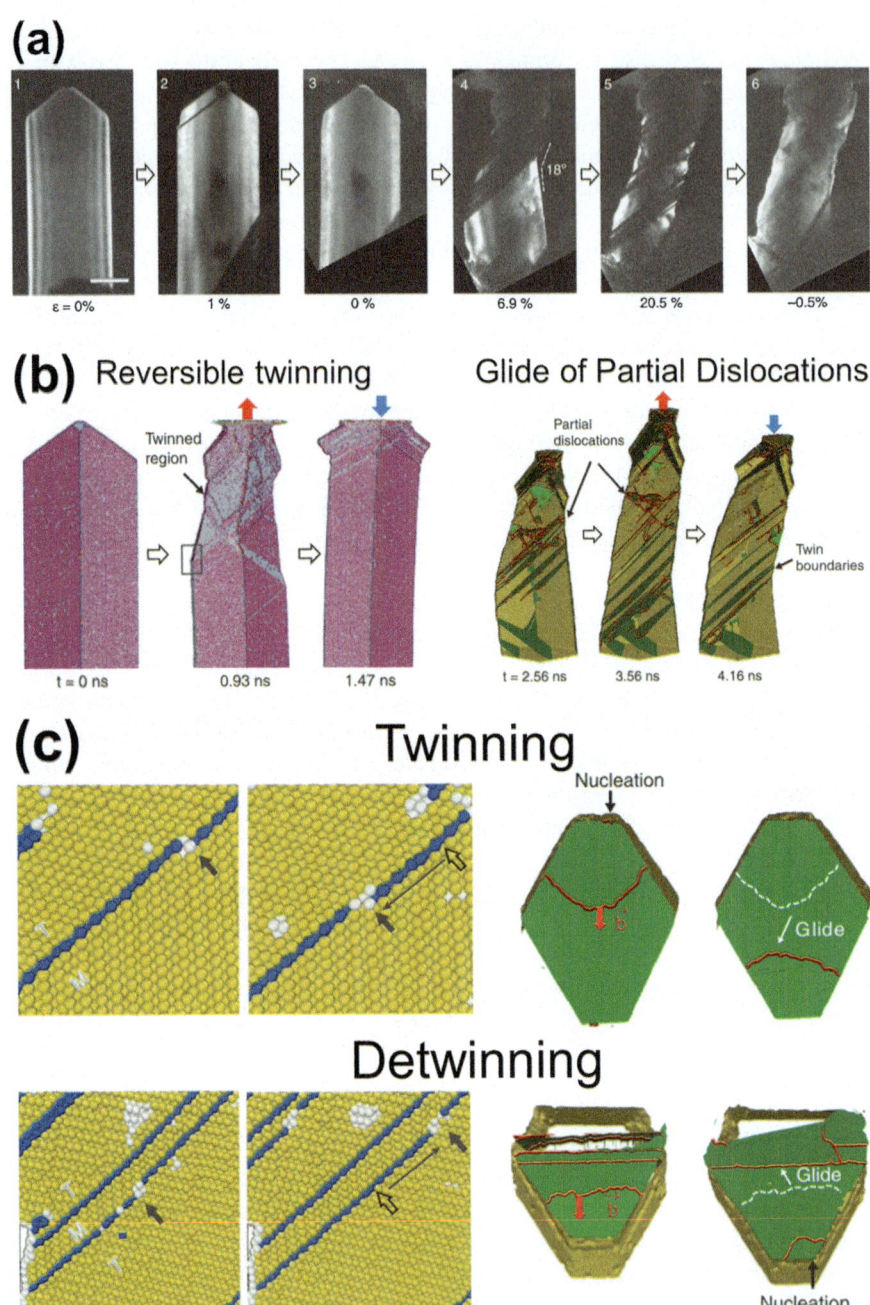

Fig. 6.10 In-situ imaging and MD simulations to study cyclic deformation in gold nanowire: (**a**) in-situ imaging during uniaxial tension-compression cycle, (**b**) MD simulations showing twinning-detwinning during cyclic loading and the glide of partial dislocations as an underlying mechanism, and (**c**) atomic-scale simulations showing layer-by-layer growth of twins (during tension) and layer-by-layer removal (during compression). (Reproduced/adapted with permission from Lee et al. (2014))

6.3 Summary

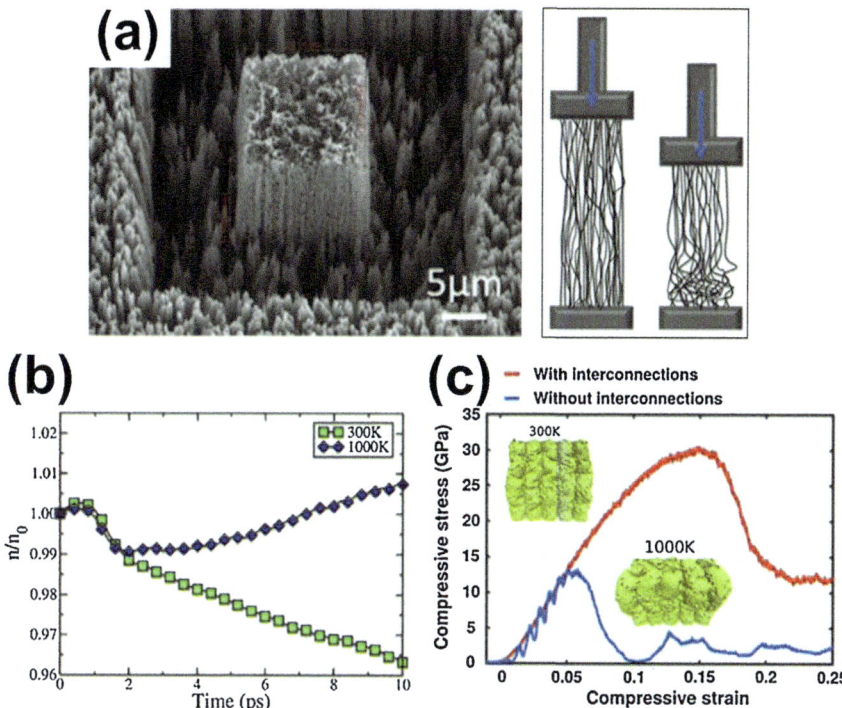

Fig. 6.11 (**a**) SEM micrograph of CNT pillars and schematic illustration of compression of the pillar, (**b**) MD simulation comparing the change in bond number at room temperature and 1000 K, and (**c**) MD simulation results comparing the stress–strain response of CNT pillars with and without covalently bonded interconnections. (Reproduced/adapted with permission from Bhowmick et al. (2018))

multicomponent microstructures. Application of DIC analysis to determine local failure strains associated with crack initiation and propagation was discussed in the chapter. Additionally, strains can be computed along different orientations (with respect to loading direction), providing a quantitative estimation of the degree of isotropy/anisotropy in the microstructure's mechanical response. DIC analysis of in-situ videos is beneficial to understand strain dispersion or stress transfer in 3D architectures, which is crucial to establish architecture-mechanics correlation. The application of the DIC technique to understand the effect of processing conditions on the mechanical response at microstructural length scales was discussed through a case study on cold-sprayed Al alloy. Understanding of processing-microstructure-mechanics correlations for emerging manufacturing techniques and material classes is important from the application standpoint. DIC mapping during small-scale mechanical testing, such as for micropillars, can be useful to examine crystallographic slip and lattice rotation. Therefore, DIC analysis with in-situ imaging provides mechanistic insights into the deformation of materials cutting across the length scales. The chapter introduced molecular dynamic investigations into the mechanics of materials, which can complement in-situ characterization. MD simulations provide insights into plastic deformation, revealing atomic-scale mechanisms, such as dislocation nucleation, stacking fault formation, and twinning-detwinning. The application of MD simulations to understand the effect of flaws or stress concentrators on mechanical properties was discussed. The computational investigations are beneficial to decipher the mechanical response when competing phenomena are active in the microstructure undergoing deformation. In-situ imaging in conjunction with simulations is vital to confirm the validity of the model or to provide guidance to improvise the model to match the real observations. The development of accurate models is beneficial for engineering microstructures with a desirable and predictable mechanical response. The case studies presented in the chapter compared in-situ observations with MD findings.

Questions and Assignments

1. Which of the following mechanical properties can be obtained from DIC analysis of real-time videos?

    (a) Failure strains.
    (b) Local strains in specific microstructure features.
    (c) Elastic modulus.
    (d) Failure strength.

2. Which of the following statement(s) hold true for DIC analysis?

    (a) DIC analysis provides information about the degree of isotropy/anisotropy in mechanical response.
    (b) DIC analysis reveals strain localization and redistribution in the microstructure during mechanical loading.
    (c) DIC analysis of in-situ SEM video provides lattice rotation angles.
    (d) None of the above.

3. MD simulations can provide additional insights into deformation mechanisms observed by in-situ imaging. Explain, with examples, how MD simulations can be useful to investigate:

    (a) The role of external flaws/stress concentrators on failure.
    (b) The role of intrinsic defects (twins/dislocations) on deformation behavior.
    (c) The effect of temperature on deformation characteristics.

4. The resolution of local strains which can be determined by DIC analysis is dependent on the choice of imaging tool. List the appropriate imaging method(s) to study strains in each of the following cases:

    (a) Strains experienced by a nanowire (diameter ~ 10 nm) under tension.
    (b) Strain localization during necking of a metallic rod (diameter ~ 10 cm) prior to failure.

(c) Strain evolution at porous sites (pore sizes ~10–100 μm) in a sintered ceramic pellet under compression loading.

5. Strain gages are typically used to determine strains in a sample under tensile testing. What kind of samples are not suitable for strain measurements using strain gages? Why? How can DIC analysis be used instead of physical strain gages? What could be the potential limitations/challenges associated with DIC strain measurements during a tensile test?

References

Bhowmick S, Ozden S, Bizão RA et al (2018) High temperature quasistatic and dynamic mechanical behavior of interconnected 3D carbon nanotube structures. Carbon N Y 142:291–299. https://doi.org/10.1016/j.carbon.2018.09.075

Brault R, Germaneau A, Dupré JC et al (2013) In-situ analysis of laminated composite materials by X-ray micro-computed tomography and digital volume correlation. Exp Mech 53:1143–1151. https://doi.org/10.1007/s11340-013-9730-9

Bufford D, Liu Y, Wang J et al (2014) In situ nanoindentation study on plasticity and work hardening in aluminium with incoherent twin boundaries. Nat Commun 5:4864. https://doi.org/10.1038/ncomms5864

Casillas G, Palomares-Báez J, Rodríguez-López J et al (2012) In situ TEM study of mechanical behaviour of twinned nanoparticles. Philos Mag 92:4437–4453. https://doi.org/10.1080/14786435.2012.709951

Christen D, Levchuk A, Schori S et al (2012) Deformable image registration and 3D strain mapping for the quantitative assessment of cortical bone microdamage. J Mech Behav Biomed Mater 8:184–193. https://doi.org/10.1016/j.jmbbm.2011.12.009

Di Gioacchino F, Clegg WJ (2014) Mapping deformation in small-scale testing. Acta Mater 78:103–113. https://doi.org/10.1016/j.actamat.2014.06.033

Embrey L, Nautiyal P, Loganathan A et al (2017) Three-dimensional graphene foam induces multifunctionality in epoxy nanocomposites by simultaneous improvement in mechanical, thermal, and electrical properties. ACS Appl Mater Interfaces 9:39717–39727. https://doi.org/10.1021/acsami.7b14078

Forsberg F, Sjödahl M, Mooser R et al (2010) Full three-dimensional strain measurements on wood exposed to three-point bending: analysis by use of digital volume correlation applied to synchrotron radiation micro-computed tomography image data. Strain 46:47–60. https://doi.org/10.1111/j.1475-1305.2009.00687.x

Gu XW, Wu Z, Zhang Y-W et al (2013) Microstructure versus flaw: mechanisms of failure and strength in nanostructures. Nano Lett 13:5703–5709. https://doi.org/10.1021/nl403453h

Huang M, Xu C, Fan G et al (2018) Role of layered structure in ductility improvement of layered Ti-Al metal composite. Acta Mater 153:235–249. https://doi.org/10.1016/j.actamat.2018.05.005

Lee S, Im J, Yoo Y et al (2014) Reversible cyclic deformation mechanism of gold nanowires by twinning-detwinning transition evidenced from in situ TEM. Nat Commun 5:3033. https://doi.org/10.1038/ncomms4033

Nautiyal P, Mujawar M, Boesl B, Agarwal A (2018a) In-situ mechanics of 3D graphene foam based ultra-stiff and flexible metallic metamaterial. Carbon N Y 137:502–510. https://doi.org/10.1016/j.carbon.2018.05.063

Nautiyal P, Zhang C, Champagne VK et al (2018b) In-situ mechanical investigation of the deformation of splat interfaces in cold-sprayed aluminum alloy. Mater Sci Eng A 737:297–309. https://doi.org/10.1016/j.msea.2018.09.065

Patterson BM, Cordes NL, Henderson K et al (2016) In situ laboratory-based transmission X-ray microscopy and tomography of material deformation at the nanoscale. Exp Mech 56:1585–1597. https://doi.org/10.1007/s11340-016-0197-3

Saucedo-Mora L, Lowe T, Zhao S et al (2016) In situ observation of mechanical damage within a SiC-SiC ceramic matrix composite. J Nucl Mater 481:13–23. https://doi.org/10.1016/j.jnucmat.2016.09.007

Chapter 7
Challenges During In-Situ Mechanical Testing: Some Practical Considerations and Limitations

The previous chapters discussed multifarious applications and advantages of the in-situ mechanical characterization approach. However, there are several challenges associated with the in-situ technique which should be taken into consideration. In-situ, high-resolution electron microscope-based measurement often requires specialized sample preparation. Alignment, placement, and clamping of miniature samples in electron microscopes for in-situ testing are time consuming. Introducing modifications, such as creating convenient surface topography on the measurement device and the use of alternative sample fixing techniques can be beneficial for miniature sample testing. The ultrahigh vacuum environment in electron microscope chambers is also known to alter the deformation characteristics. This is particularly true for samples prone to oxidation since the vacuum environment is not as reactive as the normal atmosphere. The absence of sufficient moisture in the electron microscope chambers is an impediment for the in-situ measurement of biological samples, which display mechanical properties dependent on the state of hydration. In-situ optical techniques are more desirable for moisture-sensitive biological samples, although the resolution of real-time images is not as superior as the electron microscopes. During in-situ imaging, electron beam irradiation can alter the mechanical response of materials. Localized heating can cause effects, such as the creation of artifacts or defects in the sample, abnormally accentuated dislocation activity and stress relaxation. These effects can be mitigated to some extent by minimizing beam exposure, the use of high thermal conductivity probes and sample stages to promote heat dissipation, and careful selection of acceleration voltage during imaging. Different materials respond differently to beam exposure. Low melting point and low thermal conductivity samples are expected to be more sensitive to the electron beam. The case studies presented in this chapter seek to inform the users of the in-situ characterization approach to carefully plan their experiments and understand the limitations of the techniques they intend to employ.

7.1 Sample Misalignment and Slippage

Mechanical measurements are sensitive to the attachment and alignment of samples prior to the testing. Sample and setup preparation during in-situ testing requires precautions. Poorly gripped samples can undergo slippage during tensile loading, which influences the mechanical reading as well as deformation mechanisms observed via real-time imaging. On the other hand, excess tightening can induce stress concentration and failure at the grip. Since in-situ mechanical testing is often utilized for miniature samples (such as nanowires, thin films, etc.), conventional sample placement and mechanical gripping are not appropriate. Nanosoldering by focused ion beam or focused electron beam-induced deposition is used for gripping the samples. Pick-and-place using a nanomanipulator is the common approach for securing miniature samples on the mechanical device for in-situ tensile testing. A major concern during manipulation is the possibility of a crash as the tip approaches the substrate. Accurate sample alignment by pick-and-place is rather challenging due to the limited freedom of motion available for sample manipulation inside the microscope chamber.

To overcome the challenge, the surface topography of the mechanical devices are modified. Fabrication of pillars or trenches on the surface is useful as the picked specimens can be dropped/released onto the sidewalls of these surface features. An example of nanowire placement and attachment on the MEMS mechanical testing device is shown in Fig. 7.1 (Zhang et al. 2009). The picking of a specimen is achieved by creating a joint between the manipulator tip and the specimen of interest by deposition of carbonaceous material. The deposition process for the nanowire shown in the figure took about 1–2 min. The deposition step can possibly lead to a change in sample diameter. In the MEMS device used in this case study, Young's modulus was determined by the relationship:

$$E = \frac{4}{\pi} \frac{k_s (x_C - x_A)}{D^2} \frac{l_0}{x_B - x_A} \qquad (7.1)$$

where D is nanowire diameter, k_S is sensor stiffness, l_0 is the initial length of the nanowire, $x_C - x_A$ represents sensor deformation, and $x_B - x_A$ denotes nanowire elongation. An error in the measurement of nanowire diameter (D) can result in erroneous modulus prediction. Therefore, variation in nanowire diameter due to focused ion beam-induced deposition should be taken into consideration. From the equation, it can also be seen that the estimation of modulus is highly sensitive to measured sample elongation ($x_B - x_A$). The secure clamping of the nanowire is critical to avoid sample slippage, which can cause the overestimation of elongation. In this case study, a carbon-copper composite was deposited to fix/clamp the nanowire with the mechanical stage. The focused ion beam-induced deposition is an effective technique for clamping miniature specimens for in-situ characterization.

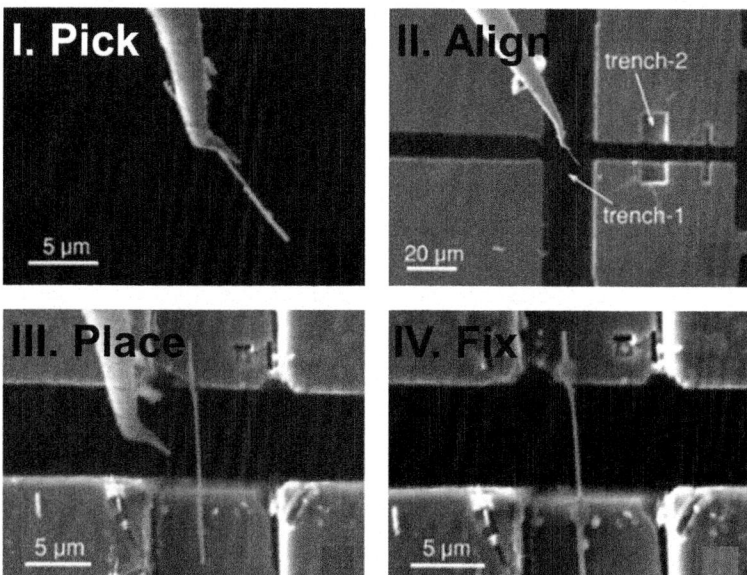

Fig. 7.1 Placement and clamping of miniature samples for in-situ tensile testing. A nanomanipulator is used to pick and place a nanowire specimen on to the MEMS device inside SEM. Focused ion beam-induced deposition was used to fix/clamp the nanowire. (Adapted with permission from Zhang et al. (2009))

7.2 Effect of Environment

The mechanical properties of many materials are a function of the environmental conditions. In-situ testing in an electron microscope requires a vacuum environment. The properties and deformation characteristics in the vacuum can deviate from the mechanical response of materials in the air. A case study on the comparison of the fatigue response of submicron thick Cu films in air and vacuum is shown in Fig. 7.2 (Kondo et al. 2016). It was observed that fatigue crack growth was decelerated in the vacuum environment. Figure 7.2a shows that the number of cycles (N) at unstable fracture was significantly arrested in the vacuum. The fracture surface for air-tested specimen was characterized by flat surfaces and surface steps with sharp edges. On the other hand, the vacuum-tested sample demonstrated blunt fracture surfaces and fine roughness (Fig. 7.2b). In the air, oxidation of the freshly exposed surface took place, preventing the possibility of rewelding. Therefore, cyclic slip deformation was irreversible and consequently, fatigue crack growth was pronounced. However, surface oxidation was subdued in the vacuum environment. Prevention of oxidation created suitable conditions for re-welding of the fracture surfaces.

The absence of moisture in the vacuum environment influences the mechanical properties of materials. This is particularly a major limitation for the mechanical testing of biological samples. It has been seen that the elastic moduli of tissues are

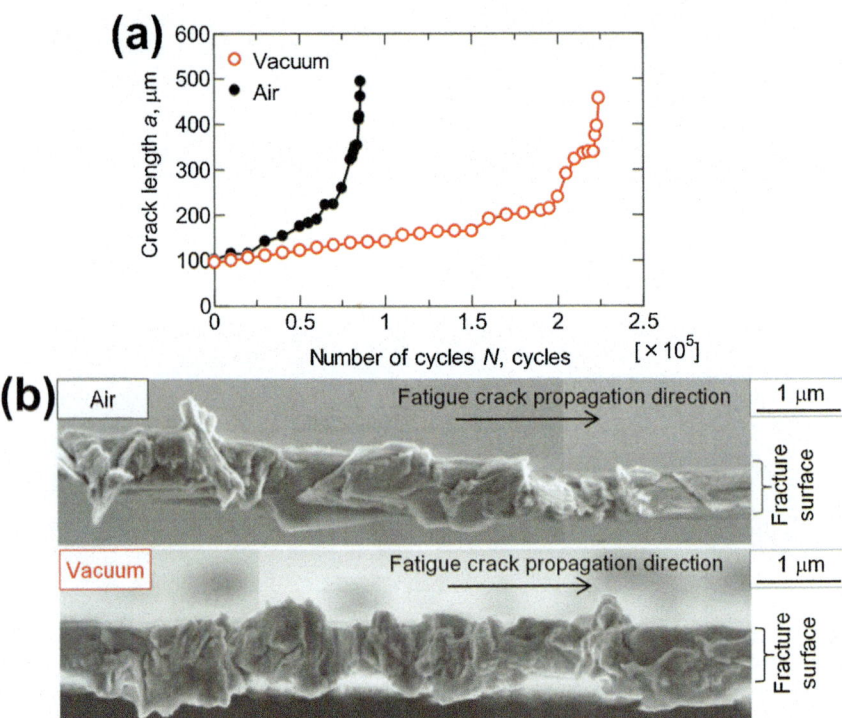

Fig. 7.2 Comparison of the fatigue response of Cu film in air and vacuum: (**a**) plot of crack length vs. the number of cycles, and (**b**) fracture surfaces for the two test conditions. (Reproduced with permission from Kondo et al. (2016))

enhanced in a dehydrated state (Ebenstein and Pruitt 2006; Nautiyal et al. 2018). Figure 7.3a illustrates nanoindentation load–displacement curves from different layers of the dactyl club, comparing its mechanical response in dry and hydrated states. The response revealed arrested deformation (depth) in the dry state. Because of this reason, prolonged dehydration of tissue specimens is avoided prior to mechanical testing. An example of nanoindentation, coupled with SPM imaging of the leaflets of the mouse tricuspid valve is shown in Fig. 7.3b (Balani et al. 2009). The tissue specimen was kept moist with phosphate-buffered saline at all times. A major limitation of high-resolution electron microscope-based in-situ characterization is that it's not possible to keep the samples hydrated. Biological samples are also time-sensitive. For instance, the tissue specimen shown in Fig. 7.3b was tested within 1 h of the heart's dissection (Balani et al. 2009). Sample preparation for SEM and waiting for a vacuum can take much longer. That's another reason why in-situ electron microscope-based techniques are rather challenging. Relatively lower magnification imaging techniques, such as optical microscopes, which do not require vacuum environments are more suitable for in-situ testing of biological samples.

7.3 Effect of Electron Beam Exposure

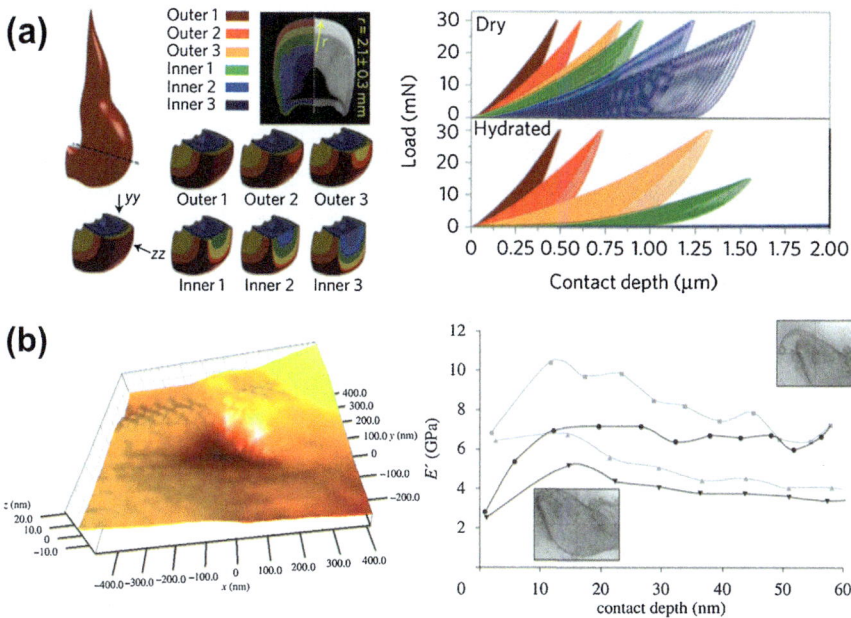

Fig. 7.3 (**a**) Comparative nanoindentation load–depth curves for dactyl club in dry and hydrated states, and (**b**) nanoindentation testing and scanning probe microscopy image of indent made on a tricuspid leaflet. (Reproduced with permissions from Amini et al. (2015). Purchased for reproduction from Balani et al. (2009))

7.3 Effect of Electron Beam Exposure

Electron beam exposure during in-situ mechanical testing influences the mechanical properties of materials. The beam irradiation can possibly introduce structural defects in the sample, which can alter the load-bearing capability of materials. A case study comparing the mechanical response of amorphous silica with and without e-beam exposure is shown in Fig. 7.4 (Zheng et al. 2010). The silica particle was subjected to loading-unloading-reloading cycles in TEM (Fig. 7.4a). The contact pressure at the end of 40% compression (with beam off) was around ~9.2 GPa. After this point, the beam was turned on and the sample was further compressed. It was observed that the forces required to deform the particle were significantly reduced (Fig. 7.4b). The contact pressure was calculated to be ~2.5 GPa (Fig. 7.4c). The electron beam can induce local damage due to bond cleavage and the creation of vacancies, resulting in lower stresses when the test is performed with the beam turned on. Additionally, there can be sample heating due to electron beam exposure. The sample heating is caused due to the energy loss of electrons passing through the sample during in-situ TEM imaging. The average energy loss (Q) of an electron is expressed as:

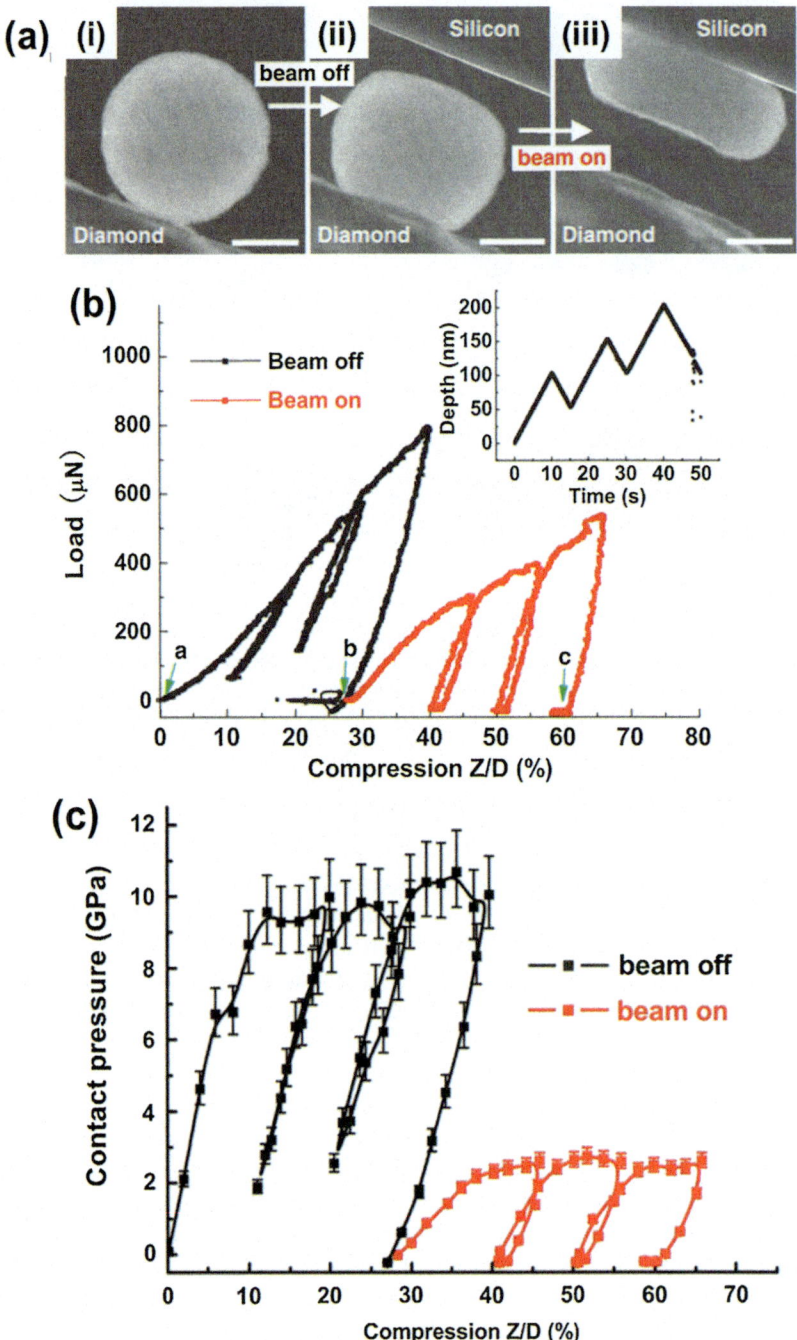

Fig. 7.4 (**a**) Compression of nanosphere under beam off and beam on conditions in TEM, and corresponding (**b**) load-compression, and (**c**) contact pressure-compression data for the cyclic testing. (Reproduced with permission from Zheng et al. (2010) CC BY-NC-SA 3.0)

7.3 Effect of Electron Beam Exposure

$$Q = Q_c + Q_r \tag{7.2}$$

where Q_c is the collision stopping power and Q_r is the radiative stopping power. The e-beam energy transferred to the sample at an energy density rate expressed by the relation:

$$H = QJ/e \tag{7.3}$$

where J is the electron current density and e is the elementary charge (1.6×10^{-19} C). The sample temperature rise due to irradiation can then be determined by the equation:

$$T_{rise} = \frac{Ht}{\rho C} \tag{7.4}$$

where t is the exposure/ irradiation time, C is the specific heat, and ρ is the mass density. From the equation, it can be deduced that longer exposure times result in significant sample heating. Sample heating can be a major issue for relatively lower melting point materials. Therefore, the exposure time for such materials should be minimized during in-situ imaging. One possible strategy is to avoid long duration tests when probing low melting point or highly temperature-sensitive materials. It is noteworthy that there is an interplay of heat accumulation-dissipation processes during in-situ testing. While electron beam irradiation causes heating, there is the dissipation of heat due to contact with the sample stage/substrate, fixtures and testing punch/probe (Fig. 7.4a). Typically, diamond probes are used for compression/indentation. Diamond has an excellent thermal conductivity of ~3320 Wm^{-1} K^{-1}. Therefore, the use of thermally conductive probes/sample holders is desirable to promote heat dissipation. Under equilibrium heat accumulation—dissipation condition, e-beam energy transfer rate (H) is correlated with temperature difference at the interface between the sample and the probe (ΔT), sample volume (V), area of contact interface (A), and contact thermal conductance (G) by the equation:

$$HV = GA\Delta T \tag{7.5}$$

Intimate contact between the sample being imaged and the high thermal conductivity probe/stage results in higher conductance, G and consequently lower ΔT. On the other hand, poor contact causes heat accumulation in the sample. By ensuring intimate contact of the specimen with the probe/base during in-situ imaging, sample heating can be avoided. It is worthy of mention that interfacial heat dissipation is effective only for samples with reasonable thermal conductivity. This can be understood by looking at the temperature field equation in a sample at equilibrium:

$$k\nabla_r^2 T = -H \tag{7.6}$$

Smaller values of conductivity, k, will result in larger value of $\nabla_r^2 T$. This implies a significant temperature gradient in the sample during in-situ imaging. The temperature around the center of the sample (away from the punch/substrate) is expected to be higher than the interfaces due to poor thermal transport. The temperature gradients created due to beam exposure can influence the mechanical response and deformation mechanisms. For instance, high-temperature conditions can induce plasticity in otherwise brittle materials. However, sample heating is expected to be minimal for materials with high thermal conductivities. In addition to conductivity, the sample size is also expected to influence heating during imaging. Thermal transport and dissipation through the interfaces are faster for smaller samples, whereas the temperature gradients are much higher for larger samples exposed to e-beam. All these aspects should be taken into consideration during in-situ mechanical testing in electron microscopes.

The effect of beam exposure on the stress response of metal films has also been reported and is ascribed to dislocation activity due to irradiation. A case study on the in-situ stress–strain response of Al thin film for different acceleration voltages in TEM is shown in Fig. 7.5 (Sarkar et al. 2015). The $\sigma_{1\%}$ was observed to decrease from 380 MPa for no beam exposure to 358 MPa when the imaging was performed

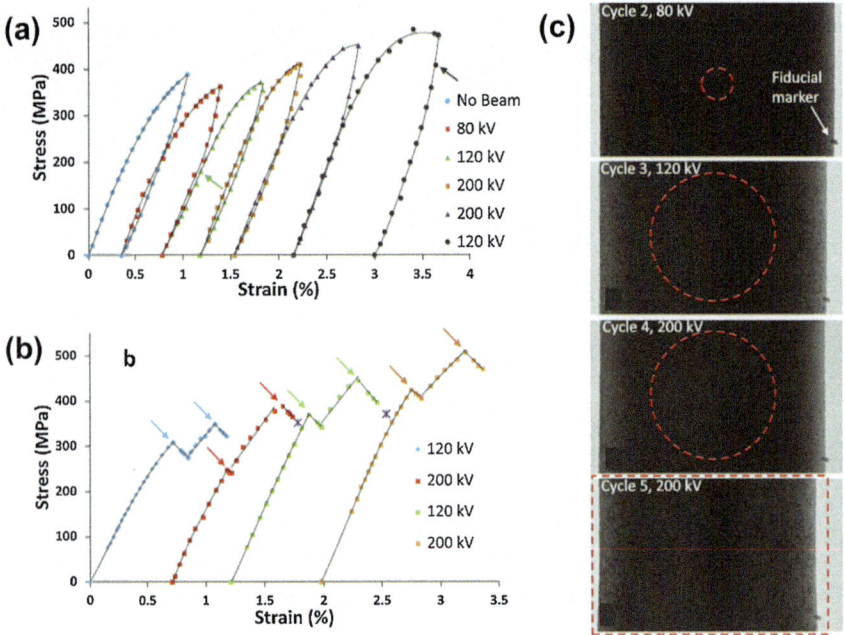

Fig. 7.5 In-situ TEM deformation of Al thin film: (**a**) effect of acceleration voltage on the stress-strain response, (**b**) effect of instantaneous beam exposures on stress relaxation (with different beam acceleration voltages), and (**c**) in-situ imaging of necking due to e-beam exposure. (Reproduced from Sarkar et al. (2015) CC BY 4.0)

with 80 kV acceleration voltage (Fig. 7.5a). Acceleration voltage influences stress relaxation in the samples during in-situ testing. In the case study discussed in Fig. 7.5b, stress relaxation was measured by performing mechanical loading without beam exposure and the beam was instantaneously turned on two times as the sample was being loaded. This instantaneous beam exposure resulted in stress drop, which was measured to compare the relaxation behavior at different voltages. The magnitude of the stress drop was higher at 120 kV than at 200 kV. Prominent dislocation activities were also noticed when the electron beam was shifted from one region of the sample to the other during mechanical loading. This dislocation activity was always accompanied with additional stress relaxation. Electron beam exposure can also promote or accentuate necking in the materials. The necking or width reduction was seen to be a function of acceleration voltage (Fig. 7.5c). The normalized width reduction is determined by the relation:

$$r = \frac{\Delta w / w}{\varepsilon_p} \tag{7.7}$$

At 120 kV, the value of r was computed to be 8.35, which is 2.5 times higher than that at 200 kV ($r = 3.28$). Additionally, shifting of e-beam to a different location in the thin film caused the formation of a new neck. This example illustrates the significance of acceleration voltage during in-situ mechanical measurements.

7.4 Summary

While the in-situ mechanical characterization approach is highly informative, there are several challenges that should be taken into consideration. This chapter discusses those issues and some practical solutions to overcome the problems. In-situ measurements are beneficial for testing miniature samples, but the placement, alignment, and clamping of ultrasmall samples can be tricky. Misalignment during sample placement can affect the determination of mechanical properties. Fabrication of mechanical measurement devices with modified topographies, such as pillars or trenches, is a helpful strategy for easy alignment. Poor gripping of the samples can lead to the overestimation of strains or underestimation of elastic modulus during tensile tests. On the other hand, miniature samples can be damaged by excess forces exerted by mechanical clamps. Focused electron and ion beam-induced deposition approaches are adopted to fix or clamp the samples. The deposition approach creates an intimate attachment of the samples to the mechanical stage while avoiding stress concentration due to mechanical fixtures. Mechanical measurements are sensitive to environmental conditions. The environment in the vacuum chamber of electron microscopes is not as reactive as air, which can influence the deformation of oxygen-sensitive materials (such as metals). The high vacuum electron microscopes are devoid of moisture, which is an impediment for testing biological samples. The hardness and stiffness of cells and tissues are much higher in the dry state than in the moist state. It is rather difficult to maintain the moist state of the samples in electron microscopes. For this reason, in-situ optical imaging/testing techniques are more suitable for biological samples. In-situ electron microscope-based mechanical testing can cause electron beam-induced sample damage. Localized sample heating, creation of defects/artifacts, dislocation activity, necking, lower strength, and stress relaxation are some of the known effects of electron beam exposure. Electron-sensitive, low melting point, and low thermal conductivity materials are more susceptible to beam irradiation. Table 7.1 below summarizes the challenges, solutions, and limitations discussed in the chapter. All these aspects should be taken into consideration during in-situ mechanical testing to improve the accuracy and reliability of results.

Table 7.1 Summary of some of the key challenges, limitations, and solutions/considerations during in-situ mechanical measurements

Challenges	Solutions and considerations	Examples
Sample alignment – Misorientation can cause error in mechanical properties	Modification of device surface topography to create release structures	Miniature samples, such as nanowires or nanotubes
Sample clamping – Poor gripping can cause slippage – Excess forces can cause stress concentration	Focused electron/ion beam-induced deposition to fix the samples	Slender and fragile samples, which cannot be clamped using mechanical fixtures
Environment sensitivity – Vacuum environment of SEM/TEM not as reactive as air – Lack of moisture impediment for testing biological samples	Use of alternate measurement techniques, such a in-situ optical/scanning probe microscopy	Tissues, cells, metallic films
Electron beam irradiation – Local sample heating – Defects/artifacts	Tweak acceleration voltages, minimize exposure time, use of thermally conductive probes/stages for heat dissipation	Low thermal conductivity materials, low melting point materials

Images reproduced with permissions from Zhang et al. (2009), Zheng et al. (2010), purchased for reproduction from Balani et al. (2009)

Questions and Assignments

1. The absence of oxidative conditions in an SEM chamber during in-situ testing can affect the mechanical response. Arrange the materials in the list below in the increasing order of their environment sensitivity:

   (a) Magnesium, Gold, Iron, Aluminum.

2. Which of the following mechanical tests is not suitable for in-situ SEM imaging?

   (b) Compression of aluminum micropillar.
   (c) Tensile testing of copper nanowire.
   (d) Nanoindentation of live cardiomyocytes.
   (e) Bending of polymer beam.

3. Out of the following four in-situ test conditions, which one is most susceptible to electron beam-induced sample damage?

   (a) A single cycle nanoindentation loading-unloading of copper in SEM.
   (b) Three-point bending fatigue testing of iron for 100 cycles in SEM.
   (c) Uniaxial tensile testing of epoxy in SEM.
   (d) Compressive fatigue testing of epoxy for 100 cycles in SEM.

4. Based on your reading of the chapter, prepare a brief report on key precautions that should be taken during in-situ SEM and TEM-based mechanical testing to avoid sample damage and minimize the effect of beam exposure on mechanical results.

References

Amini S, Tadayon M, Idapalapati S, Miserez A (2015) The role of quasi-plasticity in the extreme contact damage tolerance of the stomatopod dactyl club. Nat Mater 14:943–950. https://doi.org/10.1038/nmat4309

Balani K, Brito FC, Kos L, Agarwal A (2009) Melanocyte pigmentation stiffens murine cardiac tricuspid valve leaflet. J R Soc Interface 6:1097–1102. https://doi.org/10.1098/rsif.2009.0174

Ebenstein DM, Pruitt LA (2006) Nanoindentation of biological materials. Nano Today 1:26–33. https://doi.org/10.1016/S1748-0132(06)70077-9

Kondo T, Shin A, Hirakata H, Minoshima K (2016) Fatigue crack propagation properties of submicron-thick freestanding copper films in vacuum environment. Procedia Struct Integr 2:1359–1366. https://doi.org/10.1016/j.prostr.2016.06.173

Nautiyal P, Alam F, Balani K, Agarwal A (2018) The role of nanomechanics in healthcare. Adv Healthc Mater 7. https://doi.org/10.1002/adhm.201700793

Sarkar R, Rentenberger C, Rajagopalan J (2015) Electron beam induced artifacts during in situ TEM deformation of nanostructured metals. Sci Rep 5:16345. https://doi.org/10.1038/srep16345

Zhang D, Breguet JM, Clavel R et al (2009) In situ tensile testing of individual co nanowires inside a scanning electron microscope. Nanotechnology 20:365706. https://doi.org/10.1088/0957-4484/20/36/365706

Zheng K, Wang C, Cheng YQ et al (2010) Electron-beam-assisted superplastic shaping of nanoscale amorphous silica. Nat Commun 1:24. https://doi.org/10.1038/ncomms1021

Chapter 8
Future Outlook for In-Situ Mechanics Approach

The in-situ mechanics approach is promising for research, innovations, and product developments across the scientific and engineering disciplines. The in-situ characterization approach is expected to drive new material development, expand the application of novel manufacturing techniques to different material classes and interrogate the performance of critical engineering materials under extreme conditions. Considering the recent trend toward miniaturization, multifunctionality, and large-scale integration, the field can advance and expand by coupling in-situ mechanical testing with measurement of different functional properties. Real-time imaging can be time-consuming and expensive—interfacing in-situ experiments with machine learning can cut down the time and resources required to gain insights into the mechanics of materials and enable the development of predictive models. The chapter also presents some ideas and recommendations for developing the education curriculum on the in-situ mechanics approach relevant to the students and researchers from diverse disciplines. The development of focused courses is beneficial to prepare the future engineering workforce, popularize the in-situ approach in the industry, and to equip the researchers with knowledge set and hands-on training necessary to pursue research in the field. This book aspires to be one such comprehensive source of information for both new students and expert users alike.

8.1 Research and Innovations

This book discussed at length the importance, tools, techniques, and applications of the in-situ mechanics approach. The ability to assess mechanical properties at multiple length scales, wide-ranging load regimes, stress states and external environments makes in-situ characterization a powerful approach. The real-time imaging aspect of in-situ techniques provides the micro-to-macroscopic view of material response to mechanical loading, enabling the correlation of deformation behavior with material microstructure. As a result, in-situ characterization has strong potential for promoting research and innovations in the field of materials science, mechanical engineering, nanotechnology, biological sciences, and biomedical engineering. This section discusses some of the areas where the in-situ mechanics approach is expected to be a driving force for research and development.

8.1.1 New Material Development

Development of new materials with focused applications, such as high-strength alloys, lightweight composites, ultrahigh-temperature ceramics, oxidation-resistant polymers or multifunctional nanomaterials, requires thorough understanding of correlation between microstructure and mechanical properties. As the material microstructures become more intricate, the assessment of load transfer and deformation mechanisms becomes challenging. In-situ evaluation will allow precise estimation of load-bearing by different phases, as well as gain insights into mechanical interactions between multiple species. Real-time imaging is critical to decipher the failure in such complex material systems. Figure 8.1a shows the in-situ SEM indentation of a relatively new, MXene nanomaterial (Loganathan et al. 2019). Real-time imaging revealed indentation loading causes a reduction in inter-layer separation in MXene. It is noteworthy that materials behave differently under different loading conditions, such as tension, compression, flexure, torsion, fatigue, or creep. Performing in-situ investigations for these different loading scenarios can be useful to engineer materials with desirable mechanical characteristics. Figure 8.1b shows the in-situ testing of a novel graphene foam-metal metamaterial under tension and compression (Nautiyal et al. 2018). Real-time images captured along with load–displacement response allows to understand how the material behaves in the different load regimes, such as within the elastic limit, post-yielding, or prior to failure. These information play an important role in determining their applications. For instance, materials prone to catastrophic failure are not suited for critical load-bearing members in an aircraft. Tweaking microstructure features, such as the volume fraction of nanomaterial in a composite, alloy compositions, grain size of ceramics or the extent of cross-linking in polymers, influence the mechanical response of materials. High-resolution imaging during in-situ testing can be beneficial to capture these effects. Therefore, the in-situ approach is promising for new material development.

8.1 Research and Innovations 241

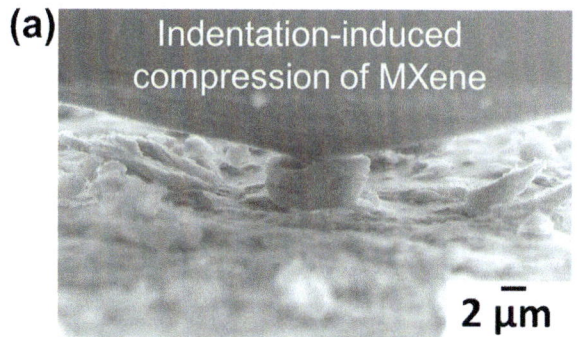

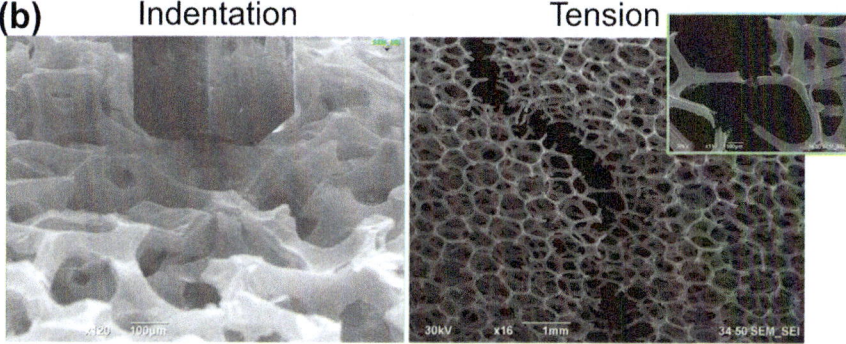

Fig. 8.1 (a) Application of in-situ indentation to characterize MXene particles, and (b) in-situ evaluation of indentation and tensile deformation of a graphene foam-metal metamaterial. (Reproduced/adapted with permissions from Loganathan et al. (2019), Nautiyal et al. (2018))

8.1.2 Novel Manufacturing Processes

There is active research interest in the development of new manufacturing processes or modification of pre-existing processes for engineering advanced materials, rapid processing, and high yield production. Some of the processes which are gaining popularity include additive manufacturing, nanomanufacturing, high-temperature/pressure processing, and sustainable manufacturing. Mechanical properties of materials are sensitive to manufacturing processes and processing parameters. Figure 8.2a shows a case study on two-photon polymerization or direct laser writing of polymer scaffolds. Multi-scale testing coupled with real-time imaging provides insights into deformation behavior of the structures fabricated by novel techniques. In addition to mechanistic insights, in-situ characterization can also be used to map processing parameters with mechanical properties. Figure 8.2b demonstrates the effect of laser scanning speed during two-photon polymerization on the elastic modulus and stiffness of the printed structure. These correlations are important for the widespread application of these emerging manufacturing techniques. Because of a

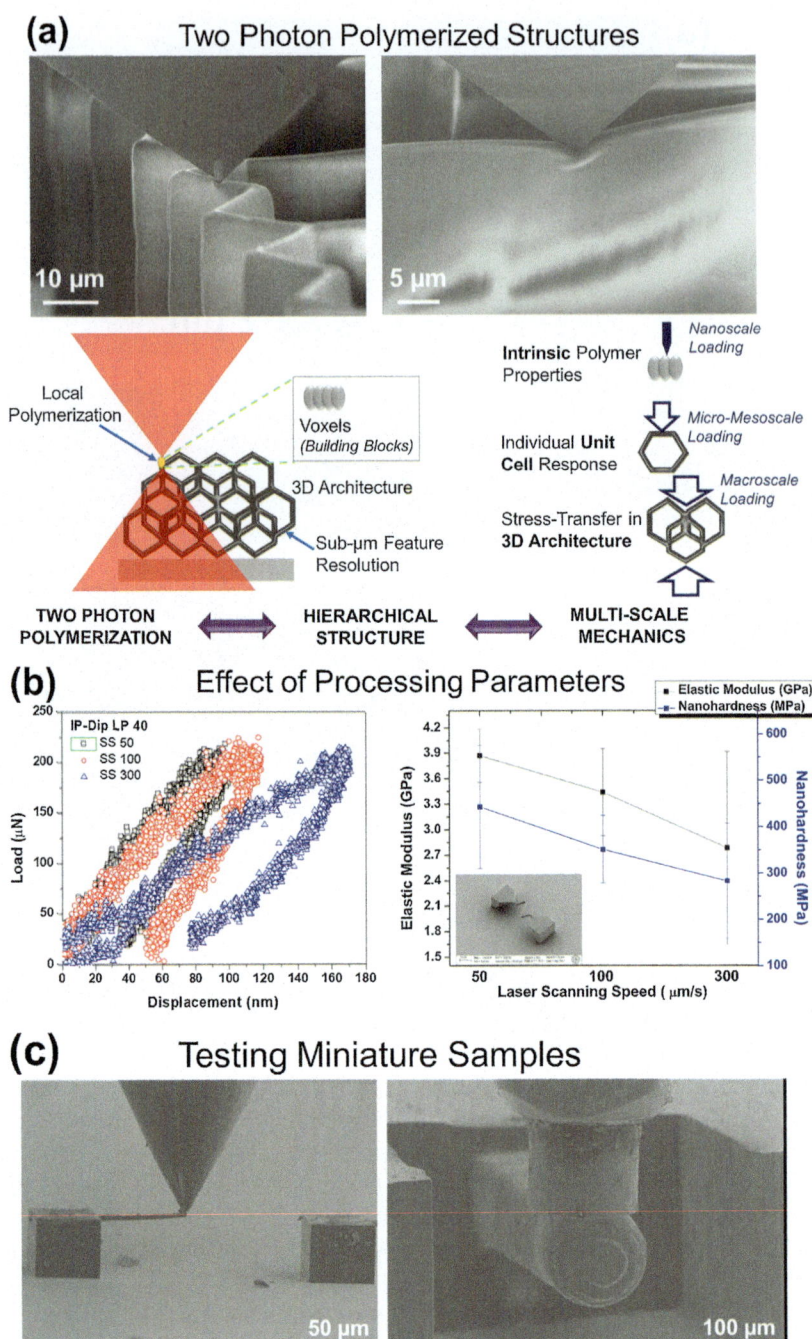

Fig. 8.2 (**a**) In-situ multi-scale mechanical characterization of two-photon polymerized scaffolds, (**b**) in-situ testing to correlate processing parameters with mechanical properties, and (**c**) application of in-situ approach for testing miniature samples. (Courtesy: Agarwal group (Authors' unpublished work) and CELL-MET NSF Engineering Research Center)

greater push toward miniaturization, nanomanufacturing is the prime focus of different disciplines in sciences and engineering. Small-scale testing techniques are essential to characterize the mechanical properties of the nano-micrometer-sized samples fabricated by nanomanufacturing approaches. Figure 8.2c shows in-situ SEM testing of polymer beams fabricated using two different techniques. The mechanical testing of these miniature samples is otherwise not possible by conventional methods. Therefore, the in-situ approach is going to play an important role in the manufacturing industry for development, improvisation, and expansion of new techniques that can be useful for fabricating a wide diversity of material classes.

8.1.3 Extreme Environment Performance

Several engineering applications require materials that can sustain and maintain their structural integrity under extreme environments. Chapter 4 discussed in detail high temperature and cryogenic temperature testing of materials. Materials can behave differently under extreme conditions. For instance, metals are known to undergo ductile-to-brittle transition at very low temperatures, resulting in susceptibility to catastrophic failure. High-temperature exposure can cause oxidation, which can be detrimental to the load-bearing capability of materials. Corrosive conditions, such as humidity or saltwater exposure also influence the mechanical performance of materials. Figure 8.3a illustrates the application of in-situ tomography to examine high humidity stress corrosion cracking in an Al alloy (Singh et al. 2014). The constant load condition resulted in the growth of cracks, as shown in the figure. X-ray tomography revealed the cracks visible on the sample surface are actually connected. It was observed the cracks grew from inside toward the surface. Another example of an environmental effect is shown in Fig. 8.3b, comparing cantilever bending in a vacuum SEM and an environmental SEM (ESEM) with 450 Pa water vapor (Deng et al. 2017a). ESEM is conducive for H-charging, resulting in embrittlement of FeAl alloy. Catastrophic failure of the cantilever beam was observed during ESEM testing, which was not seen under vacuum conditions. The loss in ductility is ascribed to the reaction of Al with water (Deng et al. 2017b):

$$2Al + 3H_2O = Al_2O_3 + 6H^+ + 6e^- \tag{8.1}$$

Extreme environment mechanical performance is vital for several applications, such as in rockets, spacecraft, submarines, nuclear reactors, and high-speed vehicles to list a few. Innovations and developments in these technologies can greatly benefit from the in-situ mechanics approach.

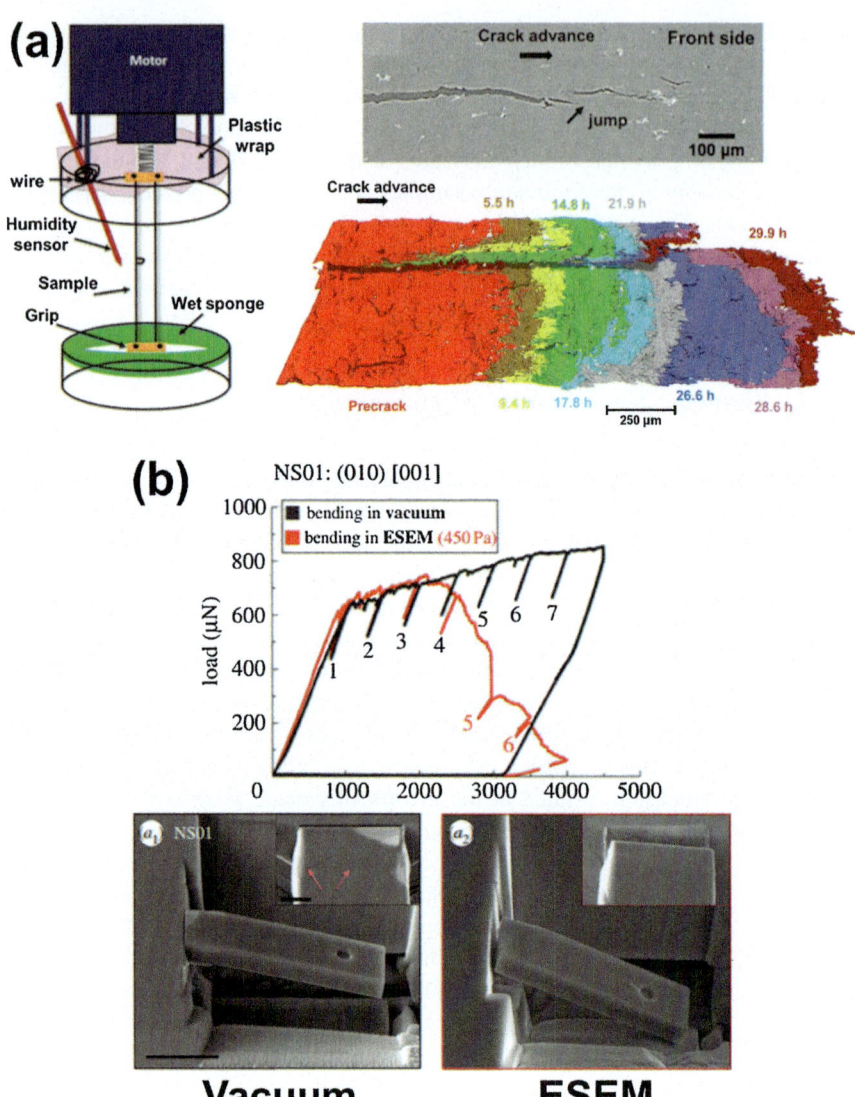

Fig. 8.3 (**a**) In-situ tomography testing under high humidity condition to image SCC crack growth in 7075 Al alloy, and (**b**) in-situ cantilever bending in vacuum vs. ESEM (water vapor) to study H-embrittlement. (Reproduced with permissions from Singh et al. (2014) CC BY 3.0. Purchased for reproduction from Deng et al. (2017a))

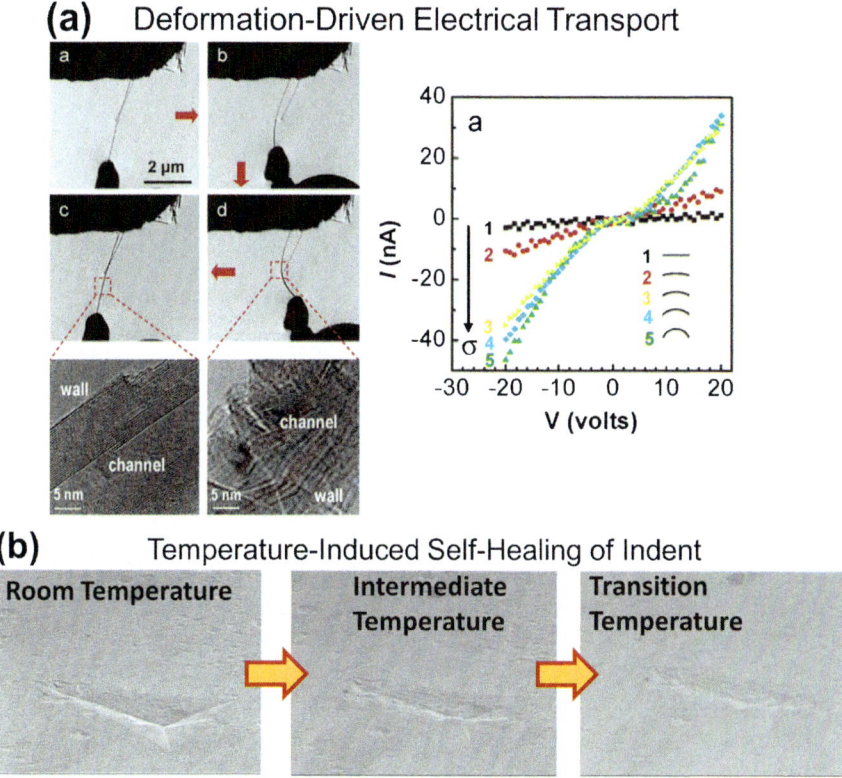

Fig. 8.4 Interfacing in-situ mechanical testing with: (**a**) electrical measurements to probe deformation-driven electrical transport in a piezoelectric nanotube[#], and (**b**) the thermal stimulus to assess self-healing/damage recovery post nanoindentation[*]. ([#]Reproduced with permission from Bai et al. (2007). *Courtesy: Agarwal group (Authors' unpublished work))

8.1.4 Coupling Mechanics with Functional Properties

Real-time imaging can be promising to interrogate the correlation between mechanical deformation and other functional properties. Figure 8.4a shows the bending deformation of boron nitride nanotubes inside TEM (Bai et al. 2007). Real-time imaging revealed the nanotubes were highly elastic. The *I–V* characteristics were recorded at different bending states. The resistance, *R* (~*dV/dI*) was observed to decrease with increasing bending curvature, signaling the transition of the nanotube from insulator to semiconductor. Based on the simultaneous in-situ mechanical and electrical interrogation, it is also possible to obtain the semiconductor parameters:

$$\ln I = \ln(SJ) + \left(\frac{q}{kT} + \frac{1}{E_0}\right) V + \ln J_S \qquad (8.1)$$

where J is the current density (through a Schottky barrier), J_S is the function of applied bias, S is the contact area (associated with the barrier). E_0 can be expressed by the relation:

$$E_0 = \frac{\hbar q}{2}\left(\frac{n}{m^*\epsilon}\right)^{1/2} \coth\left(\frac{E_{00}}{kT}\right) \tag{8.2}$$

where n is hole concentration, m^* is the effective hole mass for a deformed BNNT, and ϵ is the dielectric constant. The slope of $\ln I$ vs V provides the value of $\left(\frac{q}{kT} + \frac{1}{E_0}\right)$ term in Eq. 8.1. Therefore, the use of these equations in conjunction with I–V measurements during in-situ deformation enables the estimation of hole concentration, resistance, resistivity, and carrier mobility of the material being tested. The determination of electro-mechanical characteristics of advanced materials, like novel nanomaterials, is advantageous for engineering miniature devices.

Another application in Fig. 8.4b shows the application of in-situ mechanical testing to study shape memory effect in a polymer. An indent made on the surface was seen to recover after the polymer sample was heated to its transition temperature inside SEM. The self-healing capability is promising for engineering damage-tolerant structures or devices with potential applications in challenging environments. These two case studies highlight the future promise of coupling mechanical measurements with the interrogation of other functional properties.

8.1.5 Interfacing with Machine Learning

In-situ experimentation is time-consuming, requires highly skilled operators and demands the availability of resources, such as microscopes and mechanical instrumentation. Deep learning can be useful to minimize the number of experiments that need to be performed. Machine learning tools can be employed to make smart decisions about mechanical testing parameters, avoiding conventional trial and error approach which is time-consuming. The application of machine learning will be beneficial in scenarios where mechanical testing takes too long, such as the prediction of fatigue performance or creep deformation in materials. The machine learning approaches can provide highly reliable processing-microstructure-property correlations. These prescriptions can be useful to engineer materials with desirable microstructure, which can help save time typically spent in testing samples fabricated with multiple processing conditions. The data acquired by in-situ tests can also be used to train predictive models, minimizing the number of experiments that need to be performed to develop a comprehensive understanding of the mechanics of materials. A case study on the in-situ cyclic deflection of polymer micropillar is shown in Fig. 8.5. A machine learning model was trained using in-situ test data, and

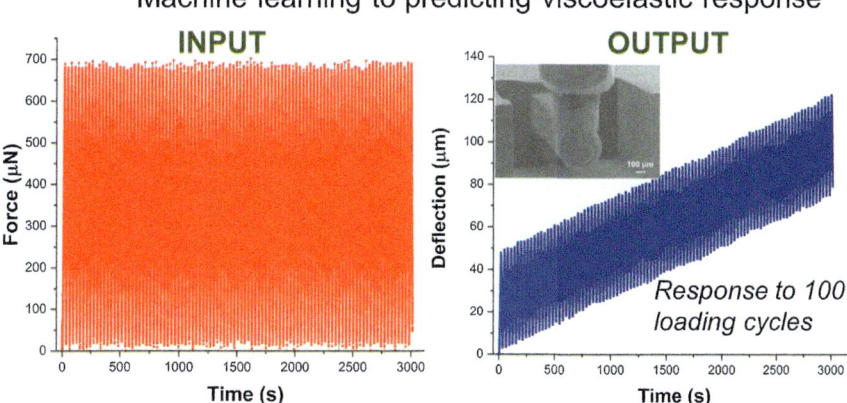

Fig. 8.5 The application of machine learning approach to predict the deformation of a viscoelastic polymer microbeam. The model was trained using in-situ test data, and the trained model was then employed for predicting material response over multiple loading-unloading-reloading cycles. (Courtesy: Agarwal group (Authors' unpublished work))

then used to predict the deflection response over 100 loading-unloading-reloading cycles. It was observed that the pillar undergoing bending displays time-dependent deformation, as peak displacement increases with the number of cycles (although the bending force is maintained constant). Therefore, the machine learning approach can complement the in-situ characterization of materials that display *viscoelastic behavior*. Machine learning can minimize the number of experiments and duration of tests required to understand the material response over long periods of time and for thousands-millions of loading cycles.

8.2 Education and Training

Considering the rising interest and expanding potential of the in-situ mechanics approach in different areas, it is high time that this technique is incorporated in education curricula (Fig. 8.6). This book is written with the intention of serving as a comprehensive source of information, providing the background, fundamental concepts, information about the tools and instrumentation, and the current state-of-the-art. In-situ mechanics is a vast field, with a multitude of applications. For effective dissemination of knowledge, courses need to be designed to cater to targeted audiences. The requirements of a mechanical engineering course will be different from a course focused on materials science. For instance, real-time testing of engineering materials, such as high-strength alloys or high-temperature composites is highly relevant in mechanical engineering. The in-situ investigation of dislocation motion and defect interactions is more relevant to material scientists and metallurgists. The main goal of designing the curricula should be to prepare the workforce

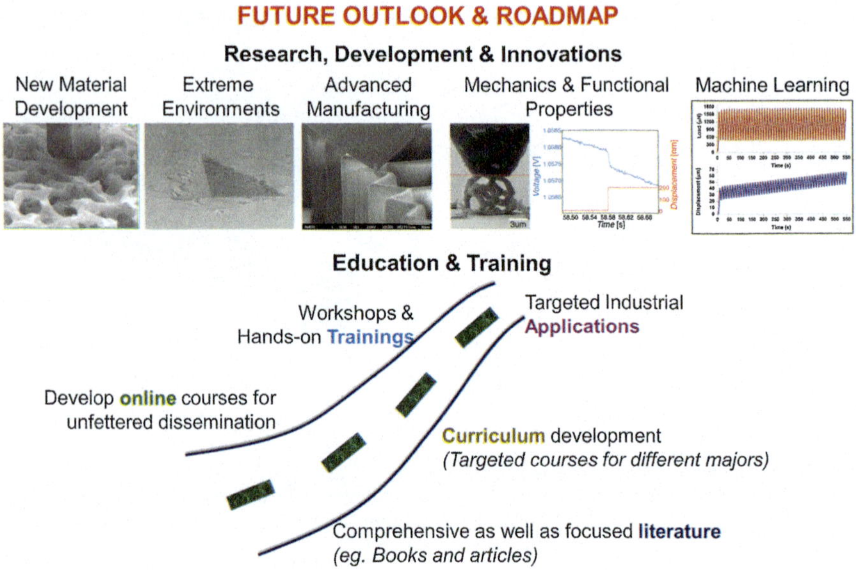

Fig. 8.6 A summary of future roadmap for the in-situ mechanics approaches in research and education. (Some of the images are reproduced with permissions from Nautiyal et al. (2018), Nautiyal et al. (2019), Yee et al. (2019), Agarwal group (FIU))

well-equipped to serve modern engineering industries. In-situ mechanics is going to play an important role in driving technological innovations, which will require skilled engineers and scientists. In order to make the students/trainees appreciate the impact and importance of the in-situ characterization approach, courses should be designed to include hands-on practical sessions (in addition to theory). Including practical sessions does not necessarily require high-end, expensive facilities. Even simple demonstration of a video camera capturing stretching and failure of different materials is good enough to provide students a practical understanding of real-time imaging approach. Research universities can leverage the advanced microscopy and mechanical characterization facilities available in their laboratories for providing practical training. It should be noted that the requirements of an expert user of in-situ techniques are altogether different from an undergraduate student who has never worked in the field. These differences in expectations should be taken into consideration while offering courses in universities or online learning platforms. Advanced level courses can also benefit and prepare graduate students who seek to carry out research in the field of in-situ mechanics. Training sessions/workshops can be organized for the students, researchers, and engineers interested in gaining hands-on expertise. Considering the rapid popularization of artificial intelligence and machine learning, it is desirable to include computational elements in the designed curricula. The curriculum development should take into consideration the educational background of the target audience. A course designed for the mechanical engineering program will not require a preliminary introduction to solid mechanics, but a curriculum for biomedical engineering students should provide mechanics background necessary to understand and appreciate the course material. Modern engineering problems demand a multidisciplinary approach, and it is the belief of the authors that advances in the field of in-situ mechanics are going to take place at the interface of disciplines. Therefore, the development of courses on in-situ mechanics should cater to students and researchers cutting across the disciplines. The research interest in the field is also expected to grow as the students, engineers and scientists are educated about the in-situ mechanics technique.

References

Bai X, Golberg D, Bando Y et al (2007) Deformation-driven electrical transport of individual boron nitride nanotubes. Nano Lett 7:632–637. https://doi.org/10.1021/nl0625401
Deng Y, Hajilou T, Barnoush A (2017a) Hydrogen-enhanced cracking revealed by in situ micro-cantilever bending test inside environmental scanning electron microscope. Philos Trans R Soc A Math Phys Eng Sci 375:20170106. https://doi.org/10.1098/rsta.2017.0106
Deng Y, Hajilou T, Wan D et al (2017b) In-situ micro-cantilever bending test in environmental scanning electron microscope: real time observation of hydrogen enhanced cracking. Scr Mater 127:19–23. https://doi.org/10.1016/j.scriptamat.2016.08.026
Loganathan A, Nautiyal P, Boesl B, Agarwal A (2019) Unraveling the multiscale damping properties of two-dimensional layered MXene. Nanomater Energy 8:84–95
Nautiyal P, Mujawar M, Boesl B, Agarwal A (2018) In-situ mechanics of 3D graphene foam based ultra-stiff and flexible metallic metamaterial. Carbon N Y 137:502–510. https://doi.org/10.1016/j.carbon.2018.05.063
Nautiyal P, Zhang C, Champagne V et al (2019) In-situ creep deformation of cold-sprayed aluminum splats at elevated temperatures. Surf Coatings Technol 372:353–360. https://doi.org/10.1016/j.surfcoat.2019.05.045
Singh SS, Williams JJ, Lin MF et al (2014) In situ investigation of high humidity stress corrosion cracking of 7075 aluminum alloy by three-dimensional (3D) X-ray synchrotron tomography. Mater Res Lett 2:217–220. https://doi.org/10.1080/21663831.2014.918907
Yee DW, Lifson ML, Edwards BW, Greer JR (2019) Additive manufacturing of 3D-architected multifunctional metal oxides. Adv Mater 31:1901345. https://doi.org/10.1002/adma.201901345

Index

A
Atomic force microscopes (AFM)
    cantilever, 57, 58
    coupling, 57
    EBID, 55
    SEM, 58
    small-scale mechanical testing, 55
    surface imaging, 55
    tensile tests, 55, 56
Atomistic molecular dynamic (MD)
    simulations, 217

B
Backscattered electrons (BSE), 39
Beam bending approach, 75
Biological materials
    bone, 190, 191
    cardiomyocytes, 192
    dragonfly wing, real-time imaging, 191, 192
    electron beam, 189
    electron microscope imaging, 192
    fracture, 190, 191
    in-situ indenter, 192, 193
    interfaces, 189
    inverted OM, 192, 194
    mechanical properties, 189, 190
    mechanics, 188
    micropillar compression, 189, 190
    nanoindentation, 189
Boron Nitride Nanotube (BNNT), 167
Brittle materials, 100
Buckling mechanism, 147
Bulk mechanical testing, 14
Bulk-scale tests, 78

C
Cantilever microbeams, 95
Carbon-linked graphene oxide nanosheets, 163
Cardiomyocytes, 192
Clamping technique, 15
Cold-spray, 210
Combined qualitative and quantitative assessment, 11
Composite nanofibers, 157
Composites
    Al-CNT, 171
    BNNT-silica interface, 168
    CNT, 165, 171
    Cu-graphene, 170
    fracture, 171
    graphene, 172
    graphene-TaC composite, 173
    in-situ approach, 165
    in-situ mechanical characterization, 169
    in-situ single nanotube pullout technique, 167, 168
    interface modification, 169
    low-dimension systems, 174
    meso-macro-scale imaging, 172
    metal/ceramic multilayer system, 175
    multilayer Al-TiN, 176
    multi-scale in-situ imaging, 174
    nanocomposites, 176, 177
    nanofillers, 165, 172
    nano-interfaces, 170
    nanotube, 165, 167, 170, 171
    PEEK-CNT, 169
    polymer-graphene foam, 174
    quantitative measurements, 175
    real-time imaging, 165, 175
    tensile stress, 169

Computed tomography (CT), 215
Contact stiffness, 76
Conventional bulk-scale mechanical testers, 14
Conventional mechanical testing, 15
Conventional optical cameras, 34
Coupling mechanical deformation, 188
Coupling mechanical measurements, 141
Crack branching, 44
Crack deflection/sword-in-sheath failure, 10
Crack initiation, 34
Crack propagation, 8, 9, 106
Crack propagation phenomena, 34
Critical stresses/strains, 11
Cryogenic cooling system, 132–134
Cryogenic indentation, 132
Cyclic in-situ indentation, 83

D
Defects, 144, 169, 190, 194, 196, 197
Deformation behavior, 13
Deformation mechanisms
   chisel-shaped stylus, 6
   mechanical loading, 3, 4
   postmortem observation, 10
   qualitative information, 1
   videos, 2
DIC mapping, 205
Different material phases, 12
Digital image correlation (DIC)
   analytical and computational investigations, 206–207
   cold-sprayed Al alloy, 210, 223
   cold-sprayed microstructure, 212
   deformation tensor, 214
   deformation, materials, 210
   determination, strains, 207
   DVC, 215, 216
   flexible foam, 210
   high-resolution in-situ video, 207
   image analysis technique, 205
   in-situ compression, Cu micropillar, 214
   in-situ imaging, 205, 223
   in-situ tensile testing, 209
   in-situ videos, 213, 223
   inter-splat sliding, 211
   local strains, 205
   manufacturing process parameters, 211
   mapping, 205
   measure strains, 209
   microstructure strain, 216
   polymer composite, 207
   polymer-graphene foam composite, 208
   qualitative assessment, 207
   real-time images, 207
   real-time imaging, 210
   real-time videos, 217, 221
   rotation values, 215
   strain contour, 208
   strain distribution, 210
   strain localization, 209
   strain maps, 207–210, 221
   strain transfer, 210
   strains, 207
   stress transfer characteristics, 3D architectures, 205
   3D graphene foam, 211
   volumetric 3D image data, 215
Digital image correlation (DIC) analysis, 66
Digital volume correlation (DVC), 215, 216
Dislocation structure evolution, 104
Dislocations motion, 8
Double cantilever testing
   brittle materials, 107
   crack growth, 105, 106
   Euler–Bernoulli theory, 106
   in-situ SEM imaging, 106
   linear elasticity solution, 106
   SiC specimen, 106
   strain rate release rate, 106
   wedge sliding, 106
Ductile-brittle transition temperature, 3
Ductile-to-brittle transition, 132
Dynamic TEM (DTEM), 45

E
Elastic theory, 153
Electrical interrogation, 245
Electron backscatter diffraction (EBSD)
   analysis, 49
   application, 48, 49
   characterization, 48
   GB misorientation, 50
   interplanar spacings, 48
   kernel average misorientation, 50
   mapping, 50, 53
   nickel pattern, 48
   orientation-slip trace relationship, 49
   surface-level information, 50
   3D HR-EBSD, 50, 52
Electron beam exposure
   acceleration voltage, 235
   beam irradiation, 231

Index 253

contact pressure, 231
irradiation, 234
mechanical properties, materials, 231
sample heating, 231, 233, 234
silica, 231
stress relaxation, 235
temperature, 234
thermal conductivity, 233
Electron beam-induced damage, 231, 236
Electron dispersive spectroscope (EDS), 4
Electron microscope
  deformation mechanisms, 39
  in-situ SEM mechanical
    characterization, 39–45
  in-situ TEM mechanical
    characterization, 45–48
  in-situ testing, 39
Electron microscope chambers, 227
Environment, 229, 230
Environmental exposure, 16
Euler–Bernoulli theory, 106
Extreme environment mechanics, 243

F
Failure strain, 207, 208, 217, 223
FIB fabrication process, 91
FIB machining, 53
FIB-milling, 26
Field number (F.N.), 32
Fluorinated nanofibers, 159
Focused electron beam-induced deposition
  (FEBID), 228, 235
Focused ion beam machining (FIB)
  application, 31
  column, 28
  in-situ mechanical testing, 31
  in-situ tensile testing, 30, 31
  in-situ testing, 26, 27
  ion irradiation, 27
  lamella, 30
  machine grips, 30, 32
  machining, 28
  milling, 26, 27, 30, 67
  miniature cantilever beam, 30
  powerful approach, 26
  pre-notches and pre-cracks, 30
  sample preparation, 27, 28
  sample/region of interest, 26
  tapered/V-shapes, 27, 28
Focused ion beam-induced deposition
  (FIBID), 228

Force–displacement profile, 86
Friction forces, 88
Fully programmable ultramicrohardness
  tester, 5
Functionalization, 158

G
GB misorientation, 50
Geometrically necessary dislocations
  (GNDs), 80
Graphene, 172
Graphene flakes, 11
Graphene nanoribbons, 162, 164
Graphene orientation, 80
Griffith's theory, 161, 162

H
Hard disc drive (HDD) film, 84
Harsh-environment mechanics, 134
High resolution-EBSD scans, 50
High strain-rate TEM holder, 45
High-load testing, 79
High-resolution imaging, 17
High-speed camera, 67
High-speed cameras, 34
High-temperature creep, 117
High-temperature exposure, 116
High-temperature indentation, 114, 115,
  118, 119
High-temperature mechanics
  components, 115
  cryogenic cooling system, 133
  indentation, 114–116, 118, 119
  mechanical testing methods, 114
  microbeam bending, 126–128
  micro-pillar compression, 119–126,
    135
  micro-pillar splitting, 124
  multilayer Cu/TiN material system,
    123
  room-temperature vs. cold-temperature
    compression, 136
  tensile testing
    macroscale sample
      characterization, 129–132
    mechanical behavior, materials, 128
    miniature samples characterization,
      128, 129
    versatile technique, 128
  working principles, 115

I

Impact-deformation of poly(L-lactide)/
  poly(ε-caprolactone) polymer, 34
Indentation-based mechanical characterization
  bulk-scale tests, 78
  contact stiffness, 76
  cyclic in-situ, 83
  dislocation nucleation phenomena, 83
  graphene orientation, 80
  high-load nanoindentation, 79, 80
  in-situ nanoindentation investigation, 78
  in-situ TEM imaging, 80–82
  in-situ testing, 78
  load–displacement curve, 76, 77
  low-load nanoindentation, 79
  multi-scale mechanics, 107
  penetration and retraction, 75
  Poisson's ratio, 76
  real-time imaging, 78, 80
  real-time TEM images, 80
  TiC sample, 80
  tip geometries, 76, 77
  YSZ, 78, 79
In-situ bending technique, 95
In-situ *frame analysis* technique, 10
In-situ image correlation, 10
In-situ imaging, 205
In-situ interface imaging, 87
In-situ manipulation, 87
In-situ measurements
  Al thin film, 234
  challenges, 236
  Cu film, air *vs.* vacuum, 230
  electron beam exposure, 231, 233–235
  environment, 229, 230
  limitations, 236
  nanosphere, 232
  sample alignment, 228
  sample clamping, 229
  sample slippage, 228
  solutions/considerations, 236
In-situ mechanics
  application, 142, 143
  concept and capabilities, 4
  crack propagation, 8, 9
  dislocations motion, 6, 8
  in-situ tools (*see* In-situ tools and
    techniques)
  low-load indentation, 5
  manufacturing techniques, 239
  microfriction, 6
  modern technological innovations, 248
  nanomaterials (*see* Nanomaterials)
  real-time imaging, 9, 10, 239
  tests methods, 75, 76
In-situ mechanics applications, 240, 243,
    246, 247
In-situ mechanics curriculum, 248
In-situ mechanics education, 247–249
In-situ mechanics research
  coupling mechanics, functional properties,
    245, 246
  extreme environment mechanics, 243
  machine learning, 246, 247
  multi-scale mechanical
    characterization, 242
  new material development, 240
  novel manufacturing processes, 241, 243
  real-time imaging, 239
In-situ mechanics training, 247–249
In-situ mechanics workshops, 248
In-situ micro-pillar compression, 119
In-situ nanoindentation investigation, 78
In-situ optical microscope
  application, 33
  conventional optical cameras, 34
  deformation mechanism, 34
  FN, 32
  folding-unfolding phenomena, 37
  glass microsphere, 36
  high-speed cameras, 34
  high-speed impact phenomena, 36
  magnification, 32
  nonequilibrium response, 37
  optical trap, 37
  optical tweezer, 37
  quantitative measurements, 33
  resolution, 32
  strain computation, 32
  stress–strain relationship, 34, 35
  tensile deformation, 33
  ultrafast in-situ imaging, 37
In-situ pillar compression, 119
In-situ scratch testing, 86
In-situ SEM mechanical characterization
  benefits, 40
  boundary-crack interaction, 43, 44
  BSE, 39
  critical mechanisms, 42
  electron-specimen interactions, 39, 40
  failure initiation mechanisms, 42
  fine-diameter nanopillars, 41
  indentation deformation mechanisms, 41
  load–displacement curves, 42
  loading cycles, 43
  nano-sized samples, 40

Index

plasticity mechanisms, 41
sample microstructure, 41
SE, 39
yield stress and ultimate tensile, 40, 41
In-situ SEM/TEM techniques, 14
In-situ single nanotube pullout technique, 167
In-situ TEM compression, 93
In-situ TEM imaging, 80
In-situ TEM mechanical characterization
    deformation mechanisms, 45
    dislocation nucleation mechanisms, 45
    DTEM, 45
    high-entropy alloy coating, 46
    high strain-rate TEM holder, 44–46
    mechanical strains, 45
    nanoindentation holders, 45
    nucleation and loss balance, 45, 47
    pillar compression, 45
    real-time TEM imaging, 45
In-situ TEM setup design, 88, 89
In-situ TEM tensile testing, 104
In-situ tensile deformation, 11
In-situ testing, 9, 16
In-situ testing micromechanical stage, 63–65
In-situ tools and techniques
    electron microscope, 39–48
    FIB (*see* Focused ion beam machining (FIB))
    in-situ optical microscope, 32–39, 48–53
    schematic representation, 25, 26
    3D internal imaging, tomography, 53–55
In-situ tribology
    applications, 84
    contact and interaction, 84
    focused ion beam milling, 87
    free-body diagram, 87, 88
    friction forces, 88
    indenter probe penetration, 87
    in-situ manipulation, 87, 88
    interfacial sliding characteristics, 87
    microstructure-level mechanistic information, 84
    MoS_2 layer beneath, 88
    multiple length scales, 84
    nanomaterial–substrate interface, 88
    nanoparticles, 88
    nanoscratch technique, 84, 85
    real-time imaging, 89
    rolling process, 89, 90
    scratch tester, 86
Instrumented indentation testing technique, 58
Interface bonding, 169
Interfacial shear strength (IFSS), 165
Inverse pole figure (IPF), 48
Inverted optical microscopy (OM), 192
Ion beam induced defects, 124
Ion beam irradiation, 194
Irradiated materials, 194–196
Irradiation, 231, 234

K
Kernel Average Misorientation (KAM), 50

L
Laboratory-tomography, 55
Lateral load sensing, 84
Liquid nitrogen-assisted cooling, 132
Load-displacement profile, 14
Load–displacement response, 2
Local strains, 205
Localized region-specific investigations, 107
Low-load indentation, 5

M
Machine learning, 246, 247
Macroscale sample characterization, 129, 130, 132
Manufacturing industry, 243
Material characterization, in situ mechanics
    aggregate material response, 4
    deformation mechanisms, 3
    ductile-brittle transition temperature, 3
    EDS, 4
    load–displacement response, 2
    material behavior, 2
    material chemistries, 3
    microscope observation, 1
    microscopes diversity, 2
    qualitative examination, 2
    quintessential step, 1
    RCC, 3
    real-time imaging, 2
    real-world engineering challenges, 3
Material microstructures, 13
Material properties, 132
Materials science, 247
Mechanical clamping, 114
Mechanics models, 9
Meso-macro-scale imaging, 172
Metamaterials, 141, 177, 178, 180, 181, 197
MgO nanocubes, 144

Microbeam bending, 126–128
Microbeam deflection
  cantilever beam, 95
  cantilever microbeams, 95
  Cu single crystal microbeams, 95
  geometries and cross sections, 94
  in-situ SEM bending, 95–97
  in-situ TEM testing, 97
  load–displacement curve, 95
  load–displacement response, 95
  SEM micrograph, 95
  stress intensity factor, 97
  three-point bending test, 95, 97, 98
  U-shaped notch, 94
Microelectromechanical systems (MEMS)
  actuator-sensor-electronics assembly, 61
  conversion factors, 62
  designs and operation principles, 60, 61
  EBID, 62, 63
  importance, 63
  in-situ application, 61
  micromechanical device, 62
  piezoelectric actuator, 62
  push-to-pull device, 62, 63
  testing platform, 63
Microfriction, 6
Micromechanical testers, 64–66
Micromechanics, 194
Micro-pillar compression, 122, 123
Micropillar compression technique, 186
Micro-pillar testing, 93
Micro-scaffolds, 13
Microstructure flaws, 218
Microstructure strains, 216
Miniature cantilever beam, 30
Miniature sample characterization, 128, 129
Miniaturization demands, 14
Misorientation, 236
Modulus mapping, 16
Moisture-sensitive mechanics, 227
Molecular dynamic (MD) simulations
  atomic-scale mechanisms, 220
  CNT pillars, 221, 223
  computational investigations, 221, 224
  cyclic loading-unloading, 220, 222
  deformation mechanisms, 220, 221
  in-situ mechanical deformation, 219
  intrinsic microstructure features, 219
  lattice dislocations, 219, 220
  microstructure flaws, 218
  nanocylinders, 218
  nanowire, 220, 222
  plastic deformation mechanisms, 224
  real-time imaging, 205
  real-time videos, 217
  strain localization, 217
  strains, 217
  stress maps, 217, 218
  temperature, 221
  twinned nanoparticle, 218, 219
  twinned nanowire, 220
Molybdenum single crystal, 41
Multiples length scales, 25
Multi-scale hierarchy, 10
Multi-scale in-situ imaging, 174
Multi-scale visualization, 2

N

Nano/microelectromechanical systems (NEMS/MEMS), 13
Nano-/micro-pillar compression approach
  complex stress-state, 90
  deformation processes and mechanisms, 91
  dislocation density, 92
  dislocation-free pillar, 92
  elastic modulus determination, 91
  FIB fabrication process, 91
  FIB-induced defects, 93
  flat-ended probe/flat punch, 90
  load–displacement readings, 90
  load–displacement response, 91
  mechanical response, 91
  misalignment, 91
  Ni nanopillars, 92
  real-time SEM micrograph, 90
  shear and plastic instability, 92
  stress–strain response, 90
  uniaxial test methods, 107
Nanocomposites, 176
Nanocrystalline, 146
Nanocubes, 145
Nanocylinders, 218
Nanofiber specimen, 100
Nanofibers, 178, 185
Nanofillers, 172
Nanoindentation, 78, 189
Nanoindentation holders, 45
Nanoindentation technique, 75
Nanoindenters, 25, 58–60, 62, 63, 67, 68
Nanomanipulator, 228
Nanomaterials, 14
  0D nanomaterials, 143, 144, 146, 147

Index

1D nanomaterials (*see* 1D nanomaterials)
2D, 160, 162, 163
in-situ testing, 143
mechanical properties, 141
microscopes, 143
strengthening mechanisms, 141
Nanoparticles, 88, 141, 143, 144, 146, 147
Nanoscratch technique, 84
Nanosheets, 141, 162
Nanotubes, 150, 165, 167, 168, 170, 184
Nanowires, 149, 152, 155, 157, 228
Need for in-situ mechanics
certainty over speculation, 10, 11
environmental effects, 16
local mechanical attributes testing, 12, 13
small-volume testing, 13–16
theories validation and development, 11
Novel manufacturing processes, 241, 243

O

1D nanomaterials
Ag nanowire, 157, 158
carbon nanofibers, 159, 160
CNT/SiC composite nanofiber, 157–159
composite nanofibers, 157
deformation mechanisms, 150
elastic modulus, 154
elastic theory, 153
fluorinated nanofibers, 159
fracture, 151, 152
functionalization, 158
GaN nanowire, 155
high-resolution imaging, 154
high-resolution in-situ mechanics, 150
in-situ technique, 155
mechanical properties, 149, 150
multiple loading-unloading-reloading cycles, 150
nanowire, 150–155, 157
Ni nanowire, 155, 156
oxide layer formation, 152
pentagonal silver nanowire, 150, 151
plasticity, 150
PTP device, 155
size effect, 157
tensile test, 149
Young's modulus, 157
ZnO nanowire, 149
Optical microscopy (OM), 2
Optical trap-based scheme, 37–39
Optical tweezer, 37, 67

P

Phase transformation, 187, 188
Piezoelectric materials, 141, 188, 197
Pile Grade-A (PGA), 194
Pillar compression, 90
Pillar compression technique, 119
Plasticity, 113, 124, 126, 134
Plasticity mechanisms, 41
Ploughing mechanism, 6
Poisson's ratio, 76
Post-buckling, 87
Post-failure microscopy, 10
Push-to-pull (PTP) micromechanical device, 155
Push-to-pull device, 63

R

Real-time imaging, 2, 17, 75, 89, 129, 141, 207, 210, 239, 240, 245
Real-time imaging analysis, 9, 10
Real-time SEM snapshots, 91
Real-time TEM images, 80
Real-time TEM imaging, 45
Reinforced carbon-carbon composite (RCC), 3

S

Sample alignment, 228
Sample clamping, 227, 229, 235
Sample heating, 129, 233, 234
Sample slippage, 228
Scanning electron microscopy (SEM), 2, 4–6, 8, 11, 16
Scratch-induced deformation, 84
Secondary electrons (SE), 39
SEM micrographs, 95
Shape memory materials, 141, 186, 187, 197
Shear strength, 5
Sliding/scratch experiments, 6
Slip dislocation, 11
Small-scale testing techniques, 243
Small-volume testing
electron-beam-induced deposition, 14
high-resolution imaging, 15
load-displacement profile, 14
miniaturization demands, 14
nanomaterials development, 14
nanometer-scale diameter, 14
nano-microelectronics, 13
nano-sized solids, 14
NEMS/MEMS, 13
stress determination, 15

Smart materials, 186–188
Smooth muscle cells (SMC), 9
Specific energy dissipation power (SEDP), 36
Sputtering, 26
Strain contour, 208, 211, 215
Strain localization, 207–209, 217
Strain mapping, 208
Strain maps, 207–210, 217, 221
Strain transfer, 210
Strain/displacement resolutions, 67
Strains, 207
Stress intensity factor, 97
Stress–strain response, 129
Stretching-relaxation curves, 37
Sub-micrometer, 28
Supplementary Video, 34
Surface-dominated geometries, 14
Sword-in-sheath mechanism, 100

T
Targeted indentation-based mechanical characterization, 13
Temperature-dependent deformation, 114, 115, 119, 122, 125, 132, 133
Temperature-dependent mechanics
   cryogenic cooling system, 132–134
   high-temperature mechanics (*see* High-temperature mechanics)
   in-situ testing, 113
   mechanical characterization, 113
   sample heating, 113
Tensile deformation, 7, 11, 12
Tensile testing
   advantage, 99
   brittle materials, 100
   challenges, 100
   characteristics, 104
   deformation mechanism/behavior, 100
   gold thin films, 130
   high-magnification imaging, 100
   in-situ SEM, 100
   in-situ TEM, 104
   linear stress–strain regime, 99
   localized mechanistic information, 101
   macroscale sample characterization, 129–132
   macro-scale samples, 131
   mechanical behavior, materials, 128
   microheater, 129
   miniature samples characterization, 128, 129
   multicomponent materials, 100
   nanofiber specimen, 100, 101
   nanoscale features, 104
   real-time TEM imaging, 103, 104
   region-specific microscopic investigations, 101
   SEM imaging, 100
   stress–strain curves, 100
   sword-in-sheath mechanism, 100
   TEM examination, 104
   uniaxial tensile loading, 99, 107
   versatile technique, 128
Thermal drift, 126, 136
Three-dimensional high-resolution EBSD (3D HR-EBSD), 50
3D architecture design, 13
3D architectures
   ceramic sponge structure, 177
   CNT structure, 184
   fracture, 184
   geometric parameters, 181
   graphene foam, 178, 179
   high-temperature in-situ investigation, 185
   in-situ characterization, 181, 184
   in-situ testing, 184
   mechanical metamaterials, 177
   mechanical properties, 180
   nanocomposites, 181
   nanofibers, 178, 185
   nanolattices, 181
   octahedron nanolattices, 181, 182
   optimal alloy coating thickness, 184
   polymer-high entropy alloy composite, 181, 183
   readers, 178
   real-time imaging, 178
   temperature, 185
3D graphene foam architecture, 11
3D graphene-ceramic metamaterial, 178
3D printing, 13
3D rendition, 55
Three-point bending test, 95, 97
Tip–surface interactions, 6
Tomography-assisted volumetric imaging, 55
Transmission electron microscopy (TEM), 2, 6, 7, 11, 16
Transmission X-ray microscopy (TXRM), 55
2D graphene, 161

Index

2D materials, 11
2D nanomaterials, 160, 162, 163
2D X-ray radiographic imaging, 54

U
Uniaxial tensile loading, 99
Universal testing machine (UTM), 65, 66

V
Volumetric imaging, 55

X
X-ray tomography, 55

Y
Young's modulus, 1
Yttria-stabilized zirconia (YSZ), 79

Z
Zero-dimensional (0D) nanomaterials, 143, 144, 146, 147
ZnO tetrakaidecahedron, 188

The manufacturer's authorised representative in the EU is Springer Nature Customer Service Centre GmbH, Europaplatz 3, 69115 Heidelberg, Germany. If you have any concerns regarding our products, please contact ProductSafety@springernature.com

Printed and bound by CPI Group (UK) Ltd, Croydon, CR0 4YY
25/03/2026
02078177-0004